Atomic Force Microscopy

Atomic Force Microscopy

Editor

Armand Vance

Atomic Force Microscopy

Edited by **Armand Vance**

ISBN: 978-1-68117-212-5
Library of Congress Control Number: 2016934757

© 2017 by
SCITUS Academics LLC,
www.scitusacademics.com
Box No. 4766, 616 Corporate Way,
Suite 2, Valley Cottage,
NY 10989

Preface

The atomic force microscope (AFM) is one kind of scanning probe microscopes (SPM). SPMs are designed to measure local properties, such as height, friction, magnetism, with a probe. To acquire an image, the SPM raster-scans the probe over a small area of the sample, measuring the local property simultaneously. The information is gathered by "feeling" or "touching" the surface with a mechanical probe. Piezoelectric elements that facilitate tiny but accurate and precise movements on (electronic) command enable very precise scanning. Compared to competitive technologies such as optical microscopy and electron microscopy, the major difference between these and the atomic-force microscope is that the latter does not use lenses or beam irradiation. Therefore, it does not suffer from a limitation of space resolution due to diffraction limit and aberration, and it is not necessary to prepare a space for guiding the beam (by creating a vacuum) or to stain the sample. Piezo-ceramics position the tip with high resolution. Piezoelectric ceramics are a class of materials that expand or contract when in the presence of a voltage gradient. Piezo-ceramics make it possible to create three-dimensional positioning devices of arbitrarily high precision. In contact mode, AFMs use feedback to regulate the force on the sample. The AFM not only measures the force on the sample but also regulates it, allowing acquisition of images at very low forces. The feedback loop consists of the tube scanner that controls the height of the tip; the cantilever and optical lever, which measures the local height of the sample; and a feedback circuit that attempts to keep the cantilever deflection constant by adjusting the voltage applied to the scanner. The atomic force microscope is a powerful tool that is invaluable if to measure incredibly small samples with a great degree of accuracy. Unlike rival technologies it does not require either a vacuum or the sample to undergo treatment that might damage it. At the limits of operation however, researchers have demonstrated atomic resolution in high vacuum and even liquid environments.The book entitled Atomic Force Microscopy covers the applications and theories of atomic force microscope.

Table of Contents

CHAPTER 1

Atomic Force Microscopy Studies on GaAs/In Bilayers Deposited on Si (100)

Miguel Ángel Venegas[1], Roberto Bernal-Correa[2], Máximo López-López[1], Álvaro Pulzara-Mora[2]

[1]Departamento de Física, Centro de Investigación y de Estudios Avanzados del I.P.N., México D.F., México
[2]Laboratorio de Nanoestructuras Semiconductoras, Universidad Nacional de Colombia, Manizales, Colombia

ABSTRACT

GaAs/In bilayers were prepared by RF Magnetron Sputtering in an Ar atmosphere on Si (100) sub- strates using high purity (95.95%) GaAs (100) and In targets. The growth temperatures were 300°C and 580°C for the high purity targets of In and GaAs, respectively. Three samples were pre- pared: the deposition time (t_d) for the GaAs layers was fixed to 30 minutes, while varied for the In layers from $t_d = 10$, 15, and 20 minutes. The morphological and optical studies of the samples were made by means of Amplitude Modulation Atomic Force Microscopy (AM-AFM). In order to analyze and correlate surface morphology and alloy composition properties, the as-prepared samples were cleaved along the [001] direction and subsequently studied by AM-AFM-micrographs. From topographic images, a statistical study of the roughness and grain size was made. Additionally, cross sectional AM-AFM-micrographs were performed for each sample, where the phase channel, which is sensitive to the material properties of the specimen, was of particular interest.

Keywords: InGaAs, Magnetron Sputtering, AFM

1. INTRODUCTION

Indium Gallium Arsenide (InGaAs) heterostructures, often regarded as GaAs and InAs binary alloys, are a III-V semiconductor compound largely used in the electronic and optoelectronic industry. The range of practical devices fabricated with this compound, grown on Si (100) substrates, due to the different band gaps related in the system (0.33 eV for InAs, 1.43 eV for GaAs and 1.12 eV for Si), varies from optical sensors, photodiodes, solar cells and many other kind of devices [1] [2] .

Among the several techniques and methods used to growth InGaAs compounds, the Radio Frequency Magnetron Sputtering technique is of special interest, due to its low cost fabrication process compared with other growing techniques. Characterization, understanding and controlling the growing process is necessary in order to obtain the required film formation that best meets the device specifications. Between the several microscopy techniques currently employed, Amplitude Modulation Atomic Force Microscopy (AM-AFM) is a powerful, flexible and economic tool that enable us to study the surface and the heterostructure morphology. Like roughness analyses [3] , grain structure [4] , cross section heterostructure features, chemical sensitivity and materials properties through the phase contrast operation mode [5] - [8] . Certainly, by AM-AFM, it is possible to exploit both signals: the topography and the phase channels. Exploring the phase signal capabilities is of particular interest because of its sensitivity to materials properties [9] [10] , which can lead to interesting and useful analysis of cross section semiconductor heterostructures. This transversal AM-AFM analysis, enable us to study not only the surface structure, but also the inner features of any multi-composed assemblage. For all this, in this work, present the results of the characterization of the GaAs/In bilayer grown by Magnetron RF Sputtering, to study the influence of the deposition time of the Indium (In) layer on morphological and optical properties of the bilayer. The AM-AFM characterizations of the three samples prepared were implemented in terms of surface morphology, cross-section imaging composition. Surface morphology was described in terms of the following statistical quantities: Maximum Height value of the grains and Root Mean Square (RMS) roughness. Grain analysis, by using the watershed algorithm [4] , was implemented to obtain the following quantities: Grain Population Density, Mean Grain Size and Total Grain Volume. The inner heterostructure features was studied in terms of cross section imaging, combining the topography features with the phase shifts signal, to observe the film thickness and the topography and composition effects on the phase channel.

The manuscript is developed as following. We will start with the experimental setup section, where a description of the surface preparation, the AM-AFM implementation is given. We will continue with the results portion, where AFM images of the three specimens are showed. After that, the discussion part will be presented, where the grain and cross sectional analysis is made, and finally the conclusion part.

2. EXPERIMENTAL SETUP

2.1. Sample Preparation and Fabrication

The In/GaAs bilayers where grown on a (100) silicon substrate by Radio Frequency Magnetron Sputtering in an Argon (Ar) atmosphere. High purity (95.95%) GaAs (100) and In targets where used. Before deposition the Si (100) surfaces were firstly degreased with acetone and methanol. Secondly, they were cleaned with hydrofluoric acid at 2%, rinsed with deionized water, and dried by flowing nitrogen. Thirdly, the substrate was introduced into the sputtering chamber to start the vacuum process. When the base pressure of 1.2×10^{-6} Torr was reached, the substrate temperature was raised up 650°C to remove the oxide on surface. The substrate temperature was lowered and stabilized at 580°C, and high purity Argon gas was introduced into the chamber until a working pressure of 1.2×10^{-3} Torr, and the plasma is turning on. Next, the GaAs target source was powered on in order to deposit a GaAs buffer layer during 30 min, after then the GaAs shutter was closed. Subsequently, the substrate temperature was lowered to 300°C and the Indium (In) source target was powered on, in order to deposit Indium during t_d = 10 minutes (sample M1), 15 minutes (sample M2) and 20 minutes (sample M3), respectively. This process is then repeated once again in order to deposit two In/GaAs films. Finally, the structure was capped with a GaAs layer (t_d = 20 min).

2.2. AFM Equipment Setup Description

The AM-AFM characterization where performed with the N8 NEOS SENTERRA system from Bruker, working at ambient conditions. This equipment combines an optics SENTERRA Raman spectrometer with the Nano's N8 NEOS Atomic Force Microscope, to allow the morphological, structural and chemical analysis of the same sample area. It is mounted on a very heavy table and everything is enclosed in a Faraday Box. The cantilever beam deflection sensor is based in a fiberoptical interferometric detection system. The cantilevers used were Point Probe Plus Non-Contact Long cantilever Reflex coated (PPP-NCLR) from Nanosensors, with resonance frequencies in vacuum of 190 Khz, and force constants around 48 N/m. The AFM machinery was set to work in AM-AFM contrast phase mode, with free oscillation amplitudes of 195 nm, set points of 50%, 65% and 80% and oscillation frequencies around 169 Khz. All the AFM images presented here shown the forward scanning direction. Free oscillation phase was set to 90°, in a range going from 0° to 180° or from 0° to −180°. The soft- ware's programs used to treat the AFM images were Gwyddion [11] , WSxM of "Nanotec Electronica" [12] and Nano Scope Analysis [13] (the sample orientation with respect to the AFM cantilever is shown in Appendix).

3. ANALYSIS AND RESULTS

Figure 1 shows the optical images obtained with a 100× lent (upper part) and the corresponding 40 × 40 μm² AM-AFM topography images (lower part), for each of the three samples: In Figure 1(a), Figure 1(d) and Figure 1(g), we have the 5 minutes double cycle of Indium deposition, corresponding here to the sample M1. The Figure 1(b), Figure 1(e) and Figure 1(h) displays the 10 minutes double cycle of indium deposition, sample M2. And Figure 1(c), Figure 1(f) and Figure 1(i) show the 15 minutes double cycle of Indium deposition, sample M3. For each sample, the optical images presented at the upper part, Figures 1(a)-(c), are marked with a blue square, which is the zone where the AFM imaging took place, Figures 1(d)-(f).

For these large scale AFM images, the corresponding statistical quantities were the following. For the sample M1 it was found a Maximum Height of 2.36 μm and the RMS roughness was 0.28 μm. For the sample M2, we found a Maximum Height of 3.18 μm and the RMS roughness was 0.38 μm. Finally, for the sample M3, we found a Maximum Height of 7.72 μm and the RMS roughness was 0.97 μm.

In order to obtain the grain discretization a 10 × 10 μm² zoom was obtained in this same area for each sample. The corresponding AFM topographic images are following shown in Figure 2.

From the Figure 2(a), sample M1, where the Indium deposition time was two cycles of five minutes, we obtained the following statistical quantities: 1.79 μm of Maximum Height and 0.24 μm of RMS roughness. The grain statistics obtained from Figure 2(a), and discretized in Figure 2(d), were: number of grains 1120, grain population density 11.2 μm^{-2}, Mean Grain Size 0.183 μm and Total Grain Volume 48 μm³. For the sample M2 the Indium deposition time was two cycles of ten minutes, and from Figure 2(b) we obtained the following statistical analysis: Maximum Height, 3.77 μm and RMS roughness, 0.51 μm. The grain statistics obtained from Figure 2(b) and discretized in Figure 2(e) were: Number of grains 171, Grain Population Density 1.71 μm^{-2}, Mean Grain Size 0.63 μm, Total Grain Volume 137 μm³. And for the sample M3, where the Indium deposition time was two cycles of fifteen, Figure 2(c), the statistical analysis were: Maximum Height, 5.44 μm and RMS roughness, 1.01 μm. The grain statistics obtained from Figure 2(c) and discretized in 2(f) were: Number of grains 73, Grain Population Density 0.73 μm^{-2}, Mean Grain Size 0.82 μm and Total Grain Volume, 227 μm³.

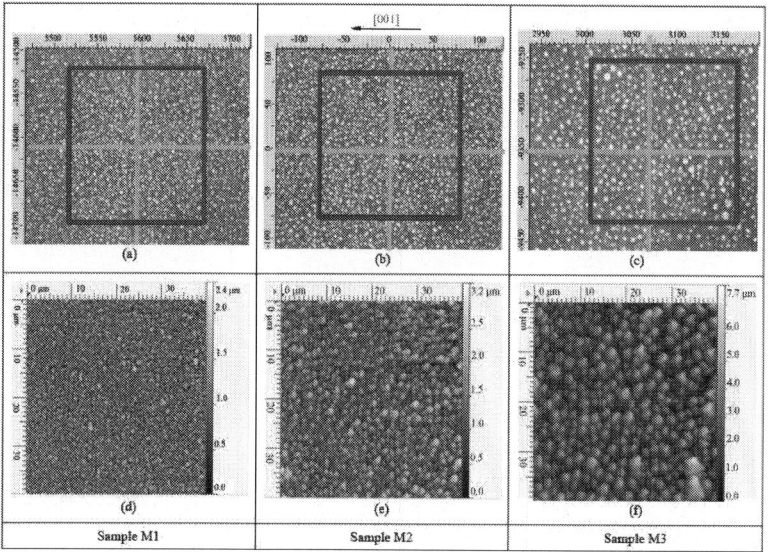

Figure 1. 100× optical image (upper part), large scale 40 × 40 µm² AM-AFM topography images (lower part). Column 1 ((a), (d)), column 2 ((b), (e)) and column 3 ((c), (f)), show the corresponding results for the sample M1, the sample M2 and the sample M3 respectively. The respective AFM parameters for the (d) were: set point (SP) = 50%, free amplitude (FA) = 195 nm, oscillation frequency (OF) = 169.12 Khz and scanning speed (SS) = 0.5 l/s. For the (e), the imaging parameters were: SP = 50%, FA = 195 nm, OF = 169.32 Khz and SS = 0.7 l/s. And for (f): SP = 50%, FA = 195 nm, OF = 169.37 Khz and SS = 0.5 l/s. The corresponding [001] directions are pointing towards the left for all the specimens.

An interesting observation presented by the AFM topography images, was the presence of linear features. The Figure 2(a) of the sample M1, presented several linear features arranged in a perpendicular way and giving rise to a repetitive pattern of corners alike forms, which apex is pointing to the [010] direction, here indicated by the green dotted lines. For the sample M2, this corner alike shapes evolved to the hexagonal alike shapes indicated by the green dotted lines of Figure 2(b). And for the sample M3, hexagonal, rectangular and polygonal alike shapes were found. Again there are indicated with the green dotted lines on Figure 2(c). Either the perpendicular features as well as the polygonal alike shapes, seems to be randomly localized. Note that the localization of these shapes is easily obtained from the grain discretized mask for the samples M2 and M3 but more difficult for the one representing the sample M1. Another interesting observation was obtained from the 3D representations, shown in Figures 2(g)-(i), for the samples M1, M2 and M3 respectively. Here, it is possible to note that the grain structures were presented in all the specimens; however a larger tilted planes with certain acute angles. This is more clearly appreciated for the samples M2 and M3, as following described.

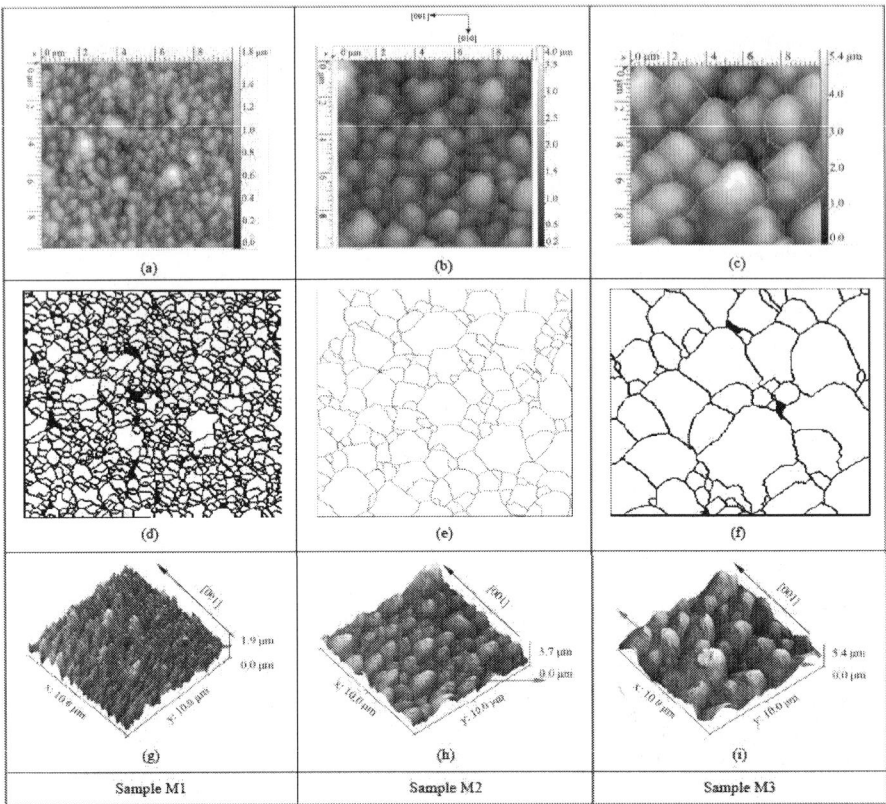

Figure 2. A 10 × 10 μm² AM-AFM topography images for (a) sample M1, (b) sample M2, and (c) sample M3. Their corresponding watershed grain discretization mask for each image is shown in the middle part ((d), (e), (f)). The AFM parameters were: (a) SP = 50%, FA = 195 nm, OF = 169.35 Khz, SS = 0.5 l/s; (b), SP = 65%, FA = 195 nm, OF = 169.27 Khz and SS = 0.6 l/s; (c) SP = 50%, FA = 195 nm, OF = 169.37 Khz and SS = 0.5 l/s; (d)-(f) We show the grain discretization for the samples M1, M2 and M3 respectively, obtained from the corresponding images (a), (b) and (c). The 3D-representations images of (a), (b) and (c) are shown in (g), (h) and (i) respectively. The corresponding [001] directions are pointing towards the left for all the samples.

The topography AFM images presented in Figure 3, shows a zoomed area of 5 × 5 μm² of the three samples. In this part of the specimens one of the larger grains found for each sample, was selected to be measured by a height profile line. These measurements showed these tilted planes (see Figure 3).

Figure 3 shows now a series of height measurements for each topography image corresponding to each sample. In this area, the highest height was 1.84

µm, 3.49 µm and 5.4 µm, for the samples M1, M2 and M3, respectively. The image shown in Figure 3(a) is marked with black dotted lines some of the linear features found on the specimen. In this case, the highest height found is the product of an island, an agglomeration of grains. As the indium quantity increased, these agglomerations leaves to the formation of bigger grains giving rise to the bigger grains presented in Figure 3(b) and Figure 2(c), for the samples M2 and M3 respectively. The profile shown in every AFM image is presented with their corresponding Height vs Distance Graph (lower part). The images of Figure 3(b) and Figure 3(c), upper part, presented the tilted plane more clearly upon subsequent Indium deposition, here marked by the black dotted line triangle. The height profile measured on the biggest grains presented by the AFM images, shown that these tilted planes presented acute angles θ, from the [011] direction, having values of about 53°, 58.5° and 61°, for the samples M1, M2 and M3 respectively. Furthermore, towards this same [011] direction, the top of the clusters seems to be smoothed upon increments of the Indium quantity. While the sample M1 presented a sharper structure on the top, the sample M2 presented a less sharp shape, with a small planar section. On the other hand, the sample M3 seemed to have small terraces separated by certain heights.

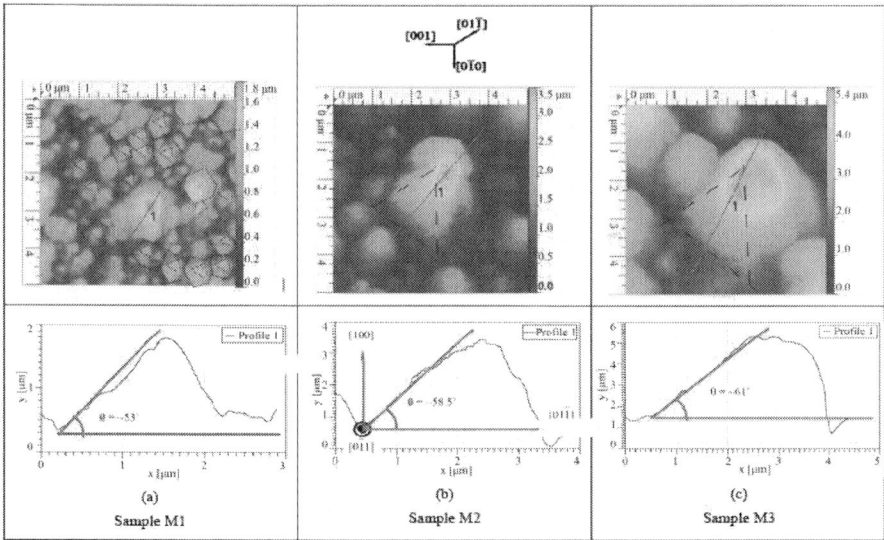

Figure 3. 5 × 5 µm² topography AFM images in warp-spectral color representation. For each sample, the images at the upper part, contains the line profile, marked with number 1. In the lower part we see the height profile for each image. The corresponding [001] direction is pointing to the left side, the [010] goes toward the bottom and the [011] is pointing to the diagonal top-right part of the image.

In the following, the cross section experiments are presented.

Cross Section Imaging: Topography and Phase Micrographs

The following images shows the cross section morphology of the samples, see Figure 4. On top, we can see the optical images with the 100× lent. Below it is shown the corresponding topography AM-AFM image of the marked zone by the green rectangle area in each of the optical images.

The optical images of Figure 4, shows the position of the In/GaAs/In/GaAs film with the red arrow, pointing toward the right side of the image, the [100] direction. The Si substrate is presented at the left side and the black color to the right is the free space after the specimen. Note that the $3 \times 10 \ \mu m^2$ AM-AFM topographic images, preserves the same orientation than that of the optical image. These AFM images with their height profiles are following compared to their corresponding phase images, presented with their corresponding Phase vs Distance Graphs, in Figure 5. A Code-V color representation for the topography and phase AM-AFM images was used. The presence of the In/GaAs/In/GaAs film is appreciated more clearly. The topographic channels are presented in the upper part, Figures 5(a)-(c) for the samples 1, 2 and 3, respectively. Here, the silicon substrate is exhibited with orange, hard-orange or red colors. The GaAs-In film is presented with yellow and/or with green color. Each image has a height profile line, which is shown below. In this particular locality, these height profiles show a film thickness of about 2.16 μm for the sample M1, 3.85 μm for the sample M2, and 6.15 μm for the sample M3. Note that these values are quite close to those observed by the height profiles of the big grains presented in Figure 3. Here the criteria followed to obtain this particular distance, was by selecting the first abrupt change in the phase channel that was followed by a different phase behavior. For the sample M1, Figure 5(a), the height profile starts with a linear tendency from the left (light orange color). Then the line profile presented a small increment until a maximal height value of 6 μm (orange color), that decreased until a height of 5.29 μm at about 7.52 μm of distance. This is what we considered to be the part of the Si substrate. After this the height value started to decrease to about 4.5 μm of height and remained there for a while (corresponding to the yellow con- trast in Figure 5(a)). Then the height decreased linearly (green contrast) until a value close to 0.73 μm of height at 9.68 μm of distance. Finally a constant value remained at this point (the blue contrast) showing that the tip is not any more in contact with the cross section of the specimen. Regarding the phase signal, Figure 5(d), it started with a linear tendency at about 80° (green contrast) starting from the left part of the image. Then the presence of some perturbations to 75° and 83° until a distance of 6.25 μm, followed by a small lobe to 85°, appeared (yellow color). After this, the phase signal decreased to 76° at a distance of 7.5 μm, (green color), and then increments and decrements to 77° and 68° are continued (green, yellow, blue colors) until a clear fallen of the phase to 31° at 9.1 μm of distance (blue color). From here the phase went up to 62° and then to 77° at 9.68 μm of distance, where it remained until the end of the profile part, right side of the image (green color) corresponding to the free oscillation of the cantilever.

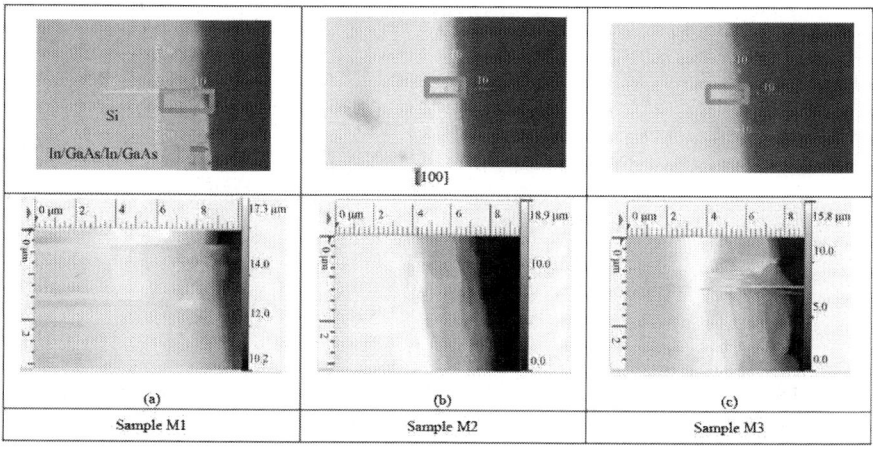

Figure 4. Optical images and the corresponding 3 × 10 μm² AM-AFM topography images of the cross section, for each of the sample. The 100× optical images are presented at the upper part, while the AM-AFM topography images are shown below. In (a) SP = 50%, OF = 169.12 Khz and SS = 0.7 l/s; In (b) SP = 80%, OF = 163.4 Khz and SS = 0.2 l/s; In (c) SP = 50%, OF = 163.27 Khz and SS = 0.2 l/s. The corresponding [100] directions for all the samples are indicated on top of the column (b).

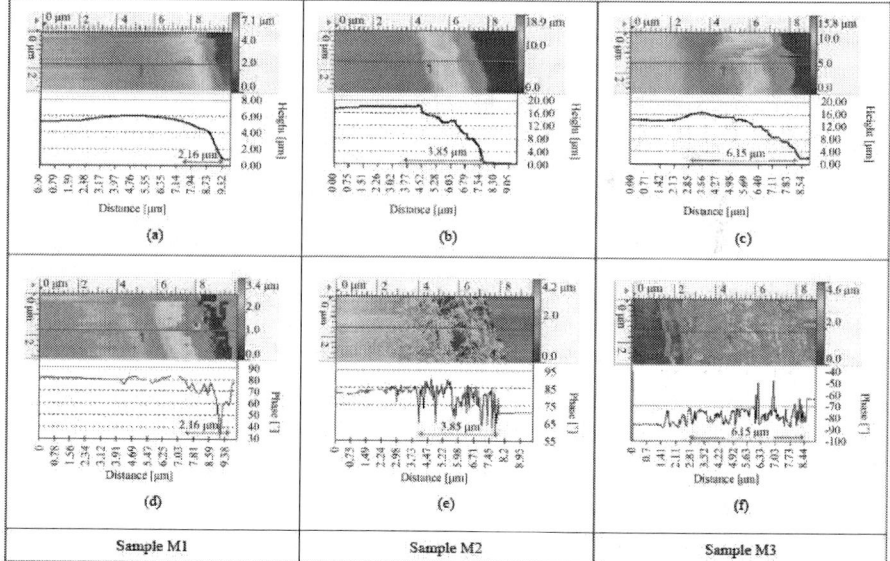

Figure 5. 3 × 10 μm² AM-AFM topography and phase images of the cross section, for each of the samples, in Code-V color representations. They are presented with their corresponding height profiles and their phase shifts vs distance curves.

For the sample M2, Figure 5(b), a similar feature is presented in the height profile measurement, but not with the same increment rate as it was presented in sample 1. Starting from the left side, a more straight linear tendency at about 17.82 μm is presented until a distance of about 4.06 μm (all the hard-orange/red color). This part is what we attributed to be the Si substrate. A small lobe is presented just before the height started to decrease. Then the profile is decreased (yellow contrast) until the profile goes to a value of about 13 μm of height and remained there for a while (green contrast). Then the height is decreased until a 0.44 μm of height is reached, at 7.91 μm of distance, where the signal remained until the end of the profile at the right part of the image (all the blue color). Here is where the cantilever is in the free space position. The phase presented values near 81°, corresponding to the orange and red contrasts, left side of Figure 5(e), and then the phase increased gradually until a close value of 86°, around a distance of 2.98 μm. Here it is presented a series of jumps near 85° until a fast fallen to 64.95° at 4.06 μm of distance. After this, increments to values close to 85° with intermittent jumps to 90° were presented (red, blue, green and orange zone) until a distance of 5.6 μm, where a noticeable decrease in the phase is observed. Here the phase presented different fallen and jumps to 75° and up to 85° (corresponding to the blue, green and yellow colors) and lasted until a distance of 7 μm. Then the phase presented jumps from 65° to 85° and 80°. These phase oscillations were attenuated until a close value to 70.23° at 7.9 μm of distance, where the attenuated phase kept a constant value (blue contrast), presented at the right part of the image. This is the part where the cantilever is not any more interacting with the specimen cross section.

Finally, for the sample 3, starting from the left part of the image, the height profile presented a constant height value, at about 13.8 μm, which lasted to a distance of 2.13 μm. Then an increment to about 14.68 μm of height started at about 2.46 μm of distance (orange and red color). This is what we considered to be the Si substrate. The profile remained for a while around 16 μm and then decreased with a logarithmic alike behavior (yellow and green colors) for all over a distance of 5.49 μm and until a value near 1.75 μm of height, at 8.52 μm of distance. From here a constant value remained all over the right side of the image (blue contrast). Regarding the phase signal of the sample 3, the image started to the left part (blue contrast), corresponding to values near −85°, Figure 5(f). At 1.41 μm of distance, a pick to −75° was observed, then returning to −85°, followed by a decrement to around −90° at 2.11 μm of distance. Two picks are then presented, one to −7.7° at 2.46 μm and the other to −71° at 2.7 μm (green and blue colors). After that, values around −75° with intermittent fallen to −80° were presented until a distance of about 6 μm (green, blue, yellow colors). Then a sharp pick to −45° with a subsequent fallen to −80° followed by another pick to −50° and a consecutive fallen to −75° were presented. The phase went down to −85° with jumps to −70° and fallen phases to −80° (green, yellow, blue, hard-orange colors) until a constant value of around −63.61° at 8.61 μm of distance. At this phase value the signal was maintained until the end of the profile, at the right part of the image (hard-orange color).

To conclude with this section, the phase signal presented a different value in the GaAs/In/GaAs/In/GaAs film compared with the silicon substrate. It is important to note that the change in phase is presented in conjunction with a change in topography; however, even though the change in topography decreased with a logarithm alike behavior, the phase shift did not. Then after the existence of the cross section, were the tip was oscillating in the free space, the phase shift was not $90°$ as it was originally when far away from the surface.

Regarding the AFM study, from the watershed algorithm of the Gwyddion software, we could obtain the discretization of the grains and the corresponding mask, containing the grain boundaries for each of the three samples studied, Figures 2(d)-(f). From these masks, and with the use of the Maximum Height statistical quantities obtained from each of the AFM images, we could observe that the grain lateral dimensions and heights, as well as the film averaged thickness, were increased upon larger Indium deposition times. This observation is consistent with the Mean Grain Size and total gran volume quantities extracted from the grain statistics see Table 1. This suggests that there exists a growing tendency produced by the number of Indium layers deposited and the grain size. The AFM images presented in Figure 2, showed the following behavior. Figure 2(a) presented a variety of very small grains that in certain zones they formed agglomerations (islands), that exhibited higher heights and greater lateral dimensions. These agglomerations were not homogeneous around the $10 \times 10 \ \mu m^2$ surface. Instead, they apparently seem to be randomly arranged. By subsequent Indium deposition, theses agglomerations collapsed and formed bigger grains, leaving the new smaller grains to accumulate around the bigger ones, see Figure 2(b). These bigger grains started to present a large tilted plane towards the [011] direction, with acute angles that increased from $53°$, $58.5°$ and $61°$ while the Indium quantity increased. Also the plane in front of this tilted plane tended to form right angles from the [011] horizontal line and the top of the grains became flatter. This suggest a pure three dimensional growth with a high tendency of expansion towards the [100] direction. This three dimensional growth comes from the lattice parameters mismatches presented by the substrates and the alloys.

Regarding the statistical quantities and grain statistics, we obtained the Table 1 and relations.

Table 1. Comparative table of the statistical quantities and grain statistics between samples.

Sample	Total time of GaAs deposition (min)	Total time of In deposition (min)	Statistical quantities		Grain statistics		
			Maximum Height (μm)	RMS roughness (μm)	Grain Population Density (μm^{-2})	Mean Grain Size (μm)	Total Grain Volume (μm^3)
M1	60	20	1.79	0.24	11.2	1.18	48
M2	60	30	3.78	0.51	1.71	0.63	137
M3	60	40	5.44	1.01	0.73	0.82	227

The increments of the In deposition time, starting from a total of 20 minutes for the sample M1, was scaledfrom this value by a factor of 2, corresponding to the 30 minutes for the sample M2, and by a factor of 3, corre- sponding to the 40 minutes for the sample M3. These scaled factors are similar to the increasing proportions found in the Maximum Height. On the other hand, the RMS roughness values measurements, presented the following features. A scale factor of 2.16 was found from the 0.237 of the sample 1, to a value of 0.512 μm of the sample 2, and a factor of 4.27 μm to 1.01 μm of the sample 3. The Grain Population Density was found to be scaled from 11.2 of the sample 1 to 1.71 of the sample 2, by a factor of 0.15, and to 0.73 of the sample 3, by a scale factor of 0.07. The Mean Grain Size was found to be scaled from 0.18 of the sample 1 by scale factors of 3.44 and 4.48 to 0.63 μm and 0.820 μm of the sample 2 and the sample 3, respectively. Finally, the Total Grain Volume was found to be scaled by 2.85 and 4.73 from 48 μm^3 of the sample 1 to 132 μm^3 of the sample 2 and 227 μm^3 of the sample 3. The height related features were found to be scaled by 2 and 3 scale factors, as well as the symmetrical factors found by the RMS roughness and Grain Population Density. However, the Grain Population Density as well as the Total Grain Volume presented asymmetrical scaled factors. From the sample 1, the Grain Population Density presented a value of 11.2 μm^{-2} and by increasing the double In quantity for the sample 2 fabrication, the Grain Population Density decreased to 1.71 μm^{-2}. Meaning that around this increment of Indium quantity a large number of clusters were collapsed into Islands and then the islands into grains. This was not the case when passing to the fifteen minutes double cycle for the fabrication of the sample M3, where the Grain Population Density was 0.73

Regarding the phases analysis, for the moment only qualitative information can be given, see Figure 6.

As shown in Figure 1, for the sample 1, this particular profile line scan speed was made at 0.7 l/s, which is faster than those speeds of 0.2 l/s used for imaging the sample 2 and sample 3. However, while the topography signal started to decrease in a logarithmic form, the phase signal presented a linear tendency composed of jumps and falls. The topography effect observed on the Phase signal was, when the oscillating tip found an abrupt fall in the topography, a fallen for the sample 1 and 2 and a jump for the sample 3. This falls and jumps are indicated by the dotted green lines in the Figures 6(a)-(c). The overall behavior of the phase signal did not show any logarithmic or exponential tendency. For example in the sample 2, when the topography signal decreased after the small lobe, at around 4.5 μm of distance, the phase fallen form a while. Then it seems to be oscillating until a more stable signal was obtained. When the topography went down to 13 μm of height, the phase suffered also a fallen. Then it came back to around 70° and even thought the topography profile, started to decrease the phase signal presented values around 70°, with intermittent falls and jumps. No exponential or logarithmic behavior was exhibited by the phase. This behavior is still more presented for the sample 3, Figure 6(c). Here, the topography profile clearly decreased logarithmically; while the phase seemed to presents a repetitive value around −75°. Again when the topography signal presented a suddenly fall, at around 6.11 µm, the phase presented a big pick to around −50°, then came back to values close the −75°. Another abrupt fall in the topography signal at 6.90 µm was presented, and again, the phase presented a big jump to around −45°, and then returned to the values (a) (b) (c)Sample M1 Sample M2 Sample M3 between −80° and −70°.

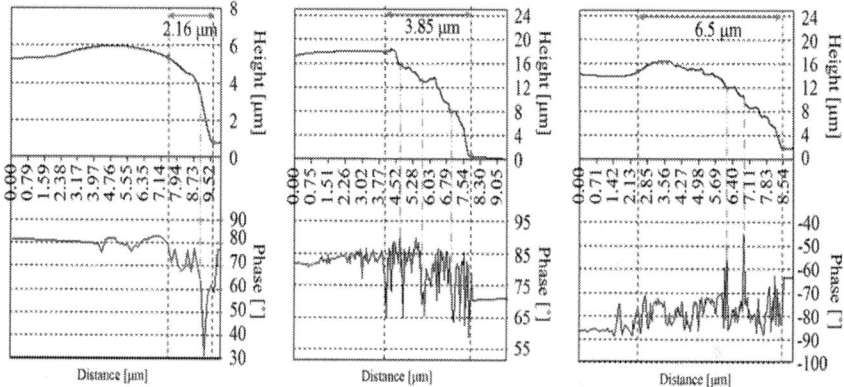

Figure 6. Qualitative comparative analysis of the topographic and phase signals. In (a) the scanning speed (SS) of the AFM image for the sample M1, was 0.7 l/s; In (b) SS = 0.2 l/s for the sample M2 and in (c) SS = 0.2 l/s, for the sample M3.

Finally, regarding the corners shapes found in the sample 1, the hexagonal shapes found in the sample 2 and sample 3 as well as the grain three dimensional structures, the growth process of In-GaAs on top of clean Si (100) substrates probably leads to the formation structures with some order that runs in certain crystallographic orientations. A subsequent analysis by electron microscopy can give us new insides about these observations as well as a detailed study of the initial growth of GaAs on Si.

4. CONCLUSIONS

Three different samples containing In/GaAs films on Si (100) substrates with different Indium quantity were successfully fabricated by RF Magnetron Sputtering. The samples were successfully characterized by AM- AFM.

The AFM was implemented in order to obtain the morphology of the surface and of the inner structure features for each of the samples. For a total time of 20 minutes of In deposition, small grains were formed. Corner alike shapes were found, and some agglomerations of grains were presented. This initial formation presented a certain growth tendency towards the [111] direction and a tilted plane towards the [011] direction could be identified. For a total of 40 minutes, this agglomerations formed bigger grains, and for a total of 30 minutes of In deposition these grains presented larger tilted planes and a flatter structure on top.

Thanks to the statistical quantities obtained, we could observe that the samples reported Maximum Heights of 1.79, 3.77 and 5.44 μm, presenting RMS roughness of 0.237, 0.512 and 1.01 μm for the sample M1, M2 and M3, respectively. Thanks to the implementation of the watershed algorithm, these grains could be discretized. From where the grain population densities of 11.2 μm^{-2}, 1.71 μm^{-2} and 0.73 μm^{-2} where obtained. The Mean Grain Size, varied from 183 nm to 630 nm and to 820 nm, and the Total Grain Volume varied from 48 μm^{-3} to 137 μm^{-3}, and then to 227 μm^{-3}. These grains formed some shapes that could be identified. The corners shapes found in the sample 1, the hexagonal shapes found in the sample 2 and polygons found in the sample 3 as well as the grain three dimensional structures, leads to the supposition that the growth process of In-GaAs on top of clean Si (100) substrates probably give rise to the formation of structures with some order that runs in certain crystallographic orientations. A subsequent analysis by electron microscopy can give us new insides about these observations as well as a detailed study of the initial growth of GaAs on Si.

Finally, the qualitative analysis observed here, the cross section phase imaging presented different phases shift for each sample, comparing the Si (100) and the In/GaAs film. The more notable result was that while the topography profile decreased logarithmically, the tendency of the phase signal did not present this behavior. The sudden falls on the topography signal, produced a sharp pick in the phase signal. As an idea, it is clear that Silicon, Gallium Arsenide and Indium Arsenide have different mechanical properties, like different harness values. It would be very interesting to try to relay any mechanical properties of the alloys and substrates compositions with the phase signal evolution. But for the moment we are not able to relate anmechanical properties with the phase shifts.

ACKNOWLEDGEMENTS

We are very much appreciated with the "Instituto de Ciencia y tecnologíadel Distrito Federal, (IcyTDF)" now the "Secretaría de Ciencia, Tecnología e Innovación del Distrito Federal" (SECITI), for the financial support. We address our most sincerely gratefulness to Prof. Jośe Trujillo and Marving Soriano. A. Pulzara Mora thanks to Secretaría de Ciencia, Tecnología e Innovación del Distrito Federal (CECITI-DF) and Centro Latinoameri- cano de Física (CLAF) for financial support.

REFERENCES

1. Borri, P., Schneider, S., Langbein, W. and Bimberg, D. (2006) Ultrafast Carrier Dynamics in InGaAs Quantum Dot Materials and Devices. Journal of Optics A: Pure and Applied Optics, 84, S33-S46.

2. Zilkie, A.J., Meier, J., Mojahedi, M., Poole, P.J., et al. (2007) Carrier Dynamics of Quantum-Dot, Quantum-Dash, and Quantum-Well Semiconductor Optical Amplifiers Operating at 1.55 mm. IEEE Journal of Quantum Electronics, 4311, 1873-1880.

3. Franta, D., Ohlídal, I., Klapetek, P., Montaigne-Ramil, A., Bonanni, A., et al. (2002) Influence of Overlayers on Deter- mination of the Optical Constants of Zn Sethin Films. Journal of Applied Physics, 92, 1873-1880.

4. Klapetek, P., Ohlídal, I., Franta, D., Montaigne-Ramil, A., Bonanni, A., et al. (2003) Atomic Force Microscopy Characterization of ZnTe Epitaxial Films. Acta Physical Slovaca, 53, 223-230.

5. Jenkins, C., Westwood, D.I., Elliott, M., Macdonald, J.E., Meaton, C., et al. (2001) Metrology of Semiconductor Device Structures by Cross-Sectional AFM. Materials Science and Engineering, 80, 138-141.

6. Fasching, G., Schrey, F.F., Roch, T., Andrews, A.M., Brezna, W., et al. (2006) Single InAs/GaAs Quantum Dots: Pho- tocurrent and Cross-Sectional AFM Analysis. Physica E: Low-Dimensional Systems and Nanostructures, 32, 183-186.

7. Noy, A., Sanders, C.H., Vezenov, D.V., Wong, S.S. and Lieber, C.M. (1998) Chemically-Sensitive Imaging in Tapping Mode by Chemical Force Microscopy: Relationship between Phase Lag and Adhesion. Langmuir, 14, 1508-1511.

8. García, R., Tamayo, J., Calleja, M. and García, F. (1998) Phase Contrast in Tapping-Mode Scanning Force Microscopy. Applied Physics A, 66, S309-S312.

9. Xu, W., Wood-Adams, P.M. and Robertson, C.G. (2006) Measuring Local Viscoelastic Properties of Complex Materials with Tapping Mode Atomic Force Microscopy. Polymer, 47, 4798-4810.

10. Magonov, S.N., Elings, V. and Whangbo, M.H. (1997) Phase Imaging and Stiffness in Tapping-Mode Atomic Force Microscopy. Surface Science, 375, L385-L391.

11. Horcas, I., Fernandez, R., Gomez-Rodriguez, J.M., Colchero, J., Gomez-Herrero, J., et al. (2007) WSXM: A Software for Scanning Probe Microscopy and a Tool for Nanotechnology. Review of Scientific Instruments, 78, Article ID: 013705.

12. http://nanoscaleworld.bruker-axs.com/nanoscaleworld/media/p/2740.aspx

CHAPTER 2

Atomic Force Microscopy Observations of the Polymer Network Structure Formed in Ferroelectric Liquid Crystals Cells

M. Petit

[1] Université 20 Août 1955-Skikda,, Algérie

1. INTRODUCTION

Polymer network stabilized liquid crystals (PSLCs) have attracted increasing interest over the past decade because of their potential applications mainly in electro-optic devices such as displays and light shutters [1-5]. The main motivation to incorporate a polymer network in liquid crystal cells was to bulk-stabilize a desired director configuration against any mechanical shock and distortions which can irreversibly alter the functionality of the cells. The PSLCs are composite materials in which a low density polymer network is dispersed within liquid crystal medium [6,7]. The polymer network is formed by chemical crosslinking of a small amount (few percent) of photo-reactive monomers dissolved in low molecular weight mesogenic material, through a polymerization reaction photochemically activated by a UV illumination. When the polymerization occurs in an aligned geometry, the resulting polymer network is roughly aligned parallel to the direction initially imposed by the liquid crystal medium in which the network has been formed [8]. Depending on the type of the reactive mesogen, the morphology of the polymer network may correspond to an open structure consisting of anisotropic fibrils [8-10]. The lateral size of fibrils is of the order of a few tenths of a micron [11-13]; their density increases with the initial reactive monomer concentration. The polymer fibrils, by creating a large internal boundary, provide a bulk anchoring mechanism which allows a control of the liquid crystal alignment in the bulk. Application of an electric field causes a distortion of the liquid crystal host, which corresponds to a field-induced director rotation, without any reorientation of the fibrils [8] considered rigid by the authors because the network is heavily cross-linked. The structure of the polymer network has been observed using different techniques. The scanning electron microscope (SEM) observations, observations between crossed polarizer after removal of the un-reacted species and observations at high temperature, all provide information on the network structure. The SEM and crossed polarizers observations give a detailed of the lateral size of fibrils [11-13] and the width of the fibers. Most of these observations have been reported in the nematic [14] or cholesteric [15] liquid crystals phases. However, a few information has been reported in the literature about the vertical distance between fibers especially when the polymer network was formed in a ordering helical ferroelectric liquid crystals (SmC*) and paraelectric SmA phases (PSFLC). The PSFLC composites have been investigated by different methods, electro optic technical [16-22] and dielectric method [23-26]. It is know that the dielectric properties of the ferroelectric liquid crystals in the ideal unbounded sample are nowadays well understood. The dielectric dispersion is essentially dominated by the dynamic of the soft mode, responsible for the phase transition from the paraelectric phase SmA to the ferroelectric phase SmC* phase, and the Goldstone mode connected with the helical structure in the ferroelectric SmC* phase [27]. The subject is already compiled in monographs [28,29].

Detailed dielectric spectroscopic study over a wide temperature and frequency ranges in the PSFLC systems reveal different molecular dynamics of this type of composites [23-25]. It was reported [23-25], that the dielectric strength of the Goldstone mode decreases with increasing polymer concentration, however, the relaxation frequency was found to be higher for the PSFLC composite film compared to that of the corresponding pure FLC. However, the behavior of the soft mode dielectric strength is not completely explained yet. Kundu et al. [25] are showed for polymer stabilized ferroelectric liquid crystal (PSFLC) systems that the soft mode dielectric strength remains unchanged when the FLC cells are stabilized by a polymer network formed from a nonmesogenic reactive monomer. The most of these studies have not given a quantitative interpretation of the dielectric response until our work published in [26].

In our previous study [26], although the range of the SmA phase is very narrow, the effect of the polymer network on the soft mode is clearly observed. By increasing the polymer concentration the dielectric strength is reduced and the relaxation frequency is increased [26].

It is knows that, in a thin enough sample when the sample thickness becomes comparable with the pitch of the helix, the entire structure in helix-free. The director twist-bend fixed by the surface polar anchoring can still exist as well as in the helicoidal samples. This spaced modulation is called a twisted structure [30]. However when the twist helical pitch is confining with the polymer network in a thin thickness, logically there is no reason that the twisted structure still remains in spite of the confinement effects. In this chapter we will illustrate that in spite of the confinement of the twist short helical structure in a polymer network the ferroelectric liquid crystal conserves the helical structure. This behavior was interpreted by the type of the morphology of the polymer network which is formed in the liquid crystals medium. After describing the dielectric responses, in sect.3-2-A we will describe in detail the effect of the polymer network density formed in a planar alignment of the ferroelectric liquid crystals cells [30] on the dielectric responses namely soft and Goldstone mode. The results of the dielectric responses were interpreted by the morphology of the polymer network using atomic force microscopy investigations. In sec.3-2-B the effect of the applied electric field during the photopolymerisation on the dielectric responses is reported, the observations of the structure of the polymer network by AFM agree well with all dielectric responses.

2. EXPERIMENTS

The liquid crystal compound used in these studies is the ROLIC 8823 (Rolic research ltd) which exhibits the following phase sequence: Crystal (Cr) -27 °C SmC* 63.5 °C SmA 65°C Isotropic (Figure 1). In the SmC* phase (Figure 2), at low temperature the helical pitch is about 0.3μm and the spontaneous polarization (P_s is close to 100 nC/cm²). To prepare the polymer network we have used a photoreactive diacrylate mesogen as photocurable monomer which presents a nematic (N) phase between Cr and Isotropic phases Cr−88 °C−N−118 °C− Isotropic. The chemical structure of the monomer is presented in figure 3.

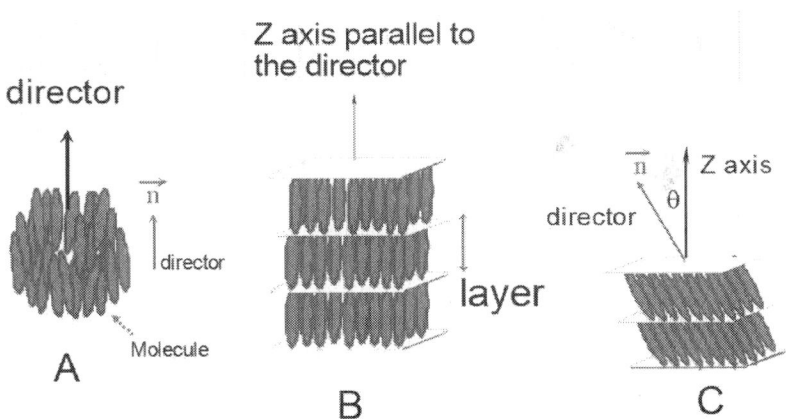

Figure 1. A schematic representation of the nematic phase (A), Paraelectric phase or smectic A (B) and Smectic C (C).

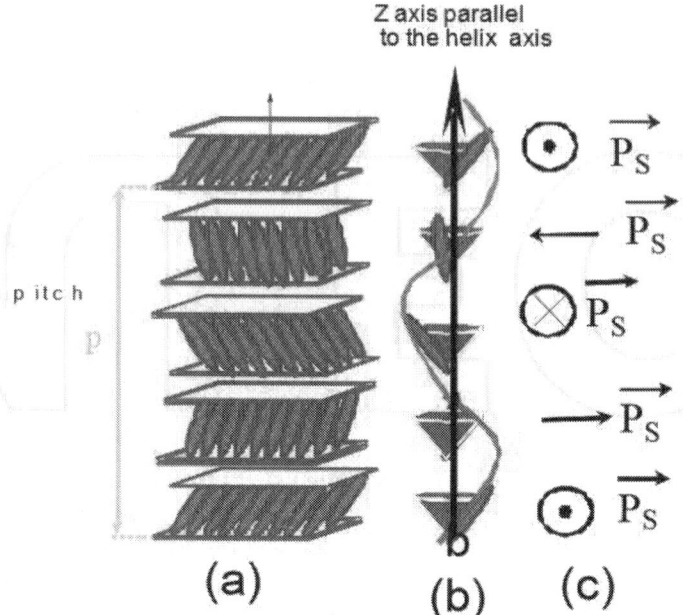

Z axis parallel
to the helix axis

pitch

p

(a)   (b)   (c)

$\vec{P_S}$

$\vec{P_S}$

$\vec{P_S}$

$\vec{P_S}$

$\vec{P_S}$

Figure 2. A schematic representation of a smectic C* phase : molecular order in the layer (a), helical structure (b) and the direction of the ferroelectric spontaneous polarization in the smectic layers planes (c).

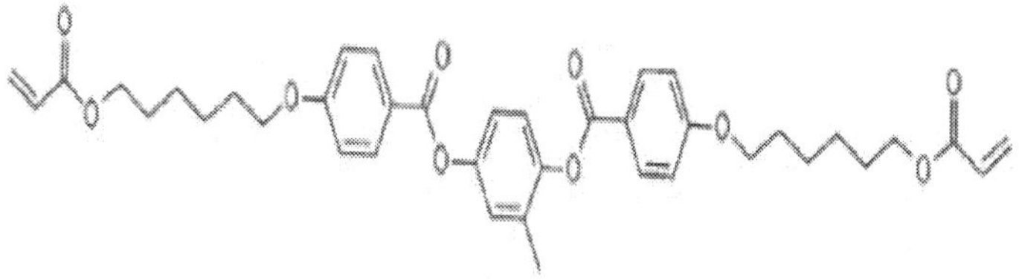

Figure 3. The molecular structure of the monomer.

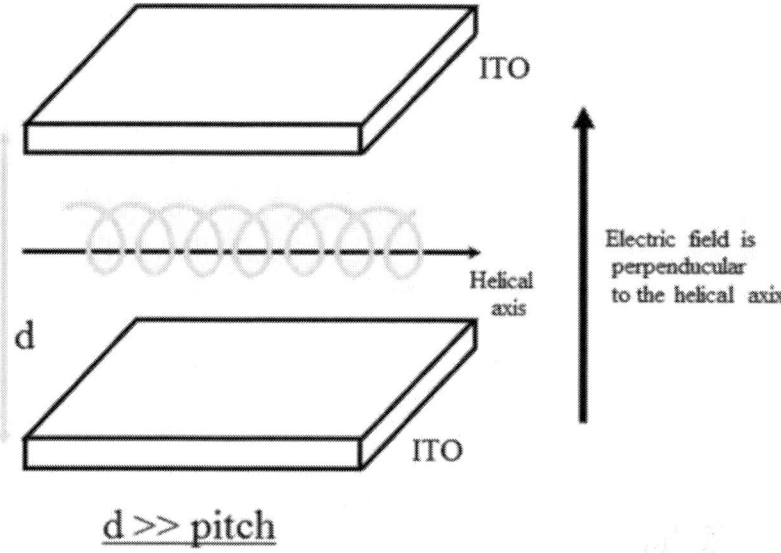

Figure 4. The Planar configuration of the helical structure in our system, the electric field is applied perpendicular to the helical axis.

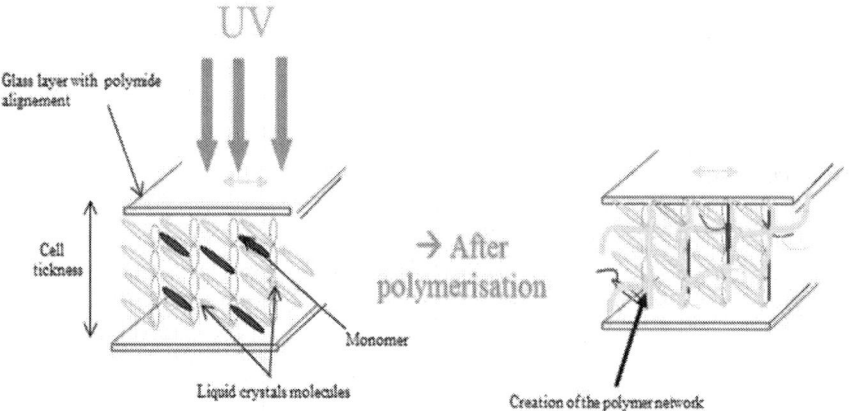

Figure 5. A schematic illustration the formation of the polymer network, before de photopolymerisation (on the left) and after the pohopolymerisation (on the right).

The PSFLC mixture was prepared by mixing the Diacrylate monomer with weight concentrations between 2 and 7%. The ferroelectric liquid crystals (FLC) compound and the diacrylate were dissolved in the isotropic phase to make a homogeneous mixture. A 5 μm thick EHC Inc, Japon-cell (two glass faces were treated with polyimide to favorite a planar alignment (Figure 4)) was filled by the mixture in its isotropic phase. In order to obtain a good alignment in the SmC* phase, the cell was slowly cooled (0.1 °C/minute) from the isotropic phase under an applied electric field (5V/μm) into the SmC* phase. For the section 3.2.A, the sample cells were then exposed to ultraviolet light (wavelength = 365nm) at 25 °C with an intensity of 5 (mW/cm²) for 30 minutes without any applied electric field. Here, polymer phase separation and network formation take place (Figure 5). During these studies, the cell was placed on a hot stage (Linkam TMS 93) for temperature control. The texture observations of the cells were carried out by means of a polarized optical microscope (POM)(LEICA DMRXP).

Dielectric measurements were performed in the frequency range of 10 Hz– 13 MHz HP 4192A. In a linear dielectric response, the time dependent polarization *P(t)* of the sample, being induced by a weak measuring electric field *E(t),* is proportional to the field :

$$P(t)= \varepsilon_0 \, (\varepsilon^*-1) \, E(t) \quad \text{where } E(t)=E_0 \exp(-i \omega t)$$

Is a sinusoidal electric field applied to the dielectric under test and * is the complex dielectric permittivity, is the angular frequency, and f is the frequency of measuring electric field displayed in the channel A of most impedance analyzers, and $_0$ is the dielectric permittivity of free space. In linear dielectric spectroscopy the amplitude of the measuring electric field should be chosen so that it does not suppress the helicoidal structure of the SmC* phase. Generally for ferroelectric liquid crystals two processes contribute to the dielectric spectra. We are interesting here only to the collective processes namely soft and Goldstone modes.

In order to obtain the characteristic dielectric strengths and relaxation frequencies of the ferroelectric relaxation modes, the dielectric spectra were fitted simultaneously by the Cole-Cole function:

$$\varepsilon^* = \varepsilon_\infty + (\Delta \varepsilon_G) \, / \, (1+jf/f_G)^{1-\alpha_G} + (\Delta \varepsilon_s) \, / \, (1+jf/f_s)^{1-\alpha_s} + (\sigma/j2\pi f \varepsilon_0)$$

Where *f* is the frequency, is the high frequency limit of the dielectric permittivity, $_G$ and $_s$ represent the dielectric strengths corresponding to Goldstone and soft modes, respectively; f_G and f_s represent the relaxation frequencies of the two modes, $_G$ and $_s$ are the distribution parameters, and is the electric conductivity. The temperature dependencies of different dielectric processes are reported and discussed below. To image the topography of polymer networks a Veeco Multimode Atomic Force Microscopy (AFM) equipped with a Nanoscope IIIa controller was used. All AFM scans were taken in tapping mode with commercially available tips made of Phosphorus doped Silicon.

3. RESULTS AND DISCUSSION

3.1. Optical observations

The first objective of this study was to investigate the effect of the applied electric field and phase order (before the polymerization) on the alignment on the polymer network formed in the FLC host. To illustrates the effect of the applied electric field during the cooling from the isotropic phase to the SmC* phase on the alignment of the SmC* layers we present on Figure 6.

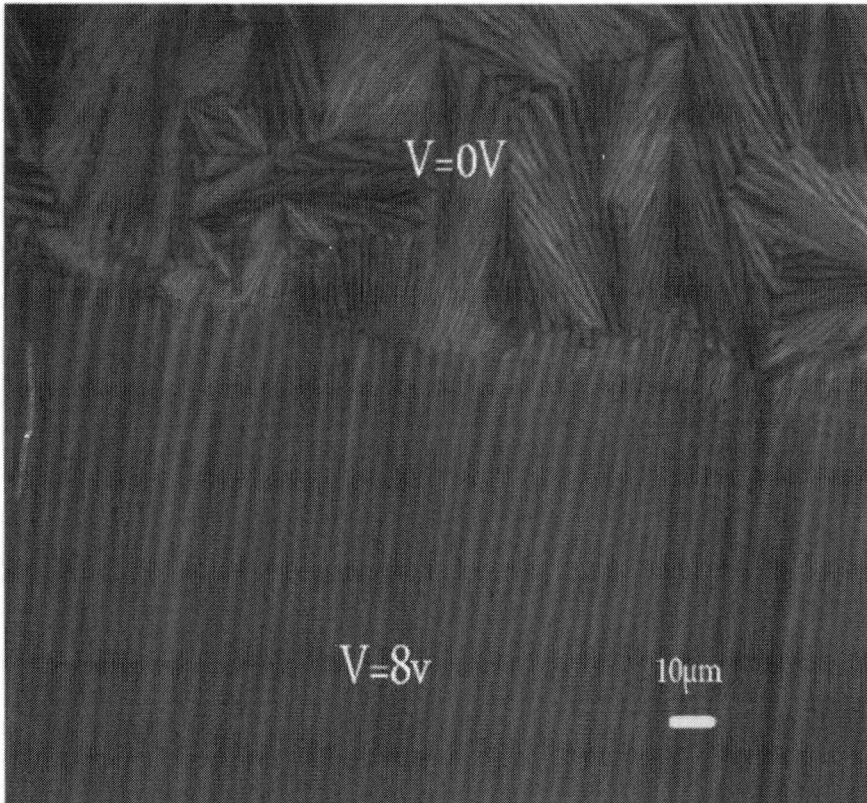

Figure 6. Optical micrographs of PSFLC samples obtained between crossed polarizers at T = 70 ∘C, for 7% polymer concentrations formed at 25°C. (V=0V) indicate the region where no field was applied and (V=8 v) indicate the region where an electric field was applied.

Figure 6 shows example of optical micrographs representing the observed textures of the PSFLC cells obtained for 7% initial monomer concentration. These micrographs were obtained at a temperature of 70 ° C (above the SmA-Isotrope transition temperature) so that only the birefringence associated to the polymer network would appear. The micrographs of Figure 6 clearly show the anisotropic structure of the polymer network. The optical observation of this anisotropic structure is due not only to the residual birefringence of polymer fibrils, but also to the remaining birefringence of the surrounding FLC molecules which are still aligned by the polymer structure. Two regions have been observed (Figure 6), the first region when no electric field has been applied (V=0V), the polymer network was randomly distributed. However, in the region where an electric field is applied (V=8V) the polymer network presents a good alignment. To illustrate the effect of the phase order on the formation of the polymer network, two cells of 7% polymer concentration were polymerized in two different temperatures. The first cell was polymerized at 25°C. The measured helical pitch at 25°C is about 0.25 µm (Figure 7). However, the second cell was polymerized at high temperature (T=58°C) where the helical pitch diverges (Figure 7). Figure 8 shows the optical micrographs representing the PSFLC cells obtained for 7% polymer concentration which are polymerized at two different temperatures. As seen in this figure, the dechiralisation lines [31] have been clearly observed on the structure of the polymer network (dechiralisation lines are perpendicular to the polymer fibers (Figure 8 (a))).

We can remember here, that the dechiralisation lines are well known lines defects can be observed in a planar samples filled by a highly twisted FLC [30-32]. In a smectic C* sample, competition between a strong surface anchoring and a helicoidal configuration in the bulk induces a double lattice of singular lines. These lines are located near both the boundary surfaces and are parallel to the plane of the layers [30-32]. Those lines are named the dechiralisation lines. Hence, these results illustrate that the polymer structure conserves locally a lines defect print. We can also indicate here that the fact that the helical structure of the FLC at T=25°C is very shorter no twisted structure on the polymer fiber has been observed by polarizing microscopy. However, when the polymer network was formed at high temperature, the helical structure of the polymer fibers is easy observed by polarizer microscopy (Figure 8 (b)). One can seen that, at high temperature, the dechiralisation lines were not printed on the polymer fibers. That illustrate that the order and structure of the liquid crystal phase are transferred onto the polymer network. It has been already shown by Archer et al [33] that the defect of the FLC twisted grain boundary were found printed on the polymer network structure. However, for short-pitch ferroelectric liquid crystals, the transfer of the lines defects has not been observed. Figure 8 shows also that the polymer network which was formed at high temperature, the dechiralisation lines were not printed on the polymer fibers. Because, when approaching the phase transition SmC*- SmA, the helical pitch diverges. The divergence of the helical pitch (unwinding of the helix) is seen as disappearance of dechiralisation lines [31].

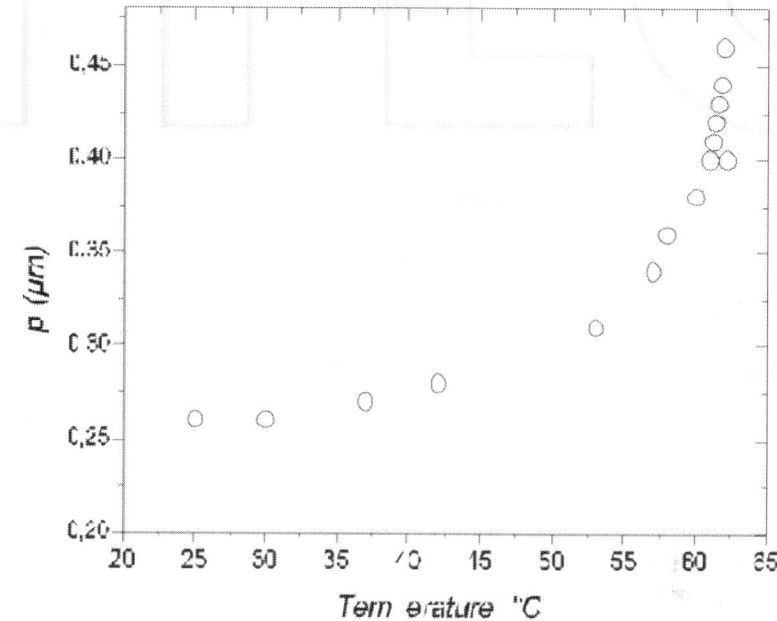

Figure 7. Temperature dependence of the helical pitch of the FLC measured by mean of Grandjean-Cano method [32].

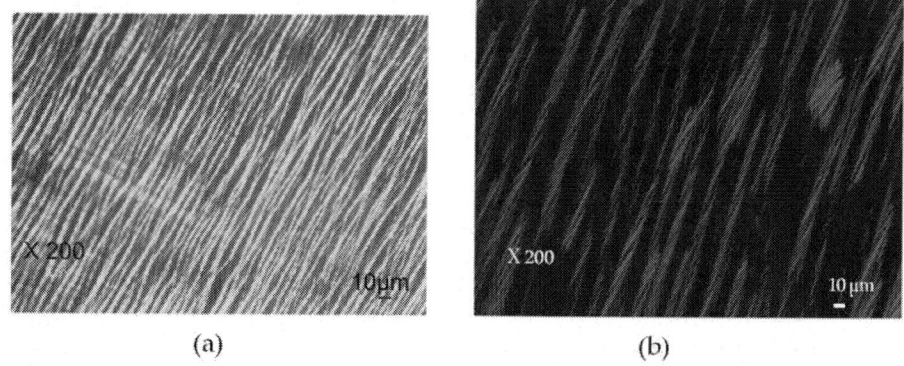

(a)   (b)

Figure 8. Optical microscopy images between crossed polarizers at T = 70 ∘C, for 7% polymer concentration. (a) polymerization at 25°C (dechiralisation lines are clearly observed and correspond to the straight lines which are perpendicular to the polymer fibers). (b) polymerization at 58°C (the polymer fibers present a helical structure).

This implies that even after polymerization, the characteristic property of the host phase in which polymerization was carried out was effectively retained.

3.2. Dielectric studies

3.2.1. Effect of the polymer network density on the dielectric responses

Goldstone. mode of the SmC*

Figures 9 (a) and (b) show examples of the dispersion, '(f) and absorption, "(f) dielectric spectra obtained in the SmC* phase at low temperatures for different polymer concentrations. For all of the concentrations studied, two relaxation mechanisms were detected. The first, at low frequencies between (1 and 3 kHz) with a high amplitude, is due to the Goldstone mode; whereas the second observed at high frequencies (> 1 MHz) with a weak amplitude is an artifact due to the indium tin oxide (ITO) conducting layers. As shown in Fig. 9(a), at low frequencies, the dielectric response shows a very strong polymer concentration dependence; at 100 Hz, for example, '(f) decreases from 100 to 30 when the polymer concentration increases from 0% to 7%.

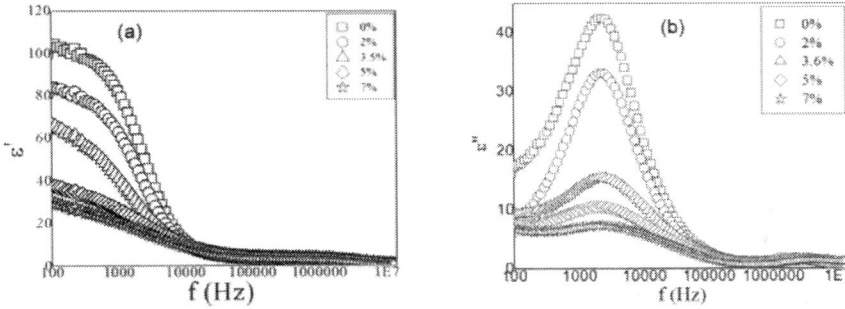

Figure 9. Frequency dependence of (a) the real and (b) the imaginary parts of complex permittivity in the smectic C* phase for different poclymer concentrations at T=25 °C.

This effect is also clearly illustrated in the behavior of the absorption peak observed in the "(f) spectra [Fig. 9 (b)]; the absorption peak strongly decreases from 48 to 8 when the polymer concentration is varied from 0% to 7%. The parameters G and f_G obtained from the curve-fit procedure are displayed in Figs. 10 (a) and (b). The behavior of G versus temperature showed the same general features for all the samples Fig.10 (a) ; G slightly increases to reach a maximum at a temperature called T_{max} 3 °C below T_c then decreases abruptly above T_{max}. The behavior of G versus temperature is dependent on that of the helical pitch of the FLC Fig. 7. Usually the maximum observed in G (T) is related to that exhibited by the helical pitch Fig. 7 at temperatures close to T_c and indicates that the helical structure of the FLC is preserved in all our PSFLC systems.

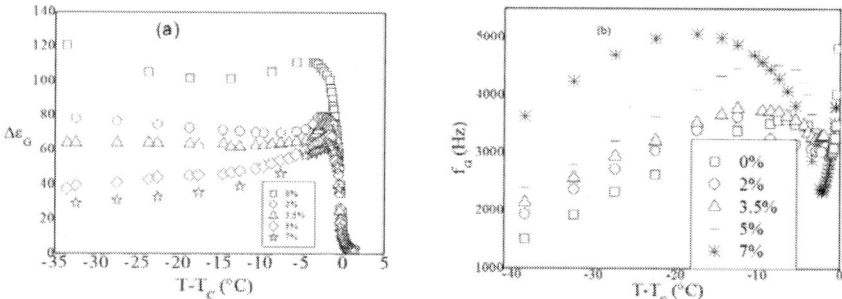

Figure 10. Temperature dependence (a) of ε_G and (b) the relaxation frequency of the Goldstone mode for different polymer concentrations.

The temperature dependence of the Goldstone relaxation frequency Fig. 10 (b) shows that f_G slightly increases with temperature, reaches a maximum, and then rapidly decreases to a minimum value at a temperature corresponding to T_{max}, after T_{max}, an abrupt increase in f_G is observed for a temperature close to T_c. Qualitatively, the thermal behavior of the Goldstone mode is not affected by the polymer network. However, quantitative differences were observed for ε_G and f_G when the polymer network density increases. To illustrate this effect, we present on Figs. 11 (a) and 11 (b) the evolution at room temperature of ε_G and f_G as a function of the polymer concentration. It can be seen from these figures that the increase in the polymer concentration from 0% to 7% leads to a breakdown of ε_G from 120 to 27 and to an increase in f_G from 1.5 to 3.5 kHz. Changes in the dynamic of the Goldstone mode have already been observed in PSFLC systems by Gasser et al. [35] and in other composite based FLC, as a random network formed from dispersions of aerosil particles within FLC media [36–38]. For aerosil/FLC composites, a decrease in ε_G and a shift of f_G toward high frequencies [37,38] with increasing the density of aerosil particles were observed.

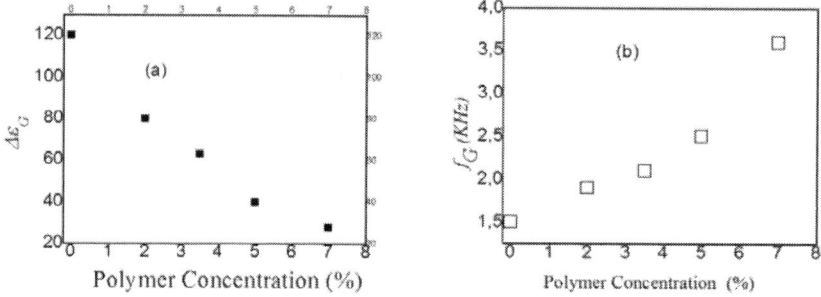

Figure 11. The dielectric strength (ε_G) (a) and relaxation frequency f_G (b) of the Goldstone mode at 25°C as function of the polymer concentration

The Goldstone mode even disappears in these systems for a sufficiently high aerosil density. The authors have interpreted the behavior of the dielectric response in these systems by size effects on smectic domains [36–38]; the reduction in the Goldstone mode strength and the increase in the relaxation frequency with increasing the concentration of aerosil particles are due, according to these authors, to the formation of smaller smectic domains where fluctuations are quenched by surface interactions, leading to a deformation of the helix. Additionally, the orientation of these smectic domains becomes randomly distributed so that fewer domains are therefore preferentially oriented in the direction of the applied electric field. We believe that the interpretation given above cannot explain the behavior of our PSFLC systems, despite similar changes in dielectric relaxation being observed. First, the polymer network in PSFLC cells are anisotropic and stabilizes the configuration of smectic domains oriented preferentially to the direction of the electric field. We think that the changes observed in dielectric response in our systems are essentially governed by elastic effects. We expect that the network-FLC interactions enhance the apparent elasticity of the PSFLC films, and accordingly, causes the increase in the relaxation frequency f_G and the reduction in dielectric strength $_G$. In fact, the dielectric strength and the relaxation frequency of the Goldstone mode in the case of a pure SmC* phase are expressed as [39]:

$$\Delta\varepsilon_G = (P_s/\theta)^2 / (2\varepsilon_0 K_\varphi q^2{}_0) \tag{2}$$

$$f_G = (K_{eff} q^2{}_0) / (2\pi\gamma_\varphi) \tag{3}$$

where is the rotational viscosity, K_{eff} is the effective elastic constant. and P_s are the tilt angle and spontaneous polarization, respectively. $q_0 = 2/p_0$ with p_0 is the helical pitch of the FLC. If we think that the Equations (2) and Equation (3) are applicable for the PSFLC composite, q_0 is considered here constant. The reduction in $_G$ as a function of polymer concentration is attributed to the variation in $(P_s/)$ and/or K_{eff}. To clarify that, we have plotted in Fig. 12 the ratio $(P_s/)$ as a function of polymer concentration. This figure shows that this ratio can be considered independent of the polymer concentration. As a consequence, the observed decrease in the $_G$ with polymer concentration can be explained rather by the increase in the effective elastic constant K_{eff}.

On the other hand, the relaxation frequency of the Goldstone mode Equation 3 is controlled both by the elastic ($K_{eff} q^2{}_0$) and viscous ($_{eff}$) forces. From the equation 3, the increase in f_G with polymer concentration can be explained by the decrease in the Goldstone rotational viscosity ($_{eff}$) and/or the increase in the effective elastic constant K_{eff}. From Eqs. 2 and 3, the effective rotational viscosity can be expressed as $(\gamma_{eff}) = (P_s \theta^{-1})^2 / 4\pi\varepsilon f_G \Delta\varepsilon_G$. According to this expression and using the experimental data of $(_G \times f_G)$ (Fig 13(b)) and $(P_s/)$ (Fig 12), $_{eff}$ was evaluated as a function of temperature for all polymer concentrations studied. Generally, $_{eff}$ of the PSFLC films increases with the polymer network density. At room temperature, for example, $_{eff}$ increases from 0.2 to 0.6 Pa. s when the polymer concentration increases from 0% to 7% (Fig. 13(b)).

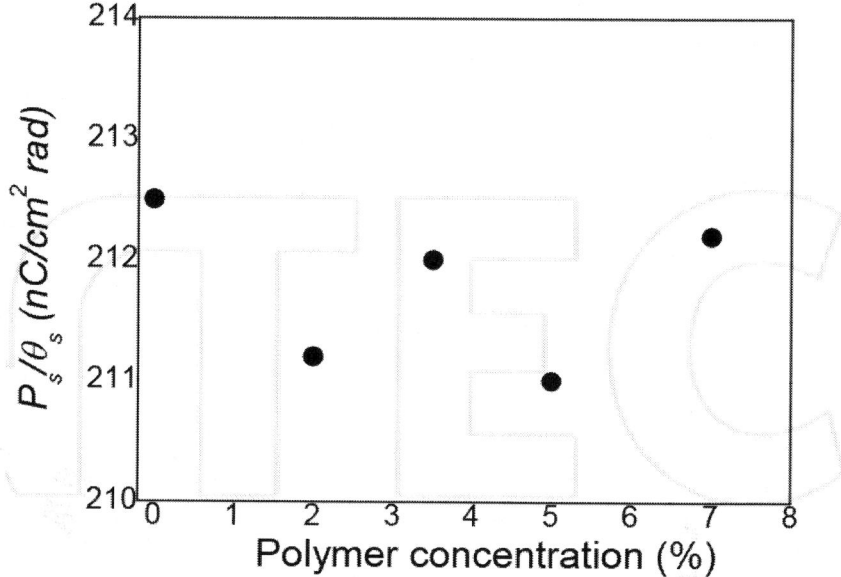

Figure 12. The ($P_s/$) ratio as function of the polymer network density

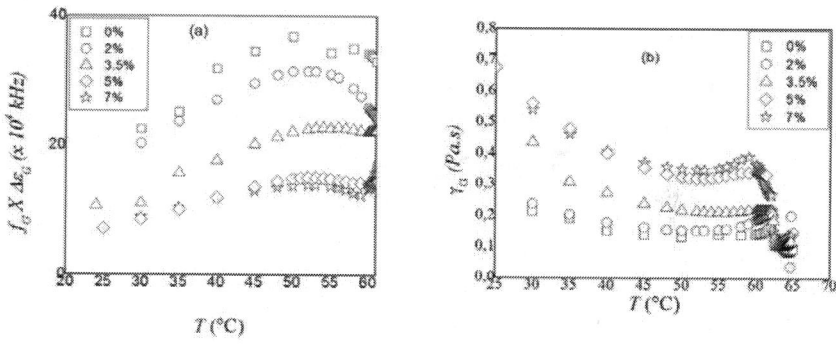

Figure 13. Temperature dependence of the Goldstone mode rotational viscosity ($_{eff}$) (b) and the product $_G$ x f_G (a) for different polymer concentrations.

Consequently, the increase in the relaxation frequency with the network density is certainly due to the increase in the effective elastic constant K_{eff}. To illustrate this, K_{eff} was evaluated at room temperature from Eqs. 2 and 3; the results are displayed in Fig. 14. This figure shows that K_{eff} linearly increases from $0.5 \ 10^{-11}$ to $2.3 10^{-11}$ N when the polymer concentration increases from 0% to 7%. The values of K_{eff} found here compare well with those obtained for the same PSFLC systems from the electro-optic measurements [18,20].

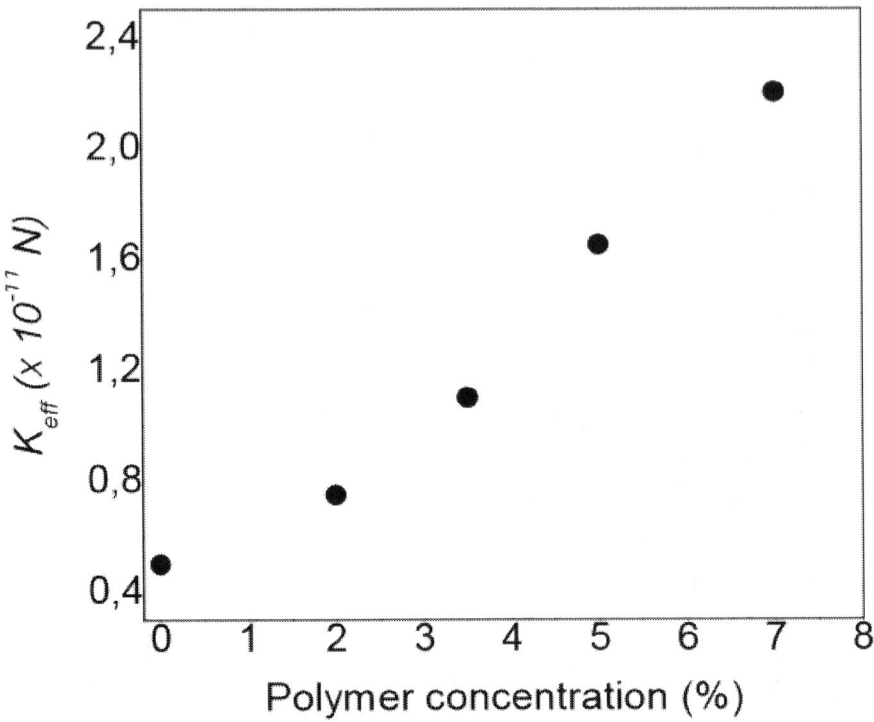

Figure 14. The effective twist elastic constant K_{eff} versus polymer concentration.

In conclusion, the increase in the relaxation frequency and the reduction in the dielectric strength of the Goldstone mode for the PSFLC films seem to be due to the increase in the twist elastic energy, resulting from the strong interaction between liquid crystal molecules and the polymer network liquid crystal molecules and the polymer network.

Polymer network morphology by AFM investigations

The principal result surprising in this party (section 3-2-A) is the role played by a polymer network to stabilize of the ferroelectric order. Indeed Bayth et al [34] which are reported that when the helical pitch of the smectic C* was confined between two parallel glass with a homogeneous planar anchoring causes an unwound of the helical structure which is proportional to the cell thickness. This transition is due to the result of the competition between the energy cost of the lines of dechiralisation slightly dependent on the thickness and that to the unwound of helix which is highly depends to the cell thickness. A similar mechanism could be compared in the case of our results by schematizing the fibril like cylinders with a homogeneous planar anchoring along their axis. In this case, the work which is reported by Baytch et al should remain valid once subsisted the thickness of the cell by the distance between cylinders. We think that the fibril seem to have a twisted morphology. In this case, these morphology could stabilizes the smectic phase. The increase of the apparent elasticity coefficients can be explained from a simple energetic argument if we take into account the polymer network morphology. One must note here that the polymer network was formed within the SmC* phase in which the director field is highly twisted in a helix. This helical structure of the FLC is transferred on the polymer network as shown in Figure 15.

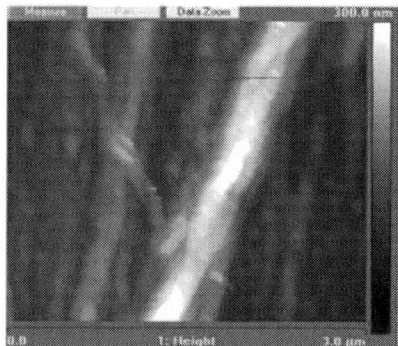 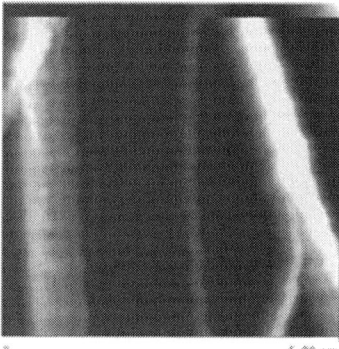

Figure 15. Tapping mode AFM height image of polymer fibers formed from a 5% polymer concentration into the short-pitch FLC (at T= 25°C). The twisted structure of fibrils is clearly observed indicating that the helical structure of the FLC from which the network was formed is printed on the polymer fibers.

For AFM experiments, the PSFLC cells were disassembled and flushed with solvent to remove the FLC. Figure 15 (on the left) presents a tapping mode AFM image (Veeco Instrument Inc.) of a 3×3 μm^2 region, and shows a fibrillar structure of the polymer network. The width of the fibers ranges from 150 to 300 nm. The AFM images clearly reveal a twisted structure of the fibers with a periodicity of about 0.24 μm which approximately corresponds to the pitch of the FLC helix used in this study. Hence, the polymer structure conserves locally a helical print which in turn stabilizes the FLC structure present during the polymerization. It has been already shown by I. Dierking et al. [12], and by G.A. Held et al.[40] that for long-pitch cholesteric liquid crystals, the helical superstaructure was transferred onto a polymer network. However, for short-pitch ferroelectric liquid crystals, this transfer has not been observed. Recently, K. Akagi et al [41] were confirmed the possibility of the printing of the short helical pitch on the polymer network. Archer et al [33] showed that the defect of the FLC twisted grain boundary were found printed on the polymer network structure. In [21] the chevron pattern has been observed on the polymer network. The printing of the helical structure on the polymer network is also reported by [42]. One could not observe any twisting of the polymer strands when polymerized in the Sm A* phase [43]. This implies that even after polymerization, the characteristic property of the host phase in which polymerisation was carried out was effectively retained. The helical aspect of the fiber structure certainly influences and explains the electro-optical: Deformed Helix ferroelectric liquid crystals (DHF) [18] and electroclinic effects [19,20].

Sof mode in the SmA phase

Figures 16 (a) and 16 (b) show examples of the dispersion, $'(f)$ and absorption, $''(f)$, dielectric spectra obtained in the SmA phase at T_c for each polymer concentration. Two relaxations mechanisms are detected. The first, at frequencies between 10 and 30 kHz with a weak strength, is attributed to the soft-mode relaxation mechanism; whereas the second observed at high frequencies 1 MHz is due to the ITO conducting layers. As shown in Fig.16, at 1 kHz frequency, the dielectric response shows a very strong polymer concentration dependence; $'$ decreases from 23 to 12 when the polymer concentration increases from 0% to 7%. This effect is also clearly demonstrated from the behavior of the absorption peak observed in Fig. 16 (b); the absorption peak decreases from 12 to 3 when the polymer concentration is varied from 0% to7%. We present in Figs. 17 (a) and 17 (b) the temperature dependence of the dielectric strength, s, and the relaxation frequency, f_s, of the soft mode. For all studied concentrations, the behavior of versus temperature shows the same general features fig. 17 (a). A rapid increase in s is observed close to T_c for all concentrations studied. The increase in s at and close to T_c is dependent on the polymer concentration. Note that s becomes relatively weakly affected by the network as temperature increases from T_c (Fig. 17 (a)). The relaxation frequency, f_s, exhibits a linear temperature dependence (Fig. 17 (b)). At T_c, f_s increases from 10 to 36 kHz when the polymer concentration increases from 0% to 7%. This effect seems to be less dependent on the polymer network density at relatively higher temperatures T_c +0.5 °C (Fig. 17 (b)). Similar behaviors have been observed in the case of dispersed silica particles on FLC matrix near the SmA-SmC phase transition [6–8]. However, Kundu et al. [25] showed for other PSFLC systems that the soft-mode dielectric strength remains unchanged when the FLC cells are stabilized by a polymer network formed from a nonmesogenic reactive monomer. These authors did not provide any indications of the network structure of their systems. However Beckel et al. [44] demonstrated that nonmesogenic monomers give rise to polymer chains which microseparate from FLC molecules in the smectic layers leading to a layer swelling. Obviously, this polymer network structure is completely different to that obtained in our systems Fig. 15. In order to examine how the polymer network influences the soft-mode dielectric strength of the PSFLC, we used the model previously developed [20] to explain the electroclinic behavior of PSFLC films.

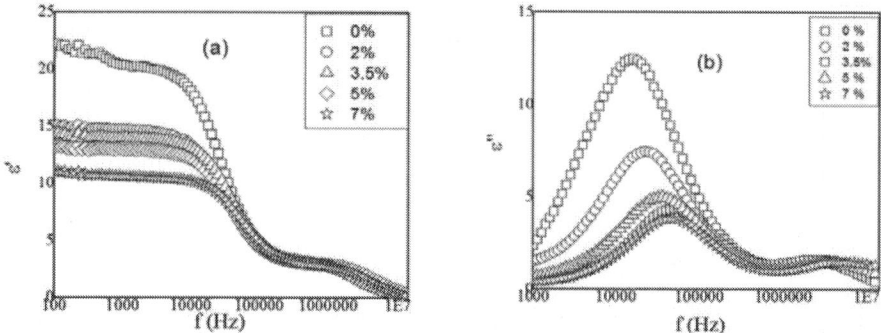

Figure 16. Frequency dependence of (a) the real and (b) the imaginary parts of the complex permittivity at the SmC-SmAphase-transition temperature, Tc, for different polymer concentrations.

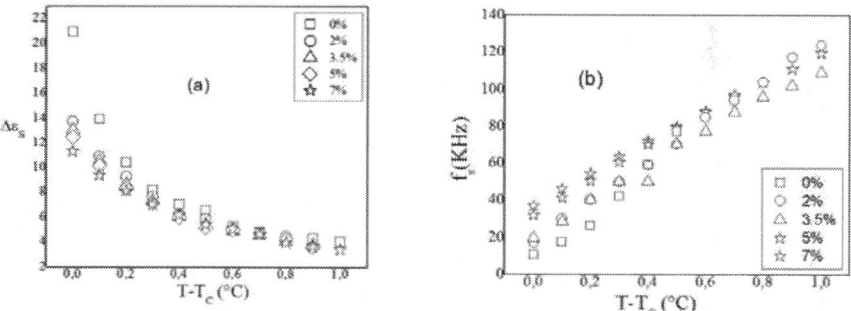

Figure 17. Temperature dependence of (a) the dielectric strength and (b) the relaxation frequency f_s of the soft mode in the smectic Aphase for different polymer concentrations.

Although polymer networks formed in liquid crystal media generally have a complex structure, they have been previously modeled as an assembly of parallel cylinders randomly distributed within the liquid crystal media [7]. The cylinders are interconnected via a chemical cross linking which ensure the network stability. In the framework of the rigid model of the network introduced by Li et al. [17], they considered that FLC molecules are affected by the bulk anchoring force from network. Li et al. interpreted this bulk anchoring force in terms of a field like effect, where the orientation of the FLC molecules was coupled to that of the anisotropic polymer network. This model gives a macroscopic description of the polymer network and was successfully applied to describe the "V-shaped" electro-optic properties of FLC gels [17]. We adopt in our theoretical approach this model structure of random polymer network of Li et al. We will consider below as a characteristic parameter of the polymer network structure the average intercylinder distance, which we call L_c, (Figure

18) and we try to analyze the effect of the applied electric field on smectic-A layers confined between two successive cylinders. The basic relation giving the free energy density of a chiral SmA phase near T_c in the presence of a small electric field E was expressed by Garoff et al. [45,46] as :

$$f_E = f_0 + (1/2)\ \alpha\ (T - T_c)\theta^2 - CP\theta + (P^2/\varepsilon_0\chi C) - PE \tag{4}$$

f_0 represents contributions to free energy density from the undistorted SmA phase. is the mean-field coefficient and C is related to the piezoelectric coupling between the polarization P and the induced electroclinic tilt . $_0$ is the dielectric constant and is the electric susceptibility. Moreover, to the free energy density term expressed in Eq.4, two other contributions are required to describe the effect of electric field on SmA blocks confined between the polymer fibers. The first one consists of the free energy density f_{ps} arising from the bulk polymer stabilization, and may be written as [17] W_p is the coupling coefficient for the interaction between the polymer network and the liquid crystal molecular director. Equation 5 is an approximative expression of f_{ps} because we are interested in low field regime where the angle is small. The second contribution comes from the elastic free energy density f_{el} arising from a director distortion upon application of E. In our system, the smectic layers are arranged perpendicular to the direction of the fibers. Between two successive groups of fibers separated by the average distance L_c (Figure18), due to anchoring forces between the liquid crystal molecules and the polymer network, the rotation of the director can be reasonably assumed not to be uniform: It is larger at or close to L_c /2 and weaker near the surface fibers. We must remark here that this theoretical approach of our PSFLC system is based on a one-dimensional model. We neglect then any splay deformation of the director, and we only consider the elastic energy arising from a twist deformation. f_{el} can then be given by the following expression:

$$f_{ps} = (1/2)\ W_p\ sin^2\theta \approx (1/2)\ W_p\ \theta^2 \tag{5}$$

K_2 is the twist elastic constant, and Z denotes the coordinate along the axis parallel to the direction of the applied electric field. The total free energy density $f_t = f_E + f_{el} + f_{ps}$ is then expressed as:

$$f_{el} = (1/2)\ K_2\ (\partial\theta/\partial z)^2 \tag{6}$$

The equilibrium values of P and are found by minimizing the free energy f_t with respect to P and , respectively. This leads to the following equations:

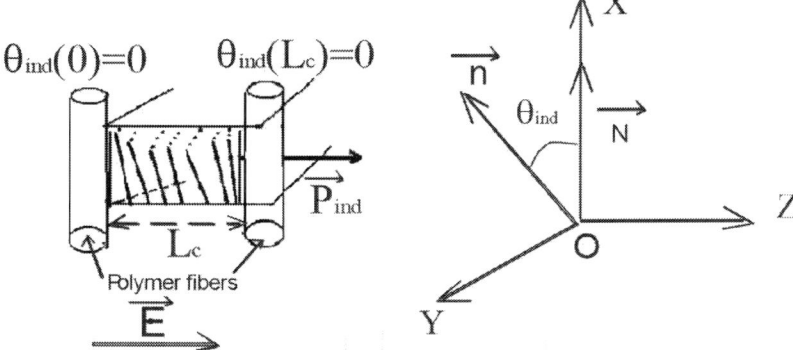

Figure 18. Tilt angle induced by electric field in the SmA* phase; distortion due to the anchorage at the polymer boundaries (on the left). L_c is the average distance between two successive polymer fibers parallel to the direction of the electric field. Cell normal and layer normal are parallel to the Z and X axis respectively.

$$f_t = f_0 + (1/2)\ \alpha\ (T-T_c)\theta^2 - CP\theta + (P^2/\varepsilon_0\chi C) - PE + (1/2)\ K_2\ (\partial\theta/\partial z)^2 + (1/2)\ W_p\ \theta^2 \tag{7}$$

$$\alpha\ (T-T_c)\theta - CP + W_p\ \theta - K_2\ (\partial^2\theta/\partial z^2) = 0 \tag{8}$$

$$C\theta - (P/\varepsilon_0\chi) + E = 0 \tag{9}$$

Inserting Eq. 9 into Eq. 8 gives

$$K_2\ (\partial^2\theta/\partial z^2) - \alpha\ (T-T'_c)\theta + C\ \varepsilon_0\chi\ E = 0 \tag{10}$$

where $T'_c = T_c - W_p\ /$ is the SmC*–SmA transition temperature of the PSFLC system; $T_c = T_0 + C^2_0\ /$ is the SmC*–SmA transition temperature of pure FLC. Equation 10 governing the director distortion induced by an applied electric field in the SmA phase the anisotropic polymer fibers in PSFLC systems constrain the molecular orientation at their surfaces as the screw dislocations at the grain boundaries in the TGBA phase do. To solve Eq. 10, we assume the anchoring of the molecules at the fiber surfaces to be rigid. The boundary conditions at these surfaces are : $\theta\ (z=0) = \theta\ (z=L_c) = 0$

Equation 10 has a solution:

$$\theta(z) = (\varepsilon_0\chi CE)(\alpha\ (T-T'_c)\ )^{-1}\ [1 - \exp(z/a)((1 - \exp(L_c/a))^{-1} - \exp(-z/a)\ (1 + \exp(-L_c/a))^{-1}] \tag{11}$$

Where

$$a = (K_2/(\alpha(T-T_c))^{1/2}$$

Averaging the (z) [15] and $P(z)$ values over the 0 z L_c domain (Figure 18) gives the expression of the mean induced tilt and polarization:

The average induced polarization can also be written as:

$$<\theta^{PSFLC}> = (\varepsilon_0 \chi CE)/\alpha(T-T'_c)[1- tanh (L_c/2a) / (L_c/2a)] \qquad (12)$$

$$<P^{PSFLC}> = (\varepsilon_0 \chi E) + (\varepsilon_0 \chi^2 C^2 E) / \alpha(T-T'_c) [1- tanh (L_c/2a) / (L_c/2a)] \qquad (13)$$

$$a= K_2 (\alpha(T-T'_c))^{-1}$$

The identification between equations (13) and (14) gives the expression of the dielectric strength of the soft mode, ^{PSFLC}, as a function of L_c:

$$<P^{PSFLC}> = (\varepsilon_0 \chi E) + (\varepsilon_0 E \Delta\varepsilon^{PSFLC}) \qquad (14)$$

$$\Delta\varepsilon^{PSFLC} \approx (\varepsilon_0 \chi^2 C^2 E) / \alpha (T-T'_c) [1-H] \qquad (15)$$

Where $H = tanh (L_c/2a) /(L_c/2a)$ which we called the elastic parameter, depends on the network density via L_c. For a same given reduced temperature, $(T-T'_c)$, the equation (10) can be expressed as:

$$\Delta\varepsilon^{PSFLC} \approx \Delta\varepsilon^{FLC} [1-H] \qquad (16)$$

^{FLC} denotes the soft mode dielectric strength of the pure FLC. Equation (16) shows that, in the SmA* phase, the main parameter that governs the soft mode dielectric strength in the PSFLC films is L_c. This means that the stored elastic energy arising from the distortion of the director upon application of electric field is as much larger than the L_c becomes smaller, which causes a decreases of the soft mode dielectric strength (equation (16)). The effect of the polymer concentration on the dielectric strength is in accordance with the theoretical results (Equation 16). According to this equation, the $_s$ is reduced by the increase of the polymer network density or by decreasing the distance L_c. To explain quantitatively the behavior of the reduction of the soft mode dielectric strength we can use the expression given by equation (16). From this equation, we conclude that the main parameter which governs the soft mode dielectric strength is the vertical distance which separated between two successive groups of polymer fibers L_c. We used the values of $8.8 \ 10^{+3} \ N \ m^{-2}K^{-1}$, $a =0.12\mu m$ [20], and the values of $_s$ presented in the Figure 17(a) at $T-T_c=0.1° \ C$, with a reasonable value of the twist elastic constant, typically, $K_2 \ 10^{-11} \ N$, the equation (16) was graphically resolved to evaluate L_c. We obtain a mean inter-fiber L_c. The results are displayed in Fig. 15. The calculated values were compared to those directly measured from the topography of the polymer networks obtained by means of AFM experiments (Fig. 19). Figure 19 shows example of AFM images and height profiles on the z direction in Fig. 18 of the polymer network. The height profile in the **z** direction indicates two successive groups of fibers. The mean distance L_c Fig. 18 between them was evaluated for each polymer concentration and displayed in Fig. 20(a). The measured values of L_c linearly decrease with the polymer density and agree well with those calculated from the model. It seems from these results that the fibrillar and anisotropic nature of the network stabilizes, at long-range scale, the SmA order and opposes the electric field effect on the deformation of the SmA director. These results are in accordance with previous works [19,20] of the electroclinic effect in the same PSFLC systems, which demonstrate that the electroclinic susceptibility of these systems is reduced with the increase in the polymer network density. It must be noted here that the average lateral separation distance (y direction in Fig.18) between polymer strands is estimated to be about 10 μm (Fig. 19), which is 1 order of magnitude higher than L_c.

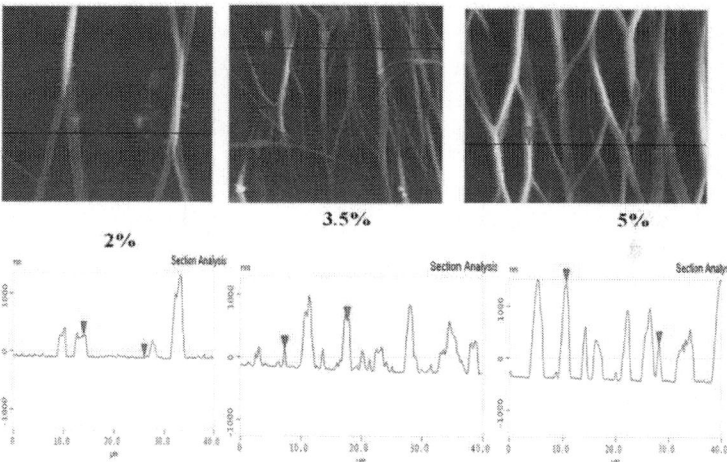

Figure 19. Tapping mode AFM images of polymer network structure of 40 x 40 μm ² (upper) and the height profile of the network structure (lower) of the 2%, 3.5% and 5%, respectively polymer concentration formed at T=25°C.

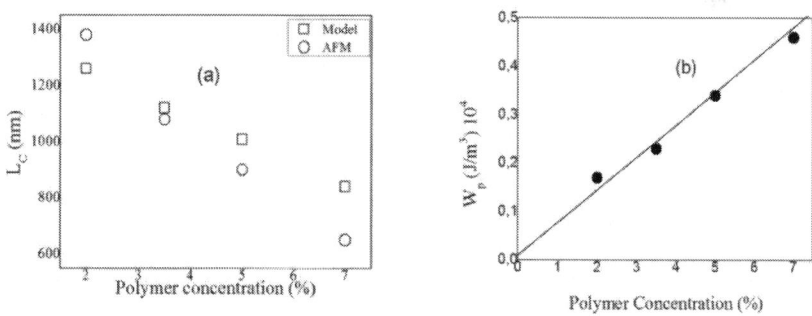

Figure 20. The average distance between two successive groups of fibers L_c as a function of polymer concentration (a) and the coupling interaction Polymer-Liquid crystals (W_p)(b).

This difference between interfiber distances in the two directions is an unexpected result. In fact, during the photopolymerization process, the mobility of the reactive monomers could be comparable within the smectic layers so that they present the same ability to come together and react to form the network. This ability could be significantly different across the smectic layers. Therefore, it would be reasonable to suspect that the average distance between polymer fibers could be of the same order of magnitude in lateral direction as well as in the z direction. The result found here is not yet clear, and the physical and chemical mechanisms governing the formation of the network could provide an explanation of our finding. This is not the aim of the work presented in this chapter. From the shift of the transition temperature, $T=T_c-T'_c = W_p/$, the coupling coefficient, W_p, characterizing the interaction energy between the FLC and the polymer network can be estimated. Tc values of 0.2 °C, 0.5 °C, 0.7 °C, an 0.9 °C were found for the polymer concentration of 2%, 3.5%, 5%, and 7%, respectively. The values of W_p are displayed in Fig. 20 (b). W_p linearly increases with the polymer concentration. The linear behavior of Wp in PSFLC system was already reported in earlier works [18-20]. The value of W_p found here are within 1 order of magnitude of those reported by Furue et al. [21] and Li et al. [17] on other PSFLC systems.

3.2.2. Effect of the polymer network morphology created under electric field in the short pitch ferroelectric liquid crystal on the dielectric responses

In this study we show that, both the dielectric intensity and the relaxation frequencies of the collective relaxations mechanisms namely soft and the Goldstone modes are strongly affected not only by the presence of the polymer network but also by the polymerization conditions.

Two samples cells with7% polymer concentration were then exposed to ultraviolet light (wavelength = 365nm) at 25 °C with an intensity of 18 (mW/cm^2) for 30 minutes. One of each is polymerized without the presence of electric field, however, the other cell is polymerized with the presence of AC electric field at 5 V μm^{-1} and for f= 1 Hz.

*Goldstone mode of the SmC**

In Figures 21 (a) and (b) the parameters G and fG obtained from the curve-fit procedure are displayed. For all studied samples, the behavior of G versus temperature shows the same general features (Figure 21 (a)). As seen in this figure, G increases slightly to reach a maximum at T_{max} (3°C below T_c). Above T_{max}, G abruptly decreases.

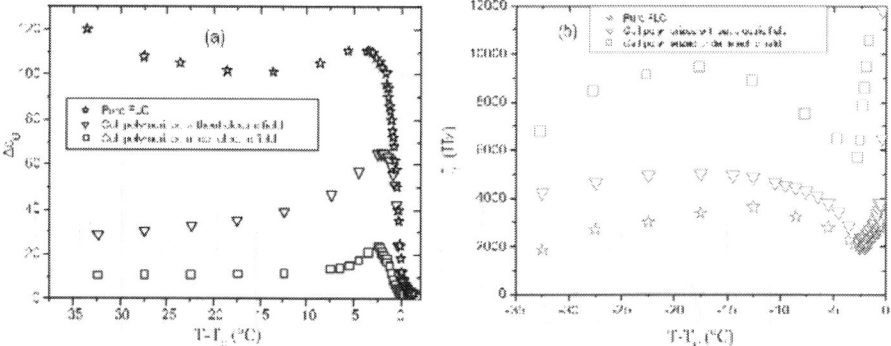

Figure 21. Temperature dependence of the Goldstone mode dielectric intensity (a) and the relaxation frequency (b).

Figure 21(b). This figure shows that the f_G slightly increases with temperature, reaches a maximum, then rapidly decreases to a minimum value at a temperature corresponding to T_{max}. It can be see from the figure 21 (b) that at room temperature, the f_G values are about 1.5 kHz, 4 kHz and 7 kHz for pure FLC, cell polymerized without and that polymerized with electric field, respectively. Significant differences were reported by Kaur et al [23] in another PSFLC systems. These authors were reported that the relaxation frequency f_G for the case when an bias field is applied during the polymerisation is very lower than in the case for sample when no bias was applied during the polymerisation. From the Equation (2) and (Equation 3), the values of K_{eff} are found about 2 10^{-11} N and 5.6 10^{-11} N for mixtures polymerized without and those polymerized with electric field, respectively. In conclusion, a lower dielectric strength (or larger relaxation frequency) could appear for a result of an increase of the elastic constant.

Soft mode of the SmA phase*

In Figures 22 (a) and (b) the temperature dependences of the dielectric strength, $_s$, and the relaxation frequencies, f_s, are shown for the soft mode detected in the SmA phase. One can see that at $(T-T_c)=0.1°$ C, an increase in the polymer concentration leads to the strong reduction of the soft mode dielectric strength. At the same temperature, $_s$ for a mixture polymerized without electric field is three times higher than those polymerized under electric field.

It can be seen that, at $T-T_c =0.1°C$, an increase of the polymer concentration from 0 to 7% leads to an increase of f_s from 10 *KHz* to 40 *KHz*, respectively. It can be seen from this figure that, at $T-T_c =0.1°$ C the cell polymerized with electric field presents a high f_s value about 80 KHz *(Figure 22(b)*. To explain quantitatively the behavior of the reduction of the soft mode dielectric strength we can use the expression given by equation (16). We used the values of 8.8 10^{+3} N $m^{-2}K^{-1}$, a =0.12µm [19,20], and the values of $_s$ presented in the Figure 22(a) at $T-T_c=0.1°$ C, with a reasonable value of the twist elastic constant, typically, K_2 10^{-11} N, the equation (16) was graphically resolved to evaluate L_c. We obtain a mean inter-fiber distance L_c 500 *nm* and 200 *nm*, for samples polymerized without and those polymerized with electric field respectively.

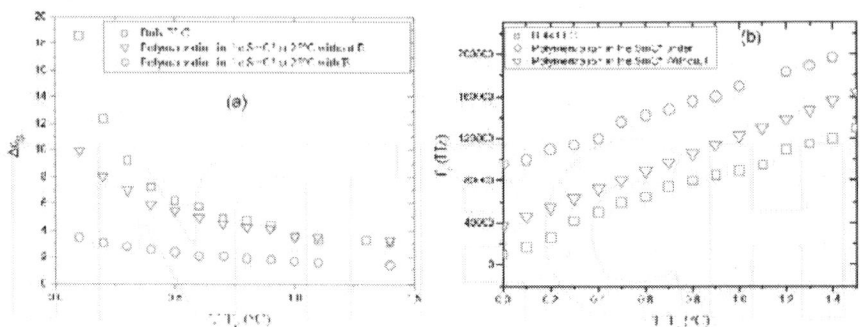

Figure 22. Temperature dependence of the dielectric intensity (a) and the relaxation frequency (b) of the soft mode in the smectic A* phase.

Atomic force microscopy observations

Figure 23 shows examples of AFM images and height profiles in the z direction of the Figure 18 of the polymer networks. These figures reveal that the separation distance between two polymer fibers (in the Y direction of the Figure 18) varies from 1 *µm* to 3 *µm* for the cell polymerized without electric field and from 300 *nm* to 500 *nm* for sample polymerized under electric field. The diameter of the fibrils was found varied between 500 *nm* and 1 *µm* for sample polymerized without electric field. However, it is found between 200 *nm* and 700 *nm* for sample polymerized with electric field.

Figures 3 and 4 (lower) represent the height profiles (on the z direction in the figure2) of the polymer networks. The height profiles show two successive groups of fibers. The mean distance, L_c between them was evaluated for each sample and are about 480 *nm* and 250 *nm* for samples when no field was applied during the polymerisation and those polymerized under electric field respectively. In conclusion, it must be noted from these observations that the polymerization under electric field is a principal factor affecting the polymer network structure. These calculated values are in accordance with those measured by the model. T_c values of the sample polymerized without Electric and those polymerized with E are 0.9°C and 1.4°C, respectively. The coupling interaction W_p values are 0.46 10^4 (J/m³) and 12,3 10^4 (J/m³) for sample polymerized without electric field and those polymerized with electric field respectively.

In conclusion, the main parameter which governs the dielectric strength and the relaxation frequency in our PSFLC systems is the distance L_c which separated between two layers of the polymer network in the Z direction parallel to the direction of the electric field. The changes observed on the dielectric response are confirmed by AFM pictures. The polymerization under electric field is another factor affecting the structure of the polymer network formed in liquid crystals.

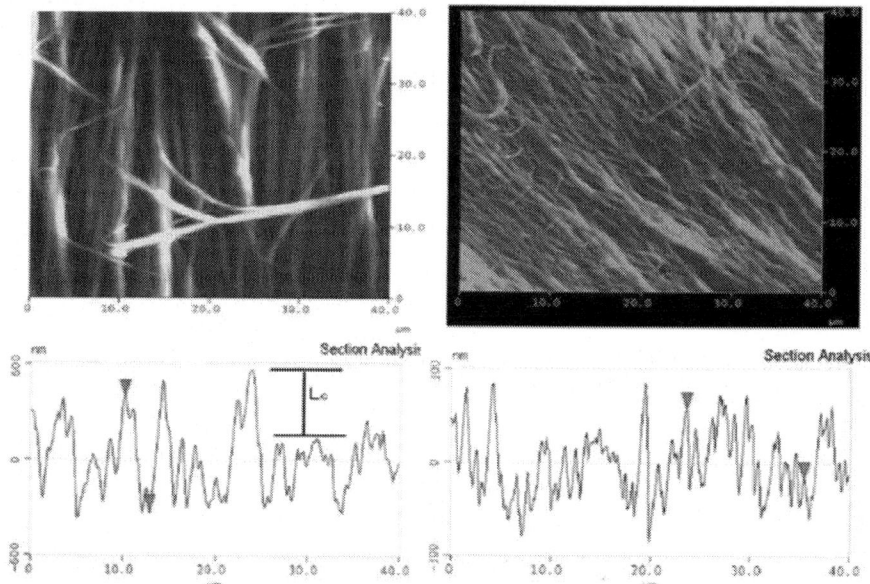

Figure 23. Tapping mode AFM images of polymer network structure of 40 x 40 μm ^{2} (upper) and the height profile of the network structure (lower) of the 7% polymer concentration (polymerization without electric field) on the left and that polymerized under electric field (on the right).

REFERENCES

1. Drzaic PS1986Polymer dispersed nematic liquid crystal for large area displays and light valves. J. Appl. Phys., 60 : 2142.
2. Fergason JL1985Polymer encapsulated nematic liquid crystals for display and light control applications. SID Dig. Tech. Pap, 16, 68.
3. Drzaic PS1995Atomic Force Microscopy Observations of the Polymer Network Structure Formed in Ferroelectric Liquid Crystals CellsWorld Scientific, NJ

4. J. W. Doane, N. A. Vaz, B. G. Wu, S. Zumer, 1986Atomic Force Microscopy Observations of the Polymer Network Structure Formed in Ferroelectric Liquid Crystals CellsAppl. Phys. Lett., 48: 296.

5. J. W. Doane, A. Golemme, J. L. West, J. B. Whitehead, B. G. Wu, 1988Atomic Force Microscopy Observations of the Polymer Network Structure Formed in Ferroelectric Liquid Crystals Cellss. Mol. Cryst. Liq. Cryst., 165: 511 EOF532 EOF

6. D. Broer, R. Gossink, R. A. M. Hikmet, 1990Atomic Force Microscopy Observations of the Polymer Network Structure Formed in Ferroelectric Liquid Crystals CellsAngew. Makromol. Chem. 183: 45.

7. G. P. Crawford, S. Zumer, 1996Atomic Force Microscopy Observations of the Polymer Network Structure Formed in Ferroelectric Liquid Crystals CellsTaylor & Francis, London).

8. Hikmet RAM, Boots HMJ,1995Atomic Force Microscopy Observations of the Polymer Network Structure Formed in Ferroelectric Liquid Crystals CellsPhys. Rev. E 5158245831

9. Chang CC, Chien LC, Meyer RB1997Atomic Force Microscopy Observations of the Polymer Network Structure Formed in Ferroelectric Liquid Crystals CellsPhys. Rev. E 56: 595 EOF599 EOF

10. Ma RQ, Yang DK2000Fréedericksz transition in polymer-stabilized nematic liquid crystals. Phys. Rev. E 6115671573

11. Y. K. Fung, D. K. Yang, Y. Sun, L. C. Chien, S. Zumer, J. W. Doane, 1995Polymer networks formed in liquid crystals. Liq. Cryst. 19797801

12. I. Dierking, L. L. Kosbar, Adakani. A. Afzali, A. C. Lowe, G. A. Held, 1997Atomic Force Microscopy Observations of the Polymer Network Structure Formed in Ferroelectric Liquid Crystals CellsJ. Appl. Phys. 81: 3007 EOF

13. Rajaram CV, Hudson SD, Chien LC, Chien. Doane, Bos. Yang-see, . chapters, . , . , 1. , 1. , . , Liquid. of, in. Crystals, Geometries. Complex, by. G. P. edited, Crawford, S. , Zumer (Taylor & Frances, London. 1996

14. Rajaram CV, Hudson SD, Chien LC1996Atomic Force Microscopy Observations of the Polymer Network Structure Formed in Ferroelectric Liquid Crystals CellsChem. Mater. 8: 2451 EOF2460 EOF

15. I. Dierking, Lagerwall. S. T. Osipov, 2000Atomic Force Microscopy Observations of the Polymer Network Structure Formed in Ferroelectric Liquid Crystals CellsEur. Phys. J. E 2: 303 EOF309 EOF

16. J. Li, X. Zhu, L. Xuan, X. Huang, 2002V-Shaped" Electro-Optic Characteristics in FLC Gels Ferroelectrics 27785105

17. M. Petit, A. Daoudi, M. Ismaili, J. M. Buisine, 2006Atomic Force Microscopy Observations of the Polymer Network Structure Formed in Ferroelectric Liquid Crystals CellsEur. Phys.J. E 20: 327 EOF333 EOF

18. M. Petit, A. Daoudi, M. Ismaili, J. M. Buisine, Costa. A. Da, 2008Effect of the Polymer Network Density Formed in Short Pitch Ferroelectric Liquid Crystal on the Electroclinic Effect Mol. Cryst. Liq. Cryst. 4876173

19. M. Petit, A. Daoudi, M. Ismaili, J. M. Buisine, 2006Electroclinic effect in a chiral smectic- A liquid crystal stabilized by an anisotropic polymer network.Phys. Rev. E 74: 061707.

20. H. Furue, T. Takahashi, S. Kobayashi, 1999Atomic Force Microscopy Observations of the Polymer Network Structure Formed in Ferroelectric Liquid Crystals CellsJpn. J. Appl. Phys., Part 1385660

21. T. Takahashi, T. Umeda, H. Furue, S. Kobayashi, 1999Atomic Force Microscopy Observations of the Polymer Network Structure Formed in Ferroelectric Liquid Crystals CellsJpn. J.Appl. Phys., Part 1385991

22. S. Kaur, I. Dierking, H. F. Gleeson, 2009Atomic Force Microscopy Observations of the Polymer Network Structure Formed in Ferroelectric Liquid Crystals CellsEur. Phys. J. E, 30265274

23. A. Mukherjee, S. S. Bhattacharyya, B. K. Chaudhuri, S. L. Wu, (200, 2009Atomic Force Microscopy Observations of the Polymer Network Structure Formed in Ferroelectric Liquid Crystals CellsJournal of Molecular Liquids127 EOF131 EOF

24. S. Kundu, T. Ray, S. K. Roy, W. Haase, R. Dabrowski, 2003Atomic Force Microscopy Observations of the Polymer Network Structure Formed in Ferroelectric Liquid Crystals CellsFerroelectrics282239248

25. M. Petit, J. Hemine, A. Daoudi, M. Ismaili, J. M. Buisine, Costa. A. Da, 2009Effect of the network density on dynamics of the soft and the Goldstone modes in short-ferroelectric liquid crystals stabilized by an anisotropic polymer network.Phys. Rev. E, 790317059

26. R. Blinc, B. Zeks, 1978Dynamics of helicoidal ferroelectric smectic-$\tilde{C}$ liquid crystals. Phys. Rev. A.18. 740745

27. Goodby JW et al.Ferroelectric Liquid Crystals : Principales, Properties and Applications.Gordon and Breach Science Publishers 1991

28. I. Musevic, R. Blinc, B. Zeks, (20, Physics. The, Ferroelectric. of, Liquid. Antiferroelectric, . Crystals, World Scientific).

29. M. Glogarova, J. Pavel, 1984The Behaviour of Thin Samples of Ferroelectric Liquid Crystals. Mol. Cryst. Liq. Cryst 114249257

30. M. Brunet, O. Parodi, (198, 1982Atomic Force Microscopy Observations of the Polymer Network Structure Formed in Ferroelectric Liquid Crystals CellsPhys. France 43515522

31. Y. Bouligand, 1975Defects and Textures in Cholesteric Analogues Given by Some Biological Systems J. Phys., 36, C1331

32. P. Archer, I. Dierking, 2009Polymer stabilisation of twisted smectic liquid crystal defect states Soft Matter, 5, 835.

33. N. Baytch, R. L. B. Selinger, J. V. Selinger, R. Shashidhar, 2003Simulations of helix unwinding in ferroelectric liquid crystals.Phys. Rev. E. 68, 041702.

34. M. Gasser, A. Gembus, D. Ganzke, I. Dierking, 2000Atomic Force Microscopy Observations of the Polymer Network Structure Formed in Ferroelectric Liquid Crystals Cells

35. Z. Kutnjak, S. Kralj, S. Žumer, (200, 2002Effect of dispersed silica particles on the smectic-A-smectic-C* phase transition.Phys. Rev. E 66, 041702.

36. S. A. Rozanski, J. Thoen, 2005Influence of dispersed aerosil particles on the collective dynamic modes in a ferroelectric liquid crystal with polarization sign reversal Non-Cryst. Solids 351, 2802.

37. S. A. Rozanski, J. Thoen, (200, dynamic. Collective, in. modes, liquid. ferroelectric, dispersions. crystal-aerosil, Liq. Cryst. 32331340

38. T. Carlsson, B. Zeks, C. Filipic, A. Levstik, 1990Atomic Force Microscopy Observations of the Polymer Network Structure Formed in Ferroelectric Liquid Crystals CellsPhys. Rev. A 42877889

39. G. A. Held, L. L. Kosbar, I. Dierking, A. C. Lowe, G. Grinstein, V. Lee, R. D. Miller, 1997Atomic Force Microscopy Observations of the Polymer Network Structure Formed in Ferroelectric Liquid Crystals CellsPhys. Rev. Lett. 7934433446

40. M. Goh, M. Kyotani, K. Akagi, 2007Atomic Force Microscopy Observations of the Polymer Network Structure Formed in Ferroelectric Liquid Crystals CellsJ. Am. Chem. Soc., 129 (27), 8519 EOF27 EOF

41. A. Hoischen, S. A. Benning, H. Kitzerow, HS(2009Atomic Force Microscopy Observations of the Polymer Network Structure Formed in Ferroelectric Liquid Crystals CellsJ. Appl. Phys. 105, 013540 EOF013540 EOF

42. S. Kaur, I. Dierking, H. Gleeson, HF(2009Eur. Phys. J. E, 2, 3, (2009).

43. E. Beckel, N. B. Cramer, A. W. Harant, C. N. Bowman, 2003Atomic Force Microscopy Observations of the Polymer Network Structure Formed in Ferroelectric Liquid Crystals CellsLiq. Cryst. 30, 1343 EOF1350 EOF

44. S. Garoff, R. B. Meyer, 1977Electroclinic Effect at the A-C Phase Change in a Chiral Smectic Liquid Crystal.Phys. Rev. Lett. 38848851

45. S. Garoff, R. B. Meyer, 1979Atomic Force Microscopy Observations of the Polymer Network Structure Formed in Ferroelectric Liquid Crystals CellsPhys. Rev. A 19338347

CHAPTER 3

Atomic Force Microscopy as a Tool Applied to Nano/Biosensors

Clarice Steffens [1,2], Fabio L. Leite [3,], Carolina C. Bueno [3], Alexandra Manzoli [1] and Paulo Sergio De Paula Herrmann [1,*]*

[1] National Nanotechnology Laboratory for Agribusiness (LNNA), Embrapa Instrumentation, P.O. Box 741, 13560-970, São Carlos, SP, Brazil

[2] Departament of Biotechnology, Federal University of São Carlos (UFSCar), 13565-905, São Carlos, SP, Brazil

[3] Department of Physics, Mathematics and Chemistry (DFMQ), Federal University of São Carlos (UFSCar), 18052-780, Sorocaba, SP, Brazil

ABSTRACT

This review article discusses and documents the basic concepts and principles of nano/biosensors. More specifically, we comment on the use of Chemical Force Microscopy (CFM) to study various aspects of architectural and chemical design details of specific molecules and polymers and its influence on the control of chemical interactions between the Atomic Force Microscopy (AFM) tip and the sample. This technique is based on the fabrication of nanomechanical cantilever sensors (NCS) and microcantilever-based biosensors (MC-B), which can provide, depending on the application, rapid, sensitive, simple and low-cost in situ detection. Besides, it can provide high repeatability and reproducibility. Here, we review the applications of CFM through some application examples which should function as methodological questions to understand and transform this tool into a reliable source of data. This section is followed by a description of the theoretical principle and usage of the functionalized NCS and MC-B technique in several fields, such as agriculture, biotechnology and immunoassay. Finally, we hope this review will help the reader to appreciate how important the tools CFM, NCS and MC-B are for characterization and understanding of systems on the atomic scale.

Keywords: atomic force spectroscopy; atomic force microscopy; nanotechnology; nanoscience; nanosensors

1. INTRODUCTION TO NANOSENSORS

In order to promote a stable adsorption of molecules on microcantilevers, the Chemical Force Microscopy (CFM) technique is used here. CFM helps the molecules to "fall into place" in a spontaneous association in such way that they form a structurally well-defined aggregate. Thus, the purpose of the use of CFM in the construction of nanosensors is to achieve the organization of the molecules, to promote an orientation of functional groups, to contribute to chemical and physical stability of adsorption molecules to turn the device reproducible and sensitive.

Recent advances in the design and development of these sensors have led to simple microelectromechanical systems that can be manufactured easily, produced on a large scale and are capable of detecting very small mechanical deflections. The spring constant of a microcantilever ranges from 10^{-3} to 10^{1} N/m, enabling it to detect very small forces (10^{-12} to 10^{-9} N) [1]. These systems allow a fast response, low cost and the construction of arrays of sensors of small dimensions, thus enabling the investigation of microenvironments [2]. Microcantilevers are usually prepared on silicon and/or silicon nitride or polymeric materials, with dimensions from 100 to 500 microns in length from 0.5 to 5 micrometers in thickness, in the shape of a "V" (triangular) or "T" (rectangular) with a needle mounted on the free end [3].

Micro Electro Mechanical Systems (MEMS) are micro-electronic systems that incorporate chemical, magnetic and radiant heat micromechanical sensors and actuators. The first developments in the area of MEMS were performed in the 60s, systems were being commercialized by the 90s [4]. MEMS represent a family of diverse designs, devices with simple cantilever configurations, as used in atomic force microscopy (AFM), that are considered especially attractive as chemical and biological sensors. The ability of microcantilevers, to change their vibrational frequency or suffer deflection upon adsorbing molecules on their surface makes them excellent probes that can act as chemical, physical or biological sensors on nanoscale. Changes in vibrational frequency of micromechanical devices can be used to measure viscosity, density and flow rates in various systems. Deflections of the cantilever are due to the stress of molecular adsorption, which can be upward or downward depending on the type of chemical bonding of the molecule. In these systems, the change in frequency of the microcantilever has been reported to be proportional to the magnitude of the adsorbed mass [5–8].

2. CHEMICAL FORCE MICROSCOPY (CFM)

The chemical characterization of surfaces developed as an important technological tool allowing goods design and fabrication processes to fulfill high standards. This was achieved thanks to scientific advances in the atomic and molecular manipulation, the understanding of pathways for molecular binding but also to the conversion of such chemical information into reliable application methods.

In this context, Atomic Force Microscopy (AFM) can provide sensitive resources to measure and to map the surface chemistry information and to quantify the adhesive or repulsive forces associated to inorganic materials and biological samples, through the control of chemical interactions between the AFM tip and the sample [9–11]. Among AFM techniques, there is a very useful tool known as Chemical Force Microscopy (CFM), which is based on AFM tips chemically modified with specific exposed functional groups, carefully architected to carry out a specific function in a system [12–17]. Noy and coworkers have pioneered this technique, utilizing Lateral Force Microscopy (LFM) [18]. Applications to this tool include titration-AFM to obtain the apparent Pka value at the surface [10,15], determination of adhesive forces and energy on a surface, finding a specific substance by measuring single intermolecular forces (host-guest interaction in a complex environment), detection of chemical groups, determining surface heterogeneity, studying surface chemical reactions on the nanoscale and in real time [19]. All these applications found use in synthetic surface chemistry and in the creation of intelligent bioarrays, sensors, chips, and micro/nanofluidic devices.

Nanosensors are called smart devices due to their recognition selectivity and sensitivity. These crucial features are also typical characteristics of Chemical Force Microscopy, precisely used to address stabilization to a specific surface, since nanomaterials tends to aggregate, react chemically, or decompose if not treated properly. Here, it is worth mentioning that this treatment provided by CFM is known in the literature under several names, such as chemical derivatization, functionalization or modification. There are many ways to chemically modify surfaces and these modification methods involve addition, subtraction, or exchange of chemical groups or even restructuration of a molecule. For instance, we can cite the biochemical modification with enzymes [20], heat treatments, polarity change with corona discharge, plasmas, UV and gamma-radiations electron or ion bombardment, ozone [20–23], silylation [24,25] and lamination [26]. In the following, we review important parameters related to the CFM applications and its implications in nanobiossensors and nanomechanical cantilever sensors.

First of all, one of the major critical tasks in CFM is to deal with several parameters that can interfere in the measurements by causing noise in the signals collected and introducing artifacts into the results. Since CFM is based on an analytical probe consisting of a functionalized AFM tip, the quality of this probe is crucial. It has been observed that low quality tips can lead to artifacts in the images and consequently, to incorrect results [9]. It is also important to check the material the probe is made of, as well as if they have a suitable nominal tip radius, spring constant and resonant frequency that fits on our aim. More accurately the choice of the tip is made, more reliable and representative will be the results and, as consequence, they will be free of the inaccuracies that might be introduced by stiffness of the cantilever, the tip radius, shape, size and its dilation in AFM/CFM measurements.

Commercial tips may come with impurities, such as dust, organic residues or other common substances. Unclean probes can lead to contamination of the material under investigation, and because of that, before an experimental part begins, a cleaning step should be carried out in a UV/ozone chamber, to remove organic contaminants. This process will help to avoid artifacts in the final images and results. As previously mentioned, the tip shape can also significantly influence the results. If the tip has many things in contact with it (owing to large probe tip radius), during the surface scan, a lot of information will be lost. This effect can be explained by the direct relation between the radius of curvature of the tip and the tip-sample contact area [16]. The sharper is the tip; the higher will be the resolution and the confidence in the results. Scanning Electron Microscope (SEM) images of the tip are suggested here to determine the tip shape.

After optimizing CFM parameters, the surface functionalization is required. Firstly it is important to identify each substance, molecule or reagent that could be involved in the functionalization; secondly, this information should allow potential reactions and binding and eventually the complete molecular architecture to functionalize the tip to be predicted. This is crucial when CFM is used to measure adhesion and force (the tip surface chemistry must be well defined). For a clear understanding, each molecule, substance or atom (depending on the system) should be thought of as a small brick that will be used to build a wall. Here, some questions arise, such as: Should we lay this brick vertically or horizontally? Or is it better to put it at a specific angle? Is angle related to a binding site (active site)? How do we bind one brick to another? Is it necessary to use "cement" with specific characteristics? Should we control the solution pH? It is not important how many bricks are set on the tip surface, but it is important to know how to put them together. Chemical functionalization is a prerequisite for the firm attachment of the "bricks" in order to withstand the lateral force exerted during the scan acquisition and force measurements [12]. On balance, the functionalization of the AFM tip must be designed to obtain stability, sensitivity, and selectivity during the scan, in order to reduce the probability of non-specific bonding and the possible agglomeration of substances on the tip.

There are several ways to functionalize the AFM tip. Two of the commonest methods are known as mixing self-assembled monolayers (SAMs) and vacuum, thermal evaporation thermal vapor deposition and even, sputtering. The method of self-assembled monolayers (SAMs) is widely used to functionalize tips and surfaces. Terminal functional groups (-COOH, -CH3) are grafted onto alkane thiols that spontaneously form monolayers, under controlled conditions, on gold surfaces [9]. Organosiloxane monolayers (silanization) can also be formed on the tip [10]. The SAMs are usually applied by dropping reagents on the tip and then rinsing it after a specific time with another reagent (or ultra-pure water), or by the immersion of the tip into a specific solution (say in a beaker, like the Layer-by-Layer (LBL) process for thin films), followed again by washing. A typical result of SAM is depicted in Figure 1. Sometimes, the design of SAMs on a tip can follow the patterns of Nature; i.e., mimicry can be an excellent source of inspiration. Self-assembled monolayers offer new opportunities to increase the fundamental understanding of self-organization structure-property relationship, and interfacial phenomena [3]. SAMs can provide the needed design flexibility, as functionalized chains of polyethylene glycol (PEGs).

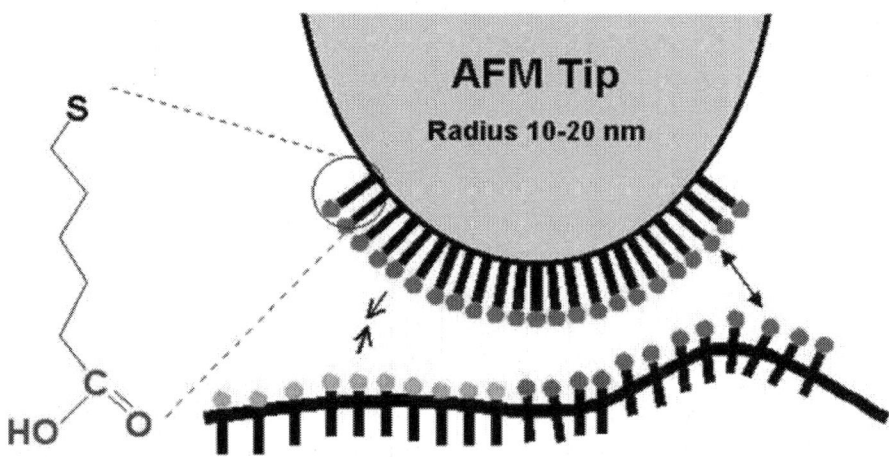

Figure 1. CFM principle: chemical tip modification with SAMs (Reprinted with permission [27]).

In vacuum/thermal vapor deposition thermal evaporation should be handled with care in order to prevent the breaking of the tip. To use this procedure, the physical and chemical conditions must be chosen carefully, as surrounding humidity and temperature, environment medium of analysis (air, liquid or vacuum) and time of incubation of reagents. Generally, this process must be done in a chamber (with vacuum or not), which chamber must be cleaned with an inert gas before the beginning of the functionalization. It is important to highlight that after the functionalization process, the probes should be gently dried in a stream of inert gas (argon is used most frequently), and stored under an atmosphere of the same gas, in a desiccator, to avoid high humidity and the oxidation of the probe. The sputtering techinique can be used to functionalize the AFM tip with a metal coating in order to induce specific properties to the tip. Some of these properties are: electric and thermal conduction, optical reflectivity and ferroelectricity [28].

Hereafter, some forms of analysis offered by CFM will be next discussed: measurement of contact angle (θ), force vs. distance curves, histograms, and adhesion maps.

Contact angle is frequently used to characterize surface energy properties across the three-phase boundaries, where liquid, gas and solid phases meet [16], and to gain a better understanding of force/adhesion events related to apolar and polar components. θ represents the ability of a liquid to spread on a plane surface and it is measured as the angle between the outline tangent of a drop deposited on a solid and the surface of this solid [29–31]. In addition, contact angle can also be related to thermodynamic concepts and are measured by fitting a mathematical expression to the shape of the drop and then calculating the slope of the tangent to the drop at the three-phase boundaries interface line. Here is important highlight the Cassie's law. This law is derived from the thermodynamic definition of contact angle and concerns to the contact angle of a macroscopic droplet on a chemically heterogeneous surface and is successfully used to explain the superhydrophobicity and self-cleaning mechanism of various natural and artificial surfaces [32–34]. Examples of such analysis can be observed in Figure 2.

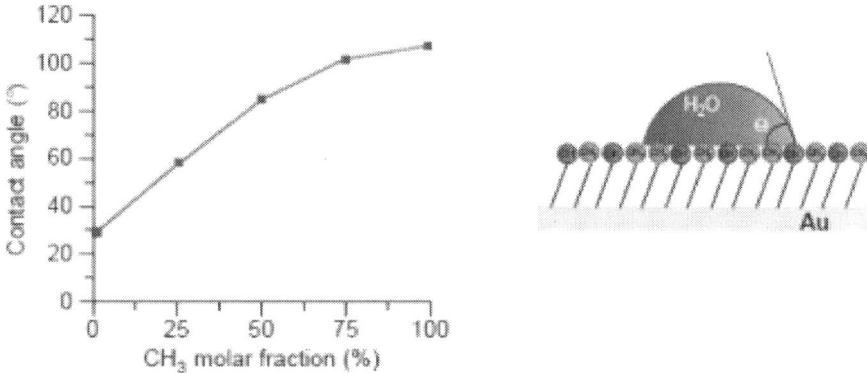

Figure 2. Chemical force microscopy (CFM): principle and application to the probing of hydrophobic forces. Water contact angle (θ) values measured on mixed self-assembled monolayers (SAMs) of CH_3- and OH-terminated alkanethiols, plotted as a function of the molar fraction of CH_3-terminated alkanethiols (reprinted from [29] with permission).

Force vs. distance curves (or force vs. displacement vertical curves) are used to collect quantitative and direct information on the force and adhesion forces between the AFM tip and the sample, including elongation, separation, elastic deformation (stretch), and binding forces. These curves can provide significant details of recognition events, homogeneity, chemical composition of the sample and biological events. Figure 3 briefly explains each part of a force vs. distance curve during the approach and retraction of the tip. Note that the force at point E is the pull-off force, defined as the highest force required separating the probe from the surface [27].

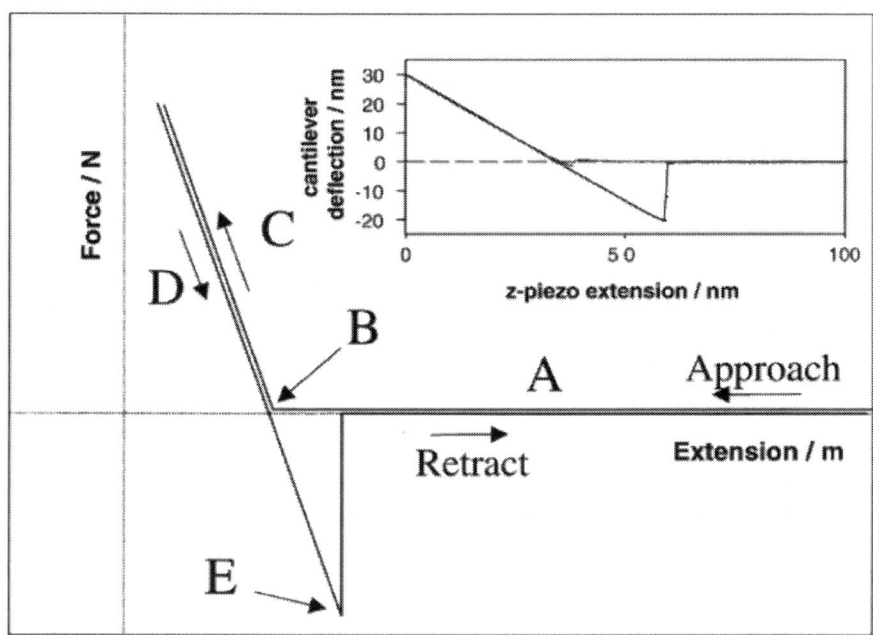

Figure 3. Force-displacement curves were formed. (**A**) tip is a far distance from the surface and there is no interaction; (**B**) the tip contacts the surface; (**C**) the cantilever is bent and a repulsive force (positive) is measured; (**D**) the cantilever holder is retracted from the surface and an adhesive (negative) interaction between the tip and surface is measured; (**E**) the pull-off point (reprinted from [9] with permission).

In order to confirm the results and quantify the differences between adhesion/repulsive force measurements and eliminate possible noise caused by artifacts, it is required to record a large number of force vs. distance curves: at least forty for every type of measurement. These data should be plotted as a histogram of the frequency (%) of the values observed for each interaction vs. the adhesion force (pN) [35]. These histograms show the statistical distribution of the adhesion forces, which are given by the vertical distance found in the force curve (point E to horizontal line).

Given that adhesive forces are directly linked to interactions between chemical functional groups on the tip and the sample surface [32], adhesion maps can give additional, and not less important, information with respect to the force vs. distance curves. As it can be seen in Figure 4, an adhesion map is a way of mapping surface forces with a very high spatial resolution (usually 10 nm) [9]. Here, the approach relies on combining information from the force vs. distance curves at each point (pixel) in an image field with simultaneously acquired topographic data to yield a high-resolution map of chemical interaction sites [9].

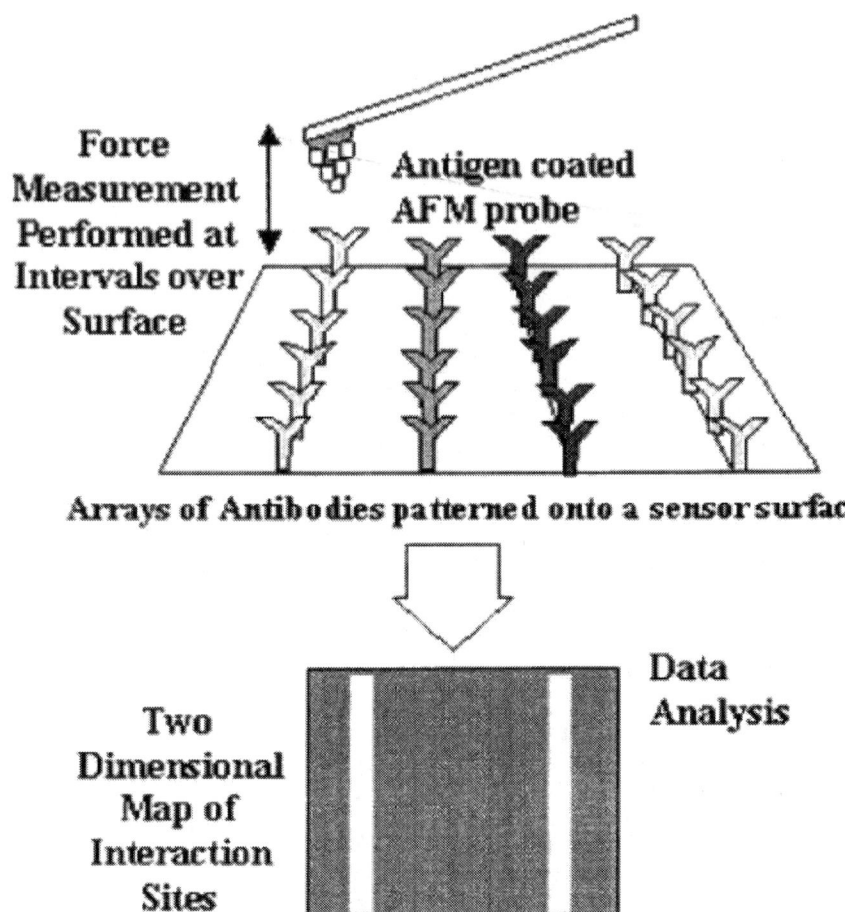

Figure 4. A schematic diagram of an antigen-coated probe mapping specific interaction sites on a substrate patterned with three antibody species (reprinted from [26] with permission).

The white lines displayed on the adhesion force map (Figure 4) are a result of the specific interaction between the antigen in the tip and its specific antibody (light grey). Thus, as may be clear, an adhesion map is built as result of repeated specific interactions between the tip and the surface. Depending on the system studied, this repetitive process can damage the functional molecules on the tip apex during mapping, which can be a hard challenge to deal with [36]. Adhesion maps have another remarkable application: establishing a fingerprint for each molecule, substance, and receptor or complexation event in order to recognize and locate specific substances on a surface [37]. This could be a useful tool that enables surface differentiation in cutting-edge technologies [38]. A typical adhesion map is formed by pixels of specific recognition (differences detected in the deflection signal) [39,40].

The interpretation of adhesion maps and force curves can occasionally be difficult because of the surface roughness and elastic behavior of the sample. Sometimes it is necessary to process the primary results with statistical programs, depending, of course, on the form in which the AFM software outputs the results.

Additional physical techniques are used to support CFM data. Usually, these are traditional methods used to analyze the molecular and elemental composition of surfaces, such as X-ray Photoelectron Spectroscopy (XPS), Fourier Transformed Infrared Spectroscopy (FTIR) and Raman Spectroscopy. For instance, Raman spectroscopy provides additional information on the chemical composition of materials and, because of that, may confirm whether the tip was indeed functionalized [41,42]. The same is valid to FTIR and XPS, which provide, respectively, the molecular composition of the surfaces of materials and the measurement of the elemental composition of the surface [43,44]. Nevertheless, it is also worth mentioning that these cited techniques provide a spatial resolution limited by diffraction and should be used just as complementary information to the CFM studies. Additionally, all of these techniques can also be compared with Dynamic Molecular Model Data too, and used to verify what is predicted by these models.

Finally in CFM studies, it is important to make the building process of the nano(bio)sensor reproducible in order to make the CFM measurements reproducible. Because of that, the majority of the parameters involved in their construction and operation should be kept as constant as possible, such as surrounding humidity and temperature, environment medium of analysis (air, liquid or vacuum) and time of incubation of reagents, as previously indicated. Even small changes in the parameters that determine the chemical force interaction between the tip and the sample can lead to enormous changes in images, curves and results [21].

3. NANOMECHANICAL CANTILEVER SENSORS (NCS)

The detection of chemical species through sensors is one of the most intensely investigated fields of science and technology, owing to the importance and variety of applications. A chemical sensor is an instrument that, when exposed to a particular type of substance (the analyte), transforms the chemical information, such as polarity or difference in concentration, for example, into an analytically measurable signal, such as electrical resistance, conductivity, potential difference or frequency. This transformation is called signal transduction and is of central importance to the working of any sensor [33].

Microcantilevers were first used in Scanning Force Microscopy. The deflection of a cantilever can be measured with an optical sensor when it is a small deflection; the vertical displacement of the tip (Z) is directly proportional to the force on it (F) and is thus a direct measurement of the strength of interaction between the tip and a surface (Figure 5). The principle of operation of a cantilever sensor is based on the adsorption of analytes at the surface of the cantilever (coated with a sensing layer), which usually leads to an induced surface stress and an increase of the apparent mass of the cantilever [8–34]. The change in mass leads to deflection of the cantilever in the Z direction.

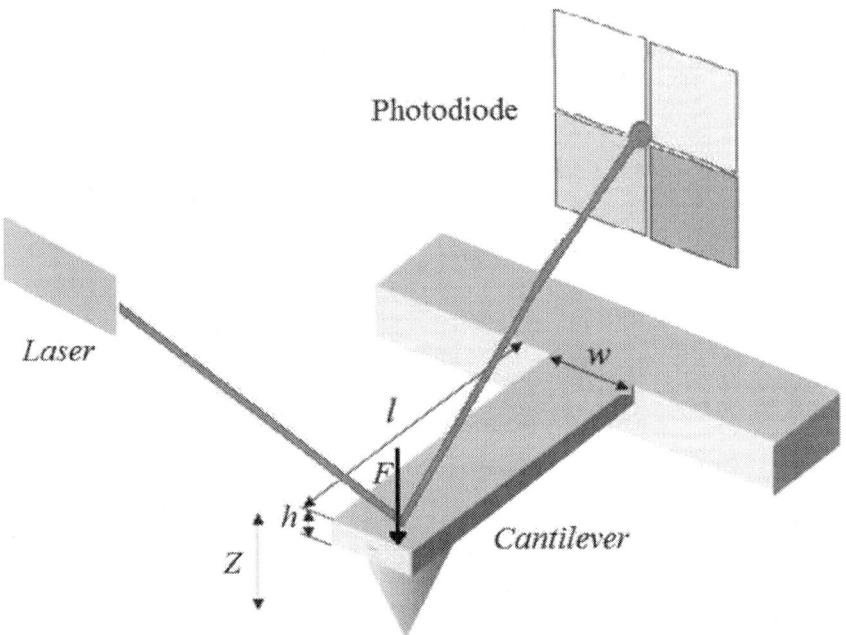

Figure 5. Illustration of the displacement of the cantilever (h, w and l, thickness, width and the length of the cantilever, respectively).

Treating the rectangular cantilever as a vibrating spring-mass system, its resonance frequency (f_{res}) can be calculated as:

$$f_{res} = \frac{1}{2\pi}\sqrt{\frac{k}{m}}$$

(1)

where k is the spring constant and m the mass.

The mass of a cantilever can be expressed as: $m = \rho.h.l.w$, where ρ is the density of the material, h its thickness, w its width and l the length.

The spring constant k is a function of the geometric and material parameters:

$$k = \frac{E * w * h^3}{4 * l^3}$$

(2)

where E is Young's modulus of the material in the cantilever. For example, for silicon (100) E = 1,3 × 10^{11} N/m². The surface stress is uniform and acts on the surface, increasing (in the case of compressive stress) or decreasing (in case of tensile stress) the surface area. If this stress is not compensated at the opposite side of a cantilever, the whole structure will bend. A change in surface stress on only one side of the cantilever will cause a static deflection and the deflection can be calculated as:

$$\Delta Z = \frac{3l^2(1-\upsilon)}{E * h^2}\Delta\sigma$$

(3)

where υ is the Poison rate (υ = 0.24 for a rectangular cantilever), $\delta\sigma = \sigma_1 - \sigma_2$ the difference in surface, σ_1 and σ_2 being the stress on the upper and lower side of the cantilever, respectively. Poisson's ratio is 0.24 for a rectangular cantilever and the difference in surface stress approximately follows Stoney's equation [45], which relates cantilever deflection (Z) to applied stress (σ).

The cantilevers can be operated in two modes static and dynamic. The static mode allows the analysis of cantilever bending in different ambient conditions to measure surface stress changes while, in the dynamic mode, the change in resonance frequency of the cantilever is monitored. Changes in surface stress result from the adsorption or electrostatic interactions between molecules on the surface, as well as changes in surface hydrophobicity and conformational changes in the adsorbed molecules [35].

When cantilevers are functionalized with sensitive materials such as metals, polymers, enzymes, thin layers, among others, the sensitive coating may interact with analyte molecules that selectively adsorb or bind by chemical affinity, converting the cantilever into a selective and sensitive sensor [36], which responds to specific substances or groups of substances.

The interaction between the analyte and the surface layer on the sensor can be reversible or irreversible. In the reversible case, the analyte interacts with the surface layer of the sensor to produce a response and when the analyte molecules are removed, the senor returns to its original state. In the irreversible case, the analyte undergoes a chemical reaction at the sensor surface catalyzed by the sensor material. Here, the analyte is consumed in the sensing process, although the number of molecules reacting is often a small proportion of the total number within the sample.

The method used in the functionalization has a strong influence on the sensitivity, because this depends on a number of parameters such as uniformity of the coating and the possibility of molecular reorganization by changing interactions due to external stimuli or analytes. The cantilever selectivity depends on the detection layer, which may be built according to principles of molecular recognition.

Several methodologies have been described in the literature for microcantilever surface modification with organic [37] or inorganic layers [39,40,46], some of which are described below. Functionalized cantilever bending may be due to the adsorption of gas, if the expansion coefficients of the materials on the two sides are different. The absorption of water vapor into an inorganic sensing layer has been investigated by Thundat et al. [39]. They coated one side of the silicon cantilever with a thin film of gelatin or phosphoric acid, as hygroscopic materials. The functionalization was carried out by sliding the cantilever into the solution until completely covered. The resonance frequency was measured in different conditions of relative humidity. The cantilevers functionalized with phosphoric acid (H_3PO_4) showed a decrease in the resonant frequency with the reduction of relative humidity due to the increasing effective mass, while the cantilevers coated with gelatin film showed an increase in the resonant frequency. Thundat et al. [40] also reported the deflection of commercially cantilevers, using a position sensitive detector. Silicon nitride cantilevers were supplied with 4 nm chromium and 40 nm gold layers and other cantilevers with 5–13 nm of aluminum on one surface. For the gold coated cantilever the deflection varied almost linearly and reversibly with changes in relative humidity, while in the uncoated cantilever, the deflection was negligible. The cantilevers with 5–13 nm of aluminum were very sensitive to changes in relative humidity. Pinnaduwage et al. [46] reported the detection of 10–30 parts per trillion of pentaerythritoltetranitrate and hexahydro-1,3,5-triazine within 20 s of contact with a silicon microcantilever whose gold surface had been modified by immersing the cantilever into a 6×10^{-3} M solution of 4-mercaptobenzoic acid, to build a self-assembled monolayer. The monolayer coating was shown to be quite stable for several months under normal operating conditions.

Functionalizing the cantilever with polymer coating renders possible highly sensitive identification of gases and volatile organic compounds. The possibility of using different polymers allows one to functionalize the cantilevers with a selective coating to analyze mixtures of volatile organic compounds [47]. When cantilevers are functionalized with polymeric layers, they can absorb molecules of an analyte, causing a swelling of the polymer matrix and thus resulting in a differential cantilever stress. This expansion induced cantilever surface stress can occur in two different ways: target molecules can be adsorbed on to the functionalized surface (Figure 6(a)) or penetrate the sensing layer deposited on the surface of the cantilever (Figure 6(b)) [38]. Spin-coating of the polymers usually results in uniform films with controlled thickness, which is easily varied by changing the spin speed or switching to a different viscosity. However, in the literature there are divergences over the technique of spin-coating deposition.

Spin-coating with polymer layers usually causes an unwanted deposition on the passive side of the microcantilever. On the other hand, microcantilevers were coated on one side by using a spin-coating technique by Betts et al. [48]. The difference in the resonance frequency between the coated and uncoated microcantilever is related to the change in the mass of the resonating structure. The results showed that the selectivity, as indicated by differences in relative responses to the test analytes, was different for the solvents phases which differed significantly in polarity (Figure 7).

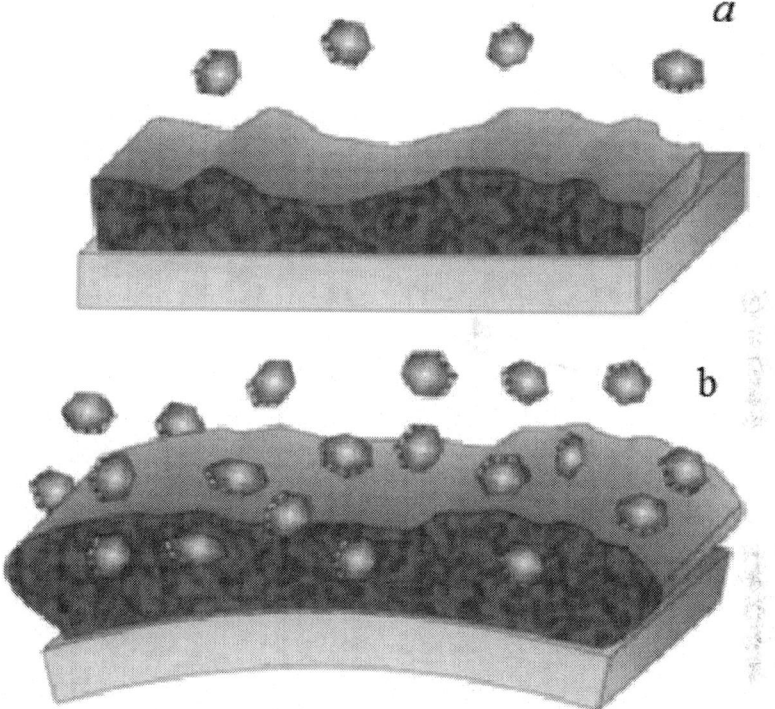

Figure 6. Sketch of the absorption-induced surface stress at the surface of cantilevers: (**a**) absorption and (**b**) penetration of target molecules into the sensing layer (reprinted from [38] with permission).

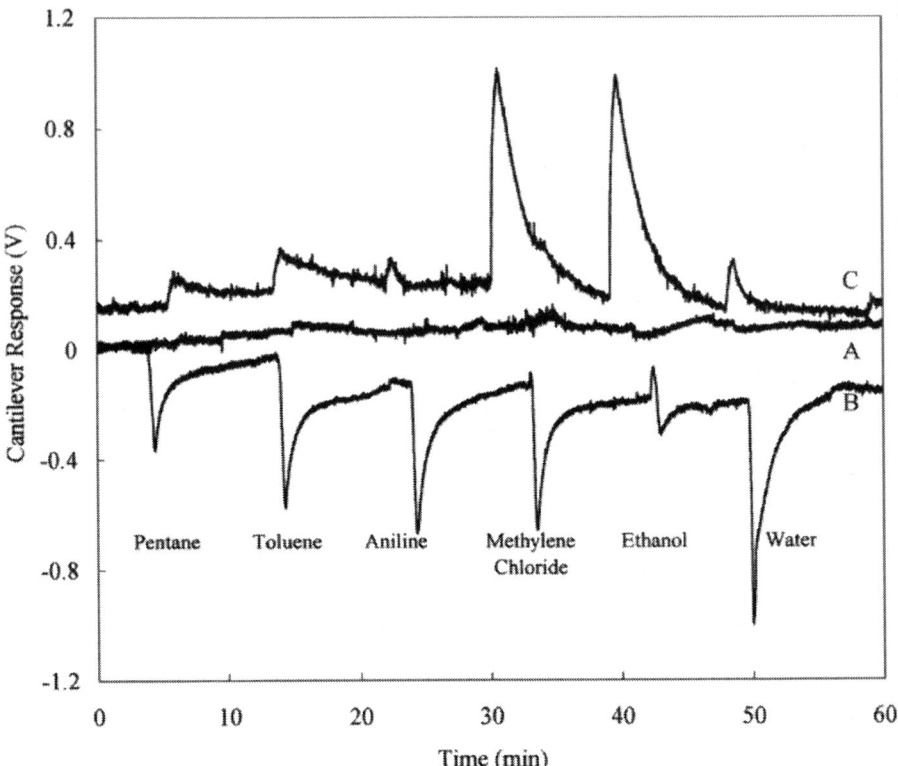

Figure 7. Cantilevers (**A**) treated with Aqua regia; (**B**) coated with gold, removed from one side by a focused ion beam and (**C**) coated with gold and thin films of polymeric chromatographic stationary phases, removed from one side a focused ion beam (reprinted with permission [48]).

Boisen et al. [49] have developed integrated piezoresistive read-out cantilevers and investigated their use as humidity and alcohol sensors. The cantilevers are silicon/silicon oxide layers with integrated polysilicon resistors. For application of cantilever based-sensors as humidity and alcohol sensors in water it was necessary coating one of the sides of the cantilevers with a water absorbing polymer. The humidity was controlled by mixing dry and wet nitrogen gas (2 to 60% of relative humidity) into a chamber. Although the response of the cantilever sensor to humidity change was not linear, it was reproducible. On the other hand, the alcohol sensor in water responded successfully to different amounts of alcohol, being the cantilever deflection proportional to alcohol entering the polymer and expelling water from the film. So, this sensor can be used to detect alcohol in water. Goericke and King [50] have reported finite element simulations of piezoresistive cantilevers. These authors investigated the sensitivity of piezoresistive cantilevers with respect to changes in cantilever length, width and thickness, and piezoresistor size, location, and depth. In the piezoresistive cantilever sensor, the cantilever width is an important parameter, which should be maximized for optimal sensitivity. The cantilever length, however, is not critical for high sensitivity. The sensitivity of piezoresistive cantilevers increases with decreasing piezoresistor thickness. The piezoresistor width need only be kept below 60% of the cantilever width for good sensitivity and the piezoresistor length should be minimized to reduce overall resistance and increase device sensitivity.

Then et al. [47] used a system of six differently polymer-coated cantilevers to analyze quantitatively and qualitatively a series of volatile organic compounds and a mixture of three different components (1-butanol, toluene and n-octane). The polymers were chosen to cover a wide range of polarities and different chemical behavior. The coating of the six different polymers (polydimethylsiloxane, polyetherurethane, poly(cyanopropylmethyldimethyl siloxane), phenylvinyl-polydimethylsiloxane, poly(phenylmethyldimethy lsiloxane) and polyethyleneglycol) on the cantilevers was performed by micromanipulator, patch-clamp pipettes and spray-coating. The array of the six differently coated cantilevers together with principal component regression was capable of quantitative analysis of complex gas mixtures as showed for ternary mixture of 1-butanol, toluene, and n-octane. The interactions of a polar polymer with a polar analyte are much stronger and therefore the partition coefficient increases. Partition coefficient is the ratio of concentrations of a compound in the two phases of a mixture of two immiscible solvents at equilibrium. The polar 1-butanol with the ability of additional H-bonding shows up with the lowest detection limit (LOD), toluene which can be polarized as an aromatic system ranges in the middle and the non-polar n-octane has the highest LOD.

Dong et al. [51] assessed three polymer layers (polyethyleneoxide, polyvinylalcohol and polyethylenevinylacetate) on resonant microcantilever for the detection of volatile organic compounds. They coated the surface of the microcantilever with an air brush and pipettor. The polarity of the volatile compound strongly influenced its diffusivity in the polymer layer. It was observed that the most hydrophilic (polyvinylalcohol) and hydrophobic (polyethylenevinylacetate) polymers showed the lowest sensitivity to the least polar solvents (hexane and octane) and most polar solvents (ethanol and acetone), respectively.

Silicon microcantilevers for humidity detection were studied by Singamaneni et al. [52]. The cantilever with a low flexural rigidity was coated with a plasma-polymerized methacrylonitrile monolayer [Figure 8(a,b)] and it showed for instantaneous changes of humidity a response time of 9.5 ms in the range from 10 to 90% of humidity [Figure 8(c)]. It was observed a linear response in the ramps of humidification and desiccation (1% of relative humidity per second) showed in [Figure 8(d)], with resolution of ±10 ppb of water vapor. The authors suggested that the desirable responsive behavior of the bimaterial cantilever is associated with the integration of a crosslinked polymer, high internal stresses, and firm adhesion to the silicon.

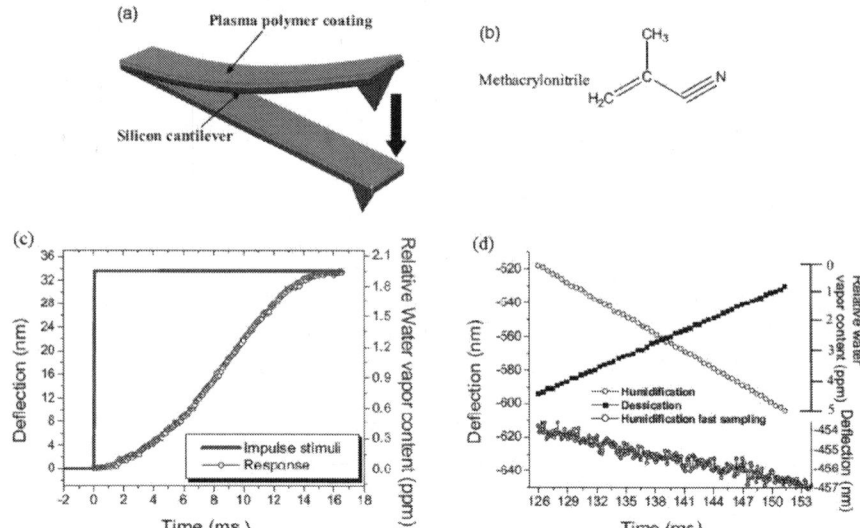

Figure 8. (a) Simplified diagram of the functionalization of the silicon cantilever coated with polymer; (b) Chemical structure of methacrylonitrile; (c) Deflection of cantilever in response to sudden humidity change and (d) Dynamic sampling during linear humidification and desiccation (reprinted with permission [52]).

Sub-ppm sensitivity and the response of piezoresistive cantilevers arrays coated with different polymers [polyvinyl alcohol (PVA), polyethylene imine, polyacryl amide and polyvinyl pyrrolidone] to detect various alkanes were investigates [53]. It was demonstrated that this array of sensors coated with different polymers had the selectivity of discriminated individual alkanes in a homologous series using principal component analysis of the pattern formed from the responses of polymer-coated cantilevers. Accordingly, the sensitivity of the piezoresistive cantilever was determined to be 0.05–0.13 ppm. The adsorption of analytes on the polymer layers induces surface stress due to swelling of the polymers, resulting in the bending of the cantilevers. The adsorption of water molecules on the surface of the cantilever increases its mass and thereby decreases its resonance frequency.

Electronic noses with arrays of cantilever sensors that respond individually to vapors can produce a distinguishable response pattern for each separate type of analyte or mixture. Their use opens up applications in many fields such as quality and process control, medical analysis, fragrance, oenology, and as sensing devices for volatiles compounds [54].

Baller et al. [54] used an artificial nose based on a microfabricated array of eight silicon cantilevers to detect analyte vapors. Each cantilever of the array was functionalized with a different polymer (polyvinylpyridine, polyurethane, polystyrene and polymethylmethacrylate). To observe the transduction of physical and chemical processes into deflection, the swelling of a polymer layer on the cantilever was monitored during exposure to the analyte. It was observed that the swelling process is related to the vapor pressure and the solubility characteristics of the analyte in the polymer. A standard recognition patterns was created upon exposure to known gaseous analytes, so that alcohols, acetone, dichloromethane, toluene and heptane could be identified.

Lang et al. [55] also evaluated the use of an artificial nose with a microcantilever sensor array to characterize solvent vapors and fragrances. All eight cantilevers were functionalized with a polymer layer by the inkjet spotting technique. The deflection was measured with a surface emitting laser and a position-sensitive detector received the optical beam. The cantilever deflection in response to solvent vapors and fragrances was caused by the diffusion of the molecules, resulting in a swelling of the polymer layers. When the cantilever was exposed to dry nitrogen, the beam returned to its initial position, demonstrating that the sensing process is reversible and reproducible.

4. MICROCANTILEVER-BASED BIOSENSORS (MC-B)

The great interest in developing microcantilever-based biosensors stems from the binding specificity of biomolecules for analytical sensing, combined with the unlimited ability of this device to transduce signals between the biomolecule and an electronic device. According to the International Union of Pure and Applied Chemistry (IUPAC) definition [56], a biosensor is a sensor possessing biological components (antibodies, enzymes, DNA), which detects their interaction with the analyte of interest by a dedicated transduction mechanism. The transducer is the device capable of converting the physical (e.g., resistance, voltage, conductivity) or chemical signals (electron transfer from a chemical reaction) into measurable, usually electronic, signals whose magnitude is proportional to the concentration of the species or chemical grouping.

The advantages of microcantilever-based biosensors are small size, label-free detection, and a potential for arrayed operation [57]. Instrumentation based on microcantilever arrays would be very attractive because it could detect multiple biomarkers simultaneously with high sensitivity and selectivity in small sample volumes [58].

Microcantilevers are micromechanical devices with dimensions on the order of 100 μm × 30 μm × 0.6 μm [6]. They can readily be fabricated on silicon wafers and other materials, but silicon microcantilevers are the most commonly used in biosensors. They are physical sensors that respond to surface stress changes due to chemical or biological processes [59]. These sensors can measure forces and stresses with extremely high sensitivity when fabricated with very small force constants [58]. Adsorption of molecules on one of the surfaces of the biomaterial cantilever results in a differential surface stress due to adsorption-induced forces, which is manifested as a deflection [58,60]. In addition to cantilever bending, the resonant frequency of the cantilever can be changed by mass loading [58,61]. These two types signal, namely the adsorption-induced cantilever bending when adsorption is confined to one side of the cantilever and the adsorption-induced vibrating frequency change due to mass loading, can be monitored simultaneously [40]. Resonant cantilevers immersed in liquid suffer from high damping losses and reduced sensitivity [62]. Thus, the resonant frequency operating mode of the cantilever is normally limited to the detection of gas-phase samples, while the bending mode can be used for both gas and liquid samples [57,62].

The commonest method used to measure cantilever displacement is the optical lever (see Figure 5) [63,64]. A focused laser beam is reflected off the cantilever surface, and captured by a PSD (position sensitive detector). The cantilever displacement causes movement of the laser spot on the PSD and hence a change in its output voltage [65]. In microcantilever biosensors, the accuracy of measurement depends strongly on accurate determination of the surface deflections. This optical lever approach is not suitable for a cantilever array, because the response of only one device can be captured. Custom-made arrays of lasers and PSDs for use with several cantilevers in parallel lead to greatly increased instrumentation complexity and difficulty of alignment [66]. The cantilever bending may also be sensed with a deflection sensor integrated into the cantilever [38]. A major advantage of this set-up is that it allows massively parallel arrays and can be performed without the need for optics. For deflection sensing with elements integrated into the cantilever, piezoresistive sensing is the most widespread approach [67], although there are others, including piezoelectric [68] or capacitive sensors [69]. The high piezoresistive coefficient of doped single-crystal silicon makes silicon piezoresistive sensors an attractive option [70].

The sensitivity of microcantilever biosensors can be increased by optimizing the geometry of the cantilevers and immobilization techniques. For adsorption-induced cantilever deflection, longer and thinner cantilevers with small force constants show higher sensitivity. However, larger area cantilevers show faster detection of low concentrations of target molecules. Thus, the optimal cantilever dimension will depend on the dimension of the cantilever chamber and the analyte delivery system [58]. The use of a reference cantilever basically improves sensitivity. New piezoresistive cantilevers designs have been developed that show less drift and higher signal-to-noise ratios. Selectivity of detection with microcantilever sensors in complex samples still remains to be solved [71].

The functionalization of the microcantilever surface is the most important step in the development of microcantilever-based biosensors. This step is critical to cantilever sensing because it determines the surface density of receptor molecules and thus the number of binding events on the active surface. It may also affect how efficient the transduction is from the biomolecular interaction to the change in surface stress on the cantilever. Additionally, an effective passivation can significantly reduce the nonspecific binding, so that the background signal can be minimized. Depending upon the final application of the device, various types of immobilization can be used. Generally, cantilevers are coated on one side with 2–3 nm of chromium and 25–30 nm of gold. Chromium acts as an adhesive layer for the gold. When the functionalization of the microcantilever surface involves silanization, both side of the cantilever is used; and when the coating involves thiol, only the gold side of the cantilever is used. For thiol self-assembled monolayers (SAMs) and organosilane modification, dip coating is the preferred method for functionalization, to allow high density immobilization on the cantilever surface. Thiol SAMs are self-limited to coverages of a monolayer of the thiol on a gold film. Silane coating can yield multilayered films upon extended exposure to the solution. Regardless of the type of coverage, it must be prepared no more than 48 h before analysis [58].

Since 1990, the technique of sensing by means of the bending response of microcantilevers is being increasingly used for the detection of biomacromolecules, rather than small molecules. A. Subramanian et al. [6] described a glucose oxidase-coated microcantilever for the specific detection and quantitation of glucose and they assessed the signal transduction mechanism. The enzyme was immobilized on one side of a silicon cantilever, coated with gold. The magnitude of the bending response of the glucose oxidase-derivatized microcantilever was concentration dependent over a large range of glucose concentrations.

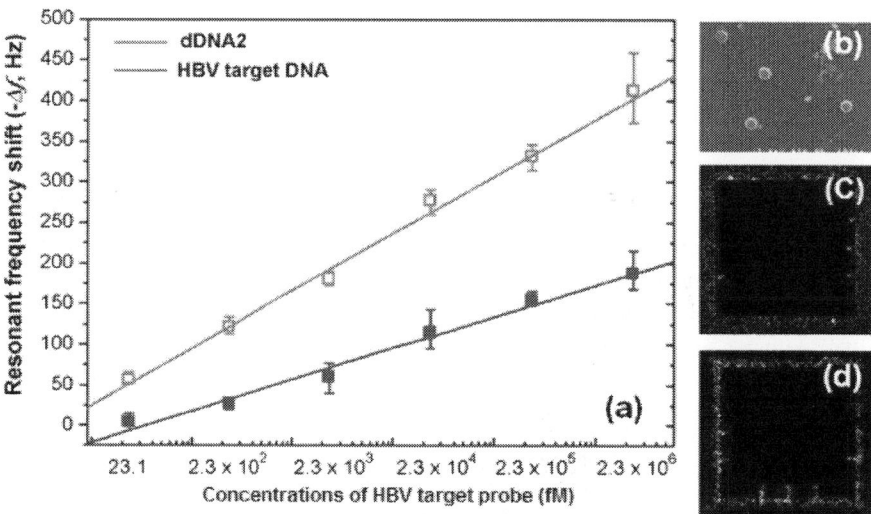

Figure 9. The results of the HBV DNA assay with silicon-nanoparticle (SiNP) enhanced dynamic microcantilevers: (**a**) plots of the resonant frequency shifts acquired from the HBV DNA assay and the SiNP enhanced HBV DNA assay; (**b**) SEM image of the microcantilever surface with the captured SiNPs at 2.3 pM HBV target DNA, and the fluorescent images of the microcantilevers; (**c**) top side and (**d**) bottom side at 2.3 pM (reprinted from [72] with permission).

The combination of the microcantilever with a highly specific enzyme provides a unique approach to quantifying enzyme substrates without the complication of sample labeling. The detection of hepatitis B virus DNA was achieved with a silica nanoparticle-enhanced microcantilever sensor. Gold-coated microcantilevers were functionalized with thiolated cDNA. Cr/Au layers (10 nm/50 nm) were deposited on the bottom side of the microcantilever with an e-beam evaporator. The resonant frequency was measured by an impedance analyzer and the working concentration range was approximately from 23 fM to 2.3×10^6 fM (Figure 9) [72].

A class of microcantilever-based biosensors is the microcantilever immunosensors, which utilize highly selective antigen-antibody interactions. The immunosensors offer a common platform for high-throughput, multiplexed label-free analyses of biomolecules in a single step in real time, based on specific biomolecular binding, such as protein-protein binding. When antibody molecules are immobilized on one surface of a cantilever, specific binding between antibodies and antigens produces cantilever deflection. Zhao et al. [73] have developed a microcantilever immunosensor and an indirect competitive enzyme-linked immunosorbed assay (icELISA), which use a highly sensitive and specific monoclonal antibody (designated mAb6A9) against a copper-chelate complex. The half maximum inhibition concentration values (IC_{50}) and working range, based on 10–90% inhibition of binding of mAb6A9 to Cu(II)-ethylenediamine-N,N,N',N'-tetraacetic acid (EDTA) [Cu(II)-EDTA], of the icELISA were approximately 1.8 ng·mL^{-1} and 0.2–17 ng·mL^{-1}, respectively. A bending response of the microcantilever immunosensor was detectable at or below 1 ng·mL^{-1} of Cu(II)-EDTA complex. The two assays developed (microcantilever immunosensor and icELISA) were sensitive enough to monitor Cu(II) in drinking water at the legal limits set by China, the U.S. Environmental Protection Agency (EPA) and World Health Organization (WHO). The results correlated well with those obtained by graphite furnace atomic absorption spectrometry.

Among the biomolecular detection assays, the microcantilever technology can also be used for DNA-DNA hybridization detection, including accurate positive/negative detection of single base-pair mismatches [74]. This type of assay can be used to detect mutations in the DNA sequence of target DNA responsible for many cancers. The detection occurs where a single non-complementary nucleotide appears in the sequence. Single-stranded DNA (ssDNA) is immobilized on the side of the cantilever coating with gold, by means of a thiol linker bonded to one end of the ssDNA. The change in surface stress resulting from the adsorption of ssDNA on a cantilever has been found experimentally to be 30–50 mN/m. Of particular interest in current genomics research is the detection of single-nucleotide polymorphisms (SNPs). The ability to locate and characterize SNPs would aid in the early detection, diagnosis, and perhaps treatment of individuals carrying mutations causing diseases such as cystic fibrosis and thalassaemia, among others. Hansen et al. [34] have used hybridization-induced cantilever deflection to demonstrate that the number and location of DNA pair mismatches in a target 10-mer oligonucleotide can be discerned under high-stringency static and flow conditions using gold-coated silicon microcantilevers. Gold-coated silicon microcantilevers were functionalized with thiolated 20- or 25-mer probe DNA oligonucleotides. The magnitude of deflection was greater for the hybrids with two internal mismatches than for those with one internal mismatch, which indicates that the degree of repulsion increased due to additional base-pair mismatch.

Prostate-specific antigen (PSA) was assayed by optically detecting nanoscale motions of two-dimensional arrays of microcantilever beams [75]. Antibodies were used, covalently bound to one surface of the microcantilevers. The 2D cantilever array chip contained 80–120 reaction wells, where each well consisted of a microfluidic chamber containing 4–8 independent cantilever sensors. The surface of the microcantilever was coated with a 25-mm layer of gold, which served as the surface for immobilizing capture molecules (antibodies). A collimated laser light beam with an expanded spot size about the same size as the whole cantilever array illuminated the gold surface of the cantilevers from the glass side. The laser light reflected off each cantilever's end pad and was collected as an array of "spots" by a charge coupled device (CCD) camera. In this study, the authors used 2-[methoxypoly(ethyleneoxy)propyl]trimethoxysilane, a silane-conjugated polyethylene-glycol chain (henceforth referred as PEG-silane) for the effective surface passivation. Quantitative detection of PSA as low as 1 ng/mL was demonstrated with an array of 400 µm long cantilevers, which yielded 2 mJ/m^2 surface stress change due to the binding of the antibody and antigen.

There have been some efforts to modify the design of the microcantilever, with the purpose of developing microcantilever-based biosensors with greater sensitivity and signal-to-noise ratio. Godin et al. [76] showed that the bending of the cantilever depends on the surface roughness of the gold film and gold films with larger grain sizes on the cantilever showed increased bending sensitivity. Tabard-Cossa et al. [77] investigated the surface stress response of micromechanical cantilever based sensors as a function of the morphology, adhesion, and cleanliness of the gold sensing surface. The surface morphology was found to influence strongly the surface stress response for molecular adsorption where ordering of the molecules is dependent on their coverage and domain size. These authors concluded that the sensitivity of bending also depends on the uniformity of the immobilization layer and cleanliness of sensing surface [77].

Ansary and Chao [78] have investigated the deflection and vibration analysis of rectangular, triangular, and step profile microcantilevers having basic and modified shapes. The surface stress induced deflection in the microcantilever is modeled by an equivalent in-plane tensile force acting on the top surface of the cantilever, in the length direction. To increase the overall sensitivity of microcantilever biosensors, both the deflection and the resonant frequency of the cantilever should be increased at the same time. The triangular and step cantilevers have better deflection and frequency characteristics than rectangular ones. The deflections usually range a few tens to a few hundreds of nanometers. Measuring deflections of this order requires extremely sophisticated readout arrangements. As the sensitivity of a microcantilever biosensor depends on the design sensitivity of the cantilever and the measurement sensitivity of the deflection measurement system, the challenge is to increase the sensitivity of a microcantilever without increasing the complexity of the deflection detection system. These authors [79] also presented a new microcantilever design with a rectangular hole at the fixed end of the cantilever, which was about 75% more sensitive than the conventional design, and the frequency analysis showed that the natural resonant frequency of the proposed design was about half the conventional frequency, i.e., $f_{0, proposed} = 0.47 f_{0, conventional}$.

The piezoresistive cantilever based biosensors are sensitive to temperature change, owing to their relatively large surface-to-volume ratio, which can have some impact on the sensor stability [80]. The geometrical parameters of this type of sensor should be optimized for optimal sensitivity. The sensitivity increases with decreasing cantilever thickness due to the reduced stiffness of the device when the ratio of piezoresistor thickness to cantilever thickness is fixed [80].

Temperature compensation has been proposed, in which a parallel microcantilever is used, with one active cantilever and another reference cantilever for noise reduction, but the different composition of the cantilevers may lead to a serious offset [81]. Yang et al. [81] proposed a bridge circuit composed of an active cantilever, a reference cantilever and two fixed piezoresistors for in situ surface stress measurements of a biochemical reaction. This circuit proved effective in offset voltage adjustment and temperature drift compensation.

Microcantilever based-biosensors can also be used in a liquid environment. Kwon et al. [82] reported the in situ real-time monitoring of a specific protein antigen-antibody interaction, using a resonating microcantilever immersed in a viscous fluid. The precise in situ real-time monitoring of protein-protein interactions was ascribed to the high quality factor of the resonating piezoelectric thick-film microcantilever.

5. CONCLUSIONS

In this review we have attempted to show the wide field of application of the atomic force microscopy (AFM) technique. Among the AFM technique, the Chemical Force Microscopy (CFM) which is based on AFM tips chemically modified with specific exposed functional groups, carefully architected to carry out a specific function in a system, appears as a powerful tool. The possibility of controlling the chemical interactions between the AFM tip and the sample can provide sensitive resources to measure and to map the surface chemistry information and to quantify the adhesive or repulsive forces associated to inorganic materials and biological samples.

The potential use of microcantilevers in a broad range of applications is also arising as a powerful tool coming from the AFM. Depending on the specific application of microcantilevers, they may be functionalized either with biological material, which converts them into microcantilever-based biosensors (MC-B) or may be functionalized with inorganic materials, which converts them into Nanomechanical Cantilever Sensors (NCS). There are two types of signals in these devices, which can be monitored simultaneously, adsorption-induced cantilever bending when adsorption is confined to one side of the cantilever and adsorption-induced vibrating frequency change due to mass loading. Depending upon the final application of the device, variety of techniques can be used to functionalize the microcantilever surface. The method used in the functionalization has a strong influence on the sensitivity and selectivity of device since they depend on the detection layer, which may be built according to principles of molecular recognition. This technology enables the development of sensors that can detect low concentrations of analytes in very small volumes of liquid or gas, with great sensitivity, reproducibility and repeatability.

ACKNOWLEDGMENTS

The authors would like to thanks Embrapa Instrumentation and Federal University of Sao Carlos (UFSCar), CNPq (IBAS 380899/2010-8; Universal 14 proc. 483303/2011-9) and FAPESP (proc. 2010/04188-6; proc. 2009/08244-0; proc. 2007/05089-9; proc. 2008/57862-6), INCT-NAMITEC (CNPq 573738/2008-4) and AgroNano Network (10.01.001.02.01) for the infrastructure and financial support for research.

REFERENCES

1. Thundat, T., Majumdar, A., Eds.; Sensors and Sensing in Biology and Engineering; Springer-Verlag: Wien, New York, NY, USA, 2003; p. 399.
2. Fagan, B.C.; Tipple, C.A.; Xue, Z.; Sepaniak, M.J.; Datskos, P.G. Modification of micro-cantilever sensors with sol-gels to enhance performance and immobilize chemically selective phases. Talanta **2000**, 53, 599–608.
3. Carrascosa, L.G.; Moreno, M.; Alvarez, M.; Lechuga, L.M. Nanomechanical biosensors: A new sensing tool. Trac Trend Anal. Chem. **2006**, 25, 196–206.
4. Swart, J.W.; Diniz, J.A.; Frateschi, N.C.; Fruett, F.; Moshkalev, S. Nanotecologia e Semicondutores. In Visão Tecnológica e Social para o Agronegócio; Vaz, C.M.P., Junior Herrmann, P.S.P., Melo, W.L.B., Eds.; Ciclo de Colóquios da Embrapa Instrumentação Agropecuária: São Carlos, Brasil, 2008; pp. 41–64.
5. Pei, J.; Feng, T.F.; Thundat, T. Glucose biosensor based on the microcantilever. Anal. Chem. **2004**, 76, 292–297.

6. Subramanian, A.; Oden, P.I.; Kennel, S.J.; Jacobson, K.B.; Warmack, R.J.; Thundat, T.; Doktycz, M.J. Glucose biosensing using an enzyme-coated microcantilever. Appl. Phys. Lett. **2002**, 81, 385–387.

7. Hansen, K.M.; Thundat, T. Microcantilever biosensors. Methods **2005**, 37, 57–64.

8. Dong, Y.; Wei, G.; Zhou, Q.; Zheng, Y.; You, Z. Characterization of the gas sensors based on polymer-coated resonant microcantilevers for the detection of volatile organic compounds. Anal. Chim. Acta **2010**, 671, 85–91.

9. Smith, D.A.; Connell, S.D.; Robinson, C.; Kirkham, J. Chemical force microscopy: Applications in surface characterisation of natural hydroxyapatite. Anal. Chim. Acta **2003**, 479, 39–57.

10. Ito, T.; Grabowska, I.; Ibrahim, S. Chemical force microscopy for materials characterization: Investigations of host-guest interactions and polymer surface chemistry. Trends Anal. Chem. **2010**, 29, 225–233.

11. Leite, F.L.; Herrmann, P.S.P. Application of atomic force spectroscopy (AFS) to studies of adhesion phenomena: A review. J. Adhes. Sci. Technol. **2005**, 19, 365–405.

12. Alsteens, D.; Dague, E.; Verbelen, C.; Andre, G.; Dupres, V.; Dufrêne, Y.F. Nanoscale imaging of microbial pathogens using atomic force microscopy. WIREs Nanomed. Nanotechnol. **2009**, 1, 168–180.

13. Azehara, H.; Tokumoto, H. Analysis of the number of hydrogen bond groups of a multiwalled carbon nanotube probe tip for chemical force microscopy. Appl. Surf. Sci. **2009**, 256, 987–990.

14. Dordi, B.; Pickering, J.P.; Schönherr, H.; Vancso, J. Probing chemical reactions on the nanometer scale: Inverted chemical force microscopy of reactive self-assembled monolayers. Surf. Sci. **2004**, 570, 57–66.

15. Zhang, H.; He, H.-X.; Mu, T.; Liu, Z.-F. Force titration of amino group-terminated self-assembled monolayers of 4-aminothiophenol on gold using chemical force microscopy. Thin Solid Films **1998**, 327–329, 778–780.

16. Brant, J.A.; Johnson, K.M.; Childress, A.E. Characterizing NF and RO membrane surface heterogeneity using chemical force microscopy. Colloid Surf. A **2006**, 280, 45–57.

17. Kim, H.; Park, J.H.; Cho, I.-H.; Kim, S.-K.; Paek, S.-H.; Lee, H. Selective immobilization of proteins on gold dot arrays and characterization using chemical force microscopy. J. Colloid Interface Sci. **2009**, 334, 161–166.

18. Noy, A.; Vezenov, D.V.; Lieber, C.M. Chemical Force Microscopy. Annu. Rev. Mater. Sci. **1997**, 27, 381–421.

19. Dordi, B.; Pickering, J.P.; Schönherr, H.; Vancso, G.J. Inverted chemical force microscopy: Following interfacial reactions on the nanometer scale. Eur. Polym. J. **2004**, 40, 939–947.

20. Bastidas, J.C.; Venditti, R.; Pawlak, J.; Gilbert, R.; Zauscher, S.; Kadla, J.F. Chemical force microscopy of cellulosic fibers. Carbohyd. Polym. **2005**, 62, 369–378.

21. Basnar, B.; Friedbacher, G.; Brunner, H.; Vallant, T.; Mayer, U.; Hoffmann, H. Analytical evaluation of tapping mode atomic force microscopy for chemical imaging of surfaces. Appl. Surf. Sci. **2001**, 171, 213–225.

22. Kwok, D.Y.; Neumann, A.W. Contact angle measurement and contact angle interpretation. Adv. Colloid Interface Sci. **1999**, 81, 167–249.

23. Fowkes, F.M. Determination of interfacial tensions, contact angles, and dispersion forces in surfaces by assuming additivity of intermolecular interactions in surfaces. Chem. Phys. **1962**, 66, 382. [Google Scholar]

24. Patankar, N.A. On the modeling of hydrophobic contact angles on rough surfaces. Langmuir **2003**, 19, 1249–1253.

25. Okabe, Y.; Furugori, M.; Tani, Y.; Akiba, U.; Fujihira, M. Chemical force microscopy of microcontact-printed self-assembled monolayers by pulsed-force-mode atomic force microscopy. Ultramicroscopy **2000**, 82, 203–212.

26. Green, N.H.; Allen, S.; Davies, M.C.; Roberts, C.J.; Tendler, S.J.B.; Willians, P.M. Force sensing and mapping by atomic force microscopy. Trends Anal. Chem. **2002**, 21, 64–73.

27. Duwez, A.-S.; Nysten, B. Study of adhesion properties of polypropylene surfaces by atomic force microscopy using chemically modified tips: Imaging of functional groups distribution. Stud. Interface Sci. **2001**, 11, 137–150.

28. Shchukin, D.G.; Möhwald, H. Smart nanocontainers as depot media for feedback active coatings. Chem. Commun. **2011**, 47, 8730–8739.

29. Alsteens, D.; Dague, E.; Rouxhet, P.G.; Baulard, A.R.; Dufrêne, Y.F. Direct measurement of hydrophobic forces on cell surfaces using AFM. Langmuir **2007**, 23, 11977–11979.

30. Dupres, V.; Menozzi, F.D.; Locht, C.; Clare, B.H.; Abbott, N.L.; Cuenot, S.; Bompard, C.; Raze, D.; Dufrêne, Y.F. Nanoscale mapping and functional analysis of individual adhesins on living bacteria. Nat. Methods **2005**, 2, 515–520.

31. Andre, G.; Kulakauskas, S.; Chapot-Chartier, M.-P.; Navet, B.; Deghorain, M.; Bernard, E.; Hols, P.; Dufrêne, Y.F. Imaging the nanoscale organization of peptidoglycan in living Lactococcus lactis cells. Nat. Commun. **2010**, 1.

32. Dague, E.; Alsteens, D.; Latge, J.P.; Verbelen, C.; Raze, D.; Baulard, A.R.; Dufrêne, Y.F. Chemical force microscopy of single live cells. Nano Lett. **2007**, 7, 3026–3030.

33. Steffens, C.; Francheschi, E.; Corazza, M.L.; Corazza, F.C.; Oliveira, J.V.; Herrmann, P.S.P. Gas sensors development using supercritical fluid

technology to detect the ripeness of bananas. J. Food Eng. **2010**, 101, 365–369.

34. Hansen, K.M.; Ji, H.-F.; Wu, G.; Datar, R.; Cote, R.; Majumdar, A.; Thundat, T. Cantilever-based optical deflection assay for discrimination of DNA single-nucleotide mismatches. Anal. Chem. **2001**, 73, 1567–1571.

35. Vashist, K.S. A review of microcantilevers for sensing application. J. Nanotechnol. **2007**, 3, 2–15.

36. Boisen, A.; Dohn, S.; Keller, S.S.; Schmid, S.; Tenje, M. Cantilever-like micromechanical sensors. Rep. Prog. Phys. **2011**, 74, 036101.

37. Singamaneni, S.; LeMieux, M.C.; Lang, H.P.; Gerber, C.; Lam, Y.; Zauscher, S.; Datskos, P.G.; Lavrik, N.V.; Jiang, H.; Naik, R.R.; et al. Bimaterial microcantilevers as a hybrid sensing platform. Ad. Mater. **2008**, 20, 653–680.

38. Lavrik, N.V.; Sepaniak, M.J.; Datskos, P.G. Cantilever transducers as a platform for chemical and biological sensors. Rev. Sci. Instrum. **2004**, 75, 2229–2253.

39. Thundat, T.; Chen, G.Y.; Warmack, R.J.; Allison, D.P.; Wachter, E.A. Vapor detection using resonating microcantilevers. Anal. Chem. **1995**, 67, 519–521.

40. Thundat, T.; Warmack, F.L.J.; Chen, G.Y.; Allison, D.P. Thermal and ambient-induced deflections of scanning force microscope cantilevers. Appl. Phys. Lett. **1994**, 64, 2894–2896.

41. Hartschuh, A. Tip-Enhanced Near-Field Optical Microscopy. Angew. Chem. Int. Ed. **2008**, 47, 8178–8191.

42. Schmidt, U.; Hild, S.; Ibach, W.; Hollricher, O. Characterization of thin polymer films on the nanometer scale with Confocal Raman AFM. Macromol. Symp. **2005**, 230, 133–143.

43. Boerio, F.J.; Starr, M.J. AFM/FTIR: A new technique for materials characterization. J. Adhesion **2008**, 84, 874–897.

44. Baty, A.M.; Suci, P.A.; Tyler, B.J.; Geesey, G.G. Investigation of mussel adhesive protein adsorption on polystyrene and poly(octadecyl methacrylate) using angle dependent XPS, ATR-FTIR, and AFM. J. Colloid Interface Sci. **1996**, 177, 307–315.

45. Stoney, G.G. The tension of metallic films deposited by electrolysis. Proc. R. Soc. Lond. A Mater. **1909**, 82, 172–175. [Google Scholar]

46. Pinnaduwage, L.A.; Boiadjiev, V.; Hawk, J.E.; Thundat, T. Sensitive detection of plastic explosives with self-assembled monolayer-coated microcantilevers. Appl. Phys. Lett. **2003**, 7, 1471–1473.

47. Then, D.; Vidic, A.; Ziegler, C. A highly sensitive self-oscillating cantilever array for the quantitative and qualitative analysis of organic vapor mixtures. Sens. Actuator B Chem. **2006**, 117, 1–9.

48. Betts, T.A.; Tipple, C.A.; Sepaniak, M.J.; Datskos, P.G. Selectivity of chemical sensors based on micro-cantilevers coated with thin polymer films. Anal. Chim. Acta **2000**, 422, 89–99.

49. Boisen, A.; Thaysen, J.; Jensenius, H.; Hanse, O. Environmental sensors based on micromachined cantilevers with integrated read-out. Ultramicroscopy **2000**, 82, 11–16.

50. Goericke, F.T.; King, W.P. Modeling piezoresistive microcantilever sensor response to surface stress for biochemical sensors. IEEE Sens. J. **2008**, 8, 1404–1410.

51. Dong, Y.; Wei, G.; Zhou, Q.; Zheng, Y.; You, Z. Characterization of the gas sensors based on polymer-coated resonant microcantilevers for the detection of volatile organic compounds. Anal. Chim. Acta **2010**, 671, 85–91.

52. Singamaneni, S.; McConney, M.; LeMieux, M.C.; Jiang, H.; Enlow, J.; Naik, R.; Bunning, T.J.; Tsukruk, V.V. Polymer-Silicon Flexible Structures for Fast Chemical Vapor Detection. Adv. Mater. **2007**, 19, 4248–4255.

53. Yoshikawa, G.; Lang, H.-P.; Akiyama, T.; Aeschimann, L.; Staufer, U.; Vettiger, P.; Aono, M.; Sakurai, T.; Gerber, C. Sub-ppm detection of vapors using piezoresistive microcantilever array sensors. Nanotechnology **2009**, 20, 015501.

54. Baller, M.K.; Lang, H.P.; Fritz, J.; Gerber, C.H.; Gimzewski, J.K.; Drechsler, U.; Rothuizen, H.; Despont, M.; Vettiger, P.; Battiston, F.M.; et al. A cantilever array-based artificial nose. Ultramicroscopy **2000**, 82, 1–9.

55. Lang, H.P.; Ramseyer, J.P.; Grange, W.; Braun, T.; Schmid, D.; Hunziker, P.; Jung, C.; Hegner, M.; Gerber, C. An Artificial Nose Based on Microcantilever Array Sensors. J. Phys. Conf. Ser. **2007**, 61, 663–667.

56. Mcnaught, A.D.; Wilkinson, A. IUPAC Compendium of Chemical Terminology; Royal Society of Chemistry: Cambridge, UK, 1997.

57. Li, L. Recent development of micromachined biosensors. IEEE Sens. J. **2011**, 11, 305–311.

58. Datar, R.; Kim, S.; Jeon, S.; Hesketh, P.; Manalis, S.; Boisen, A.; Thundat, T. Cantilever Sensors: Nanomechanical Tools for Diagnostics. MRS Bull. **2009**, 34, 449–454.

59. Thundat, T.; Oden, P.I.; Warmack, R.J. Microcantilever sensors. Microscale Therm. Eng. **1997**, 1, 185–199.

60. Fritz, J.; Baller, M.K.; Lang, H.P.; Rothuizen, H.; Vettiger, P.; Meyer, E.; Guntherodt, H.J. Translating biomolecular recognition into nanomechanics. Science **2000**, 288, 316–318.

61. Su, M.; Li, S.; Dravid, V.P. Microcantilever resonance-based DNA detection with nanoparticle probes. Appl. Phys. Lett. **2003**, 82, 3562–3564.

62. Calleja, M.; Tamayo, J.; Johansson, A.; Rasmussen, P.; Lechuga, L.; Boisen, A. Polymeric cantilever arrays for biosensing applications. Sens. Lett. **2003**, 1, 20–24.

63. Braun, T.; Barwich, V.; Ghatkesar, M.K.; Bredekamp, A.H.; Gerber, C.; Hegner, M.; Lang, H.P. Micromechanical mass sensors for biomolecular detection in a physiological environment. Phys. Rev. E **2005**, 72, 031907.

64. Zhang, X.R.; Xu, X. Development of a biosensor based on laser-fabricated polymer microcantilevers. Appl. Phys. Lett. **2004**, 85, 2423–2425.

65. Koev, S.T.; Bentley, W.E.; Ghodssi, R. Interferometric readout of multiple cantilever sensors in liquid samples. Sens. Actuators B Chem. **2010**, 146, 245–252.

66. Lang, H.P.; Baller, M.K.; Berger, R.; Gerber, C.; Gimzewski, J.K.; Battiston, F.M.; Fornaro, P.; Ramseyer, J.P.; Meyer, E.; Guntherodt, H.J. An artificial nose based on a micromechanical cantilever array. Anal. Chim. Acta **1999**, 393, 59–65.

67. Yang, M.; Zhang, X.; Vafai, K.; Ozkan, C.S. High sensitivity piezoresistive cantilever design and optimization for analyte-receptor binding. J. Micromech. Microeng. **2003**, 13, 864–872.

68. Adams, J.D.; Parrott, G.; Bauer, C.; Sant, T.; Manning, L.; Jones, M.; Rogers, B.; McCorklr, D.; Ferrell, T.L. Nanowatt chemical vapor detection with a self-sensing, piezoelectric microcantilever array. Appl. Phys. Lett. **2003**, 83, 3428–3430.

69. Britton, C.L.; Jones, R.L.; Oden, P.I.; Hu, Z.; Warmack, R.J.; Smith, S.F.; Bryan, W.L.; Rochelle, J.M. Multiple-input microcantilever sensors. Ultramicroscopy **2000**, 82, 17–21.

70. Smith, C.S. Piezoresistance effect in germanium and silicon. Phys. Rev. **1954**, 94, 42–49.

71. Rasmussen, P.A.; Thaysen, J.; Hansen, O.; Eriksen, S.C.; Boisen, A. Optimised cantilever biosensor with piezoresistive read-out. Ultramicroscopy **2003**, 97, 371–376. [Google Scholar]

72. Cha, B.H.; Lee, S.M.; Park, J.C.; Hwanga, K.S.; Kim, S.K.; Lee, Y.S.; Ju, B.K.; Kim, T.S. Detection of Hepatitis B Virus (HBV) DNA at femtomolar concentrations using a silica nanoparticle-enhanced microcantilever sensor. Biosens. Bioelectron. **2009**, 25, 130–135.

73. Zhao, H.; Xue, C.; Nan, T.; Tan, G.; Li, Z.; Li, Q.X.; Zhang, Q.; Wang, B. Detection of copper ions using microcantilever immunosensors and enzyme-linked immunosorbent assay. Anal. Chim. Acta **2010**, 676, 81–86.

74. Wu, G.; Datar, R.H.; Hansen, K.M.; Thundat, T.; Cote, R.J.; Majumdar, A. Bioassay of prostate-specific antigen (PSA) using microcantilevers. Nat. Biotechnol. **2001**, 19, 856–860.

75. Yue, M.; Stachowiak, J.C.; Lin, H.; Datar, R.; Cote, R.; Majumdar, A. Label-free protein recognition two-dimensional array using nanomechanical sensors. Nano Lett. **2008**, 8, 520–524.

76. Godin, M.; Williams, P.J.; Tabard-Cossa, V.; Laroche, O.; Beaulieu, L.Y.; Lennox, R.B.; Grütter, P. Surface stress, kinetics, and structure of alkanethiol self-assembled monolayers. Langmuir **2004**, 20, 7090–7096.

77. Tabard-Cossa, V.; Godin, M.; Burgess, I.J.; Monga, T.; Lennox, B.; Grütter, P. Microcantilever-based sensors: Effect of morphology, adhesion, and cleanliness of the sensing surface on surface stress. Anal. Chem. **2007**, 79, 8136–8143.

78. Ansari, M.Z.; Cho, C. Deflection, Frequency, and Stress Characteristics of Rectangular, Triangular, and Step Profile Microcantilevers for Biosensors. Sensors **2009**, 9, 6046–6057.

79. Ansari, M.Z.; Cho, C. A Study on Increasing Sensitivity of Rectangular Microcantilevers Used in Biosensors. Sensors **2008**, 8, 7530–7544.

80. Khaled, A.R.A.; Vafai, K.; Yang, M.; Zhang, X.; Ozkan, C.S. Analysis, control and augmentation of microcantilever deflections in bio-sensing systems. Sens. Actuator B Chem. **2003**, 94, 103–115. [Google Scholar]

81. Yang, S.M.; Chang, C.; Yin, T.I. On the temperature compensation of parallel piezoresistive microcantilevers in CMOS biosensor. Sens. Actuator B Chem. **2008**, 129, 678–684.

82. Kwon, T.Y.; Eom, K.; Park, J.H.; Yoon, D.S.; Kim, T.S. In situ real-time monitoring of biomolecular interactions based on resonating microcantilevers immersed in a viscous fluid. Appl. Phys. Lett. **2007**, 90, 223903.

CHAPTER 4

Resolving Intra- and Inter-Molecular Structure with Non-Contact Atomic Force Microscopy

Samuel Paul Jarvis

School of Physics & Astronomy, University of Nottingham, Nottingham

ABSTRACT

A major challenge in molecular investigations at surfaces has been to image individual molecules, and the assemblies they form, with single-bond resolution. Scanning probe microscopy, with its exceptionally high resolution, is ideally suited to this goal. With the introduction of methods exploiting molecularly-terminated tips, where the apex of the probe is, for example, terminated with a single CO, Xe or H_2 molecule, scanning probe methods can now achieve higher resolution than ever before. In this review, some of the landmark results related to attaining intramolecular resolution with non-contact atomic force microscopy (NC-AFM) are summarised before focussing on recent reports probing molecular assemblies where apparent intermolecular features have been observed. Several groups have now highlighted the critical role that flexure in the tip-sample junction plays in producing the exceptionally sharp images of both intra- and apparent inter-molecular structure. In the latter case, the features have been identified as imaging artefacts, rather than real intermolecular bonds. This review discusses the potential for NC-AFM to provide exceptional resolution of supramolecular assemblies stabilised via a variety of intermolecular forces and highlights the potential challenges and pitfalls involved in interpreting bonding interactions.

Keywords: atomic force microscopy; intramolecular; intermolecular; NC-AFM; supramolecular; hydrogen; bond

1. INTRODUCTION

The ability to see a single atom, for many, was considered something of an impossible dream, with the smallest units of stable matter, that is, a unit still retaining identifiable chemical properties, remaining an abstract notion. Atoms were of course known to exist, but predominantly inferred from experimental measurements and only visualised via simulations or pictorial representations of their properties. That was until the invention of the field ion microscope (FIM) [1,2] and later the scanning tunnelling microscope (STM) [3,4,5], whose exceptional spatial resolution now lets us routinely visualise the atomic nature of materials in real space, allowing us to directly see single atoms. The development of the STM triggered a whole host of complementary methods, known under the umbrella term of scanning probe microscopy (SPM), which utilise almost every conceivable type of detectable tip-sample interaction. Arguably the most intuitive interaction of these is force, which has given us the ability to feel materials at the atomic and molecular level, much as we are familiar with interacting with everyday objects. This led to the development of one of the most challenging, yet exciting scanning probe methods for atomic scale interrogation: the non-contact atomic force microscope (NC-AFM) [6,7,8,9].

NC-AFM has provided ever more detailed insights into the atomic world and has been used to manipulate single atoms at room temperature [10,11], to yield chemical resolution [12] and to measure the force required to move atoms and molecules [13]. Arguably, however, one of NC-AFM's greatest recent achievements has been to characterise single molecules with unprecedented detail, resolving their internal structure with atomic resolution. Although STM measurements have for many years revealed submolecular features, the STM is limited in that it provides information on the electronic structure of the molecule arising from its frontier molecular orbitals, which often bear little resemblance to the real atomic geometry.

Similar to the breakthrough that STM provided for imaging single atoms of crystal surfaces, NC-AFM (and in fact, the scanning tunnelling hydrogen microscopy (STHM) [14] variant of STM, as discussed later) can now directly provide beautiful real space images of the atomic structure of individual molecules, revealing a vivid appearance sharing an amazing similarity to school textbook ball-and-stick drawings. This unparalleled capability has led to an explosion of interest, fettered only by the difficulty in instrument operation and in achieving the highly-controlled environments required.

This review summarises many of the advances NC-AFM has made in characterising molecules with submolecular resolution. It begins with a summary of the techniques available to achieve submolecular resolution, collecting together many of the key studies now reporting images resolving features relating to the intramolecular structure of single molecules. The effect of flexure in the tip-sample junction is then discussed, which turns out to be essential for enhancing the appearance of the bond structure within molecules.

The review then proceeds to discuss recent exciting work investigating assemblies of molecules stabilised via intermolecular forces, with a particular focus on hydrogen bonding, where the question is posed: can NC-AFM resolve and uniquely identify single intermolecular bonds? The many potential challenges in answering these questions are then discussed within the context of a larger number of supramolecular structures stabilised through a variety of intermolecular interactions.

2. INTRA-MOLECULAR RESOLUTION-RESOLVING INTERNAL BOND STRUCTURE WITH NC-AFM

In order to save both the reader (and the author) from a great deal of confusion, I wish to define at this point the terminology that shall be used throughout this review. From this point on, when referring to intra-molecular resolution, that is, resolving submolecular structure relating to the atoms and bonds within a single covalently bound molecule, I will instead refer to resolving the internal bond structure of the molecule. This distinction is made to avoid confusion when discussing effects arising from inter-molecular forces (that is, non-covalent interactions between molecules) and features that will often be discussed alongside internal molecular bond structure.

The internal bond structure of a molecule (see reference [15] for an overview of submolecular resolution with various SPM techniques) was first observed by Temirov et al. [14] for perylene-tetracarboxylic-dianhydride (PTCDA) and tetracene, who introduced molecular hydrogen (H_2) into an STM chamber during scanning. Due to the low temperature of the scan head (~ 10 K), the H_2 would spontaneously condense and trap itself within the tunnelling junction, significantly enhancing the resolution of the observed image (see Figure 1A). In this so-called scanning tunnelling hydrogen microscopy (STHM) mode of imaging, the position of the trapped H_2 molecule is determined by the degree of Pauli repulsion felt within the small (<1 nm) tip-sample junction [16], causing the trapped molecule to effectively act as a transducer, modulating the STM signal via changes in the degree of Pauli repulsion. As will be described in more detail below, the effect of Pauli repulsion, which is largest when the tip is directly above the atoms and bonds of the molecule, is essential for achieving internal bond resolution. In addition to a H_2 molecule, D_2, CO, Xe and CH_4 [17] have also all been shown to produce very similar results, demonstrating the general applicability of the STHM method.

Submolecular resolution in NC-AFM was first achieved using the qPlus [18] setup operated in the frequency modulation mode [19]. In the qPlus setup, the tuning fork is typically oscillated at its first eigenfrequency, usually at around 20–30 kHz. The relatively high stiffness of the tuning fork (nominally 1800 $N \cdot m^{-1}$, although measurements often vary [20,21]) enables sub-Angstrom oscillation amplitudes to be reached, which is often considered essential for atomic resolution imaging of molecules (although recent work now demonstrates that submolecular resolution is also achievable with cantilever AFM with large amplitudes up to ~17 nm [22,23]).

In their seminal report, Gross et al. [24] successfully resolved the internal bond structure of pentacene using NC-AFM, shown in Figure 1B. Although somewhat similar STHM images had been published a year earlier, Gross et al. captured the clearest real space images of the internal bond structure of a molecule to date. Moreover, they were able to collect detailed quantitative information on the interaction responsible for imaging (as NC-AFM is sensitive to the tip-sample force gradient), making the images much simpler to interpret than the earlier STHM results. The fundamental mechanisms underlying contrast formation according to Gross et al. (at least to a first approximation) have an elegant simplicity, primarily relying on two prerequisites: (1) the tip must be chemically passivated, such that it weakly interacts with the surface-adsorbed molecule; and (2) the tip must be "sharp" such that its radius is sufficiently small to resolve atomic features. These two requirements enable the scanning probe to be placed extremely close to the surface-adsorbed molecule, such that a repulsive force is felt between the scanning probe and the molecule arising from Pauli repulsion [25,26]. Due to the strong localisation of the electronic density directly above the atomic positions of the molecule, the repulsion is strongest when the probe is positioned directly over the atoms and bonds. Therefore, a sufficiently sharp atomic probe can trace the corrugations of repulsion with atomic resolution, thus producing such exceptional images.

The first prerequisite in particular, that the tip must be chemically passivated, was essential for Gross et al.'s work (although it has since been observed that reactive tips can also achieve submolecular resolution when the molecule is strongly bound to the substrate [32,33,34].) Typically, either an etched W or cut PtIr wire is glued to the free tine of the tuning fork, thus acting as the tip. Without further tip functionalisation, these metallic tip structures make submolecular imaging on weakly binding substrates unfavourable, as the tip's high reactivity usually induces lateral or vertical manipulation of the target molecule during imaging (long before the region where internal bond resolution can be achieved). To counter this problem, Gross et al. used a well-established technique from low temperature STM experiments [35,36,37] where a single small molecule is "picked up" from the surface and used to terminate the scanning probe apex.

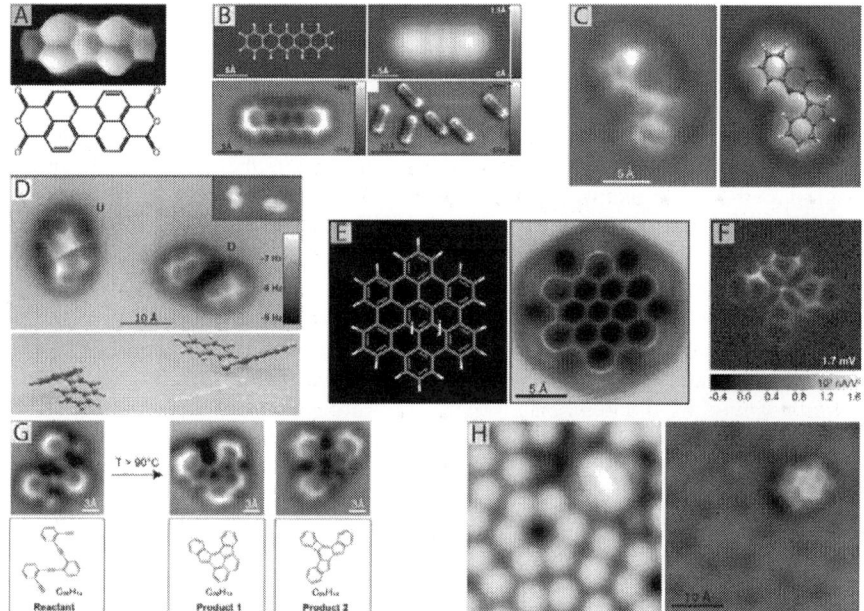

Figure 1. Imaging internal bond structure with CO-mediated non-contact atomic force microscopy (NC-AFM). (**A**) Internal bond structure of perylene-tetracarboxylic-dianhydride (PTCDA) resolved with scanning tunnelling hydrogen microscopy (STHM) using a trapped hydrogen molecule in the scanning tunnelling microscope (STM) tunnel junction [14,16]; (**B**) First images of internal bond structure resolved with NC-AFM achieved via pick-up of a single CO molecule onto the tip apex (from [24], reprinted with permission from AAAS); (**C**) NC-AFM images showing molecular structure identification of cephalandole A adsorbed on NaCl (2 ML)/Cu(111). Red and blue coloured atoms correspond to oxygen and nitrogen, respectively (reprinted by permission from Macmillan Publishers Ltd.: Nature Chemistry [27], copyright 2010); (**D**) NC-AFM reveals bistable configurations of dibenzo[a,h]thianthrene (DBTH) adopting a "butterfly" arrangement. Yellow coloured atoms correspond to sulphur (reprinted with permission from [28], copyright 2012 by the American Physical Society); (**E**) Pauling bond order discrimination in hexabenzocoronene with NC-AFM (from [29], reprinted with permission from AAAS); (**F**) Inelastic tunnelling spectroscopy (IETS) image revealing chemical bonds with a CO-terminated tip (from [30], reprinted with permission from AAAS); (**G**) NC-AFM images of the different steps of a chemical reaction with submolecular resolution (from [31], reprinted with permission from AAAS); (**H**) STM (left) and NC-AFM (right) imaging of the structure and adsorption site of naphthalene tetracarboxylic diimide (NTCDI) on the Si(111)-7×7 surface at 77 K [32,33].

Following a similar method to Bartels et al. [36,37], Gross et al. first ensured a sharp metallic tip by picking up single atoms of Au or Ag deposited onto a bilayer film of NaCl on Cu(111) (NaCl (2 ML):Cu(111)), achieved by approaching the scanning probe ~4 Å towards the metal adatom (starting at a height defined by an STM tunnelling set point of $I = 2$ pA at 200 mV applied sample bias). Subsequently, the metal tip was then either terminated with a single CO molecule using a voltage pulse of 2.5 V, a Cl ion removed from the NaCl layer or one of the surface-adsorbed pentacene molecules. Out of these four tip terminations, only the CO- and Cl-terminated tips were able to resolve the surface-adsorbed pentacene with internal bond resolution, with the CO proving to be the most effective. In the case of an Ag/Au-terminated tip, Condition 1 was not met, and the target molecule was manipulated before the repulsive regime could be reached. For the pentacene-terminated tip, Condition 2 was not met, as the pentacene renders the probe sufficiently complex that a single sharp point is no longer present.

Interestingly, it was found that CO provided much greater apparent spatial resolution compared to Cl, which can be explained not only by the smaller spatial extent of the O atom's electron density as compared to Cl, but also due to the effect of the flexibility of the CO molecule around the scanning probe, a topic discussed in much greater detail later in this review. In addition, Br and Xe tip terminations have also been tested [38], resulting in similar, albeit slightly poorer, improvements in resolution as found for CO-terminated tips. Due to the small size of the molecular termination, it is typically essential to perform experiments in a controlled environment at 5 K temperature, where diffusion can be avoided in order to maintain a stable molecular-terminated tip. Despite this, however, some studies have shown that submolecular resolution in NC-AFM can be achieved at temperatures as high as room temperature [23,39] and 77 K [22,32,40]. At 77 K , improved resolution was found to originate from spontaneous functionalisation of the scanning probe with much larger molecules, such as naphthalene tetracarboxylic diimide (NTCDI) [32,40] (where it was found that the C=O units of the NTCDI molecule behaved in a similar fashion to a single CO molecule) or the formation of TiO_2 tip clusters [22]. It has also been shown that several other spontaneously occurring tip structures may give rise to submolecular resolution, particularly in semiconductor systems [32,41]. Other notable examples observing internal molecular bond structure with non-CO terminated tips were reported for decastarphene molecules on Cu(111) [42], 4-(4-(2,3,4,5,6-pentafluorophenylethynyl)-2,3,5,6-tetrafluorophenylethynyl)phenylethynylbenzene (FFPB) on Au(110) [43] and submolecular resolution observed on C_{60} molecules [44,45,46].

Since the initial report by Gross et al. [24] there has been an explosion of interest in CO-terminated probes applied to ever increasing numbers of molecular species, revealing exciting physics and chemistry at the submolecular scale. One of the most exciting prospects of CO-mediated NC-AFM is its ability to unambiguously reveal the unknown structure of molecular species, particularly in the case where other spectroscopic measurements, such as NMR, fail to identify a unique molecular structure. This was first shown by Gross et al., who determined the structure of cephalandole A [27] adsorbed onto a thin film of NaCl, identifying a single compound out of a possible four (see Figure 1C). Later, the technique was also applied to much more complex molecules [47], breitfussin A and B, containing multiple chemical species, including I, Br, O and N. Despite the complexity of the molecular geometry, including deviations from a perfectly planar arrangement, NC-AFM images were able to resolve the detailed molecular architecture, greatly assisting in identifying a unique structural model and pointing towards the possibility of chemical identification. This highlights one of the major current challenges of the technique, as molecules deviating too far from a planar arrangement can be particularly challenging to image.

In addition to determining the internal structure of organic molecules, CO-mediated NC-AFM has shown excellent potential for determining conformational properties by directly imaging distortions of the molecular skeleton. For instance, the change in geometry of a PTCDA molecule during reversible covalent bond formation with a single gold atom [48] was observed. It was found that translation of the gold atom, initially located beside the molecule, to a position beneath it, caused a significant tilt in the molecule, clearly resolved in the NC-AFM image. Bistable configurations of dibenzo[a,h]thianthrene (DBTH) were also identified, and their adsorption site directly determined from simultaneous imaging of the molecule and surface [28,49]. Although STM measurements were able to differentiate between each conformer, only with the benefit of submolecular NC-AFM images were the exact adsorption geometries of each conformer identified (see Figure 1D). The adsorption geometries of several other molecules have also been identified [50,51], including members of the so-called olympicene family of benzopyrenes, which each showed variations in adsorption height and molecular tilt with respect to the surface plane when adsorbed on Cu(111).

Amazingly, beyond even imaging the internal bond structure of a molecule, CO-mediated NC-AFM provides important information regarding the type of atomic bond, specifically the bond order of individual carbon-carbon bonds within aromatic hydrocarbons. As demonstrated by Gross et al. in 2012 [29] variations in bond length corresponding to bond order were detected in NC-AFM images of C_{60} fullerenes and hexabenzocoronene (pictured in Figure 1E). The images revealed that carbon-carbon bonds with higher bond order consistently appeared shorter, and in some cases brighter (potentially arising from increased electronic density) than their lower bond order counterparts. As will be returned to later, the flexibility of the CO attached to the scanning probe is essential for providing such exquisite detail, as the flexible CO probe exaggerates the length of each bond, making the fractional distance between bonds of different order appear much larger in NC-AFM images than they really are.

There are several other important studies in the field worth highlighting before proceeding to discuss molecular assemblies. de Oteyza et al. have recently demonstrated NC-AFM's capability to resolve the internal bond structure of single molecules at different steps during a chemical reaction (see

Figure 1G and reference [31]), once again underlining NC-AFM's ability to provide important insights into unknown chemical structures. It has now also been shown that internal bond resolution is achievable for a variety of substrates and tip terminations, such as NTCDI molecules adsorbed on the highly-reactive Si(111)-7×7 ([32,33] and Figure 1H) and chemically-passivated Ag:Si(111)-(3√×3√) surfaces [40], both at 77 K, where spontaneous tip termination was observed to facilitate submolecular resolution. In addition, Moreno et al. [22] recently reported similar results with NC-AFM imaging of pentacene and C_{60} molecules on the (101) surface of anatase TiO_2 also at 77 K, demonstrating the wide range of substrates on which internal bond resolution is now possible. In the same work, Moreno et al. also demonstrate submolecular imaging of non-planar molecules using a novel technique incorporating imaging in and out of feedback (with the out of feedback scan following the in-feedback profile at a reduced tip-sample separation). In very recent work, submolecular resolution in NC-AFM has now even been achieved under room temperature conditions [23,39]. Finally, in addition to NC-AFM, the use of CO-terminated tips is finding use in a number of complementary SPM techniques, such as Kelvin probe force microscopy (KPFM) and inelastic tunnelling spectroscopy (IETS). The combination of CO-mediated NC-AFM/STM and KPFM now allows unprecedented detail on the internal charge distribution within single molecules to be examined [43,52,53,54], and, with IETS [30], on the electronic and vibrational properties (see Figure 1F), opening the way for ever more detailed characterisation of a whole host of molecular properties in the coming years.

3. MORE THAN JUST AN IMAGE—THE EFFECT OF TIP FLEXIBILITY

Scanning probe microscopes are microscopes like no other; there are no lenses or mirrors, as light is not used to obtain an image. SPMs instead rely on sensing the interactions between an atomically-sharp cluster of atoms (i.e., the scanning probe tip) and the surface material. SPM by its very definition is therefore invasive, and always affects the system under study in some way (although most of the time, we assume this is a minor effect). Indeed, the same interactions we exploit to acquire images are often used to manipulate those same individual atoms and molecules under study. Artefacts arising in STM and NC-AFM, therefore, unlike many optical techniques, cannot be shrugged off as unwanted effects, but are instead intrinsic to the technique and rooted in the underlying physical principles on which the microscopes operate. This provides an interesting challenge, as the simplicity and beauty of many SPM images often fails to convey the complexity of the microscope and methods required to obtain them. In other words, sometimes things are not quite as simple as just looking at an image.

This is particularly true in submolecular resolution imaging with NC-AFM. Even as early as the seminal 2009 paper by Gross et al. [24] it was noted that a Cl-terminated tip produced images where the carbon rings of pentacene appeared smaller in diameter compared with a CO-terminated tip, and did so without an asymmetry where the rings appeared elongated in one direction (particularly noticeable when imaged on NaCl (2 ML)/Cu(111)). This was followed by quantitative measurements of the molecular pair potential between two CO molecules by Sun et al. [55] who noted that the Cu-adsorbed CO molecules are far from an idealised rigid probe, and in reality show a great deal of flexibility, as shown in Figure 2A, making the point that "... chemical repulsion between the CO molecules is relaxed at the expense of weaker bonding of the CO molecules to the Cu atoms of the tip and substrate, respectively".

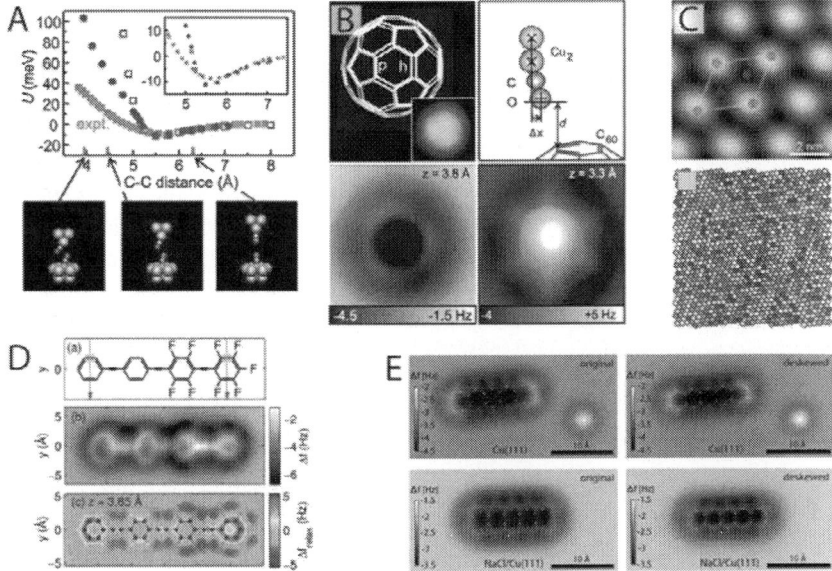

Figure 2. The effect of tip flexibility in high-resolution imaging. (**A**) CO bending observed in measurements of the CO–CO molecular pair potential (reprinted with permission from [55], copyright 2011 by the American Physical Society); (**B**) NC-AFM images showing bond length enhancement and sharpening due to flexibility at very small tip-sample distances (from [29], reprinted with permission from AAAS); (**C**) NC-AFM image showing variations in carbon ring size due to corrugation of Ir(111)-supported graphene (reprinted with permission from [56], copyright 2014 by the American Chemical Society); (**D**) Measurement and simulation of image distortions observed in NC-AFM images of 4-(4-(2,3,4,5,6-pentafluorophenylethynyl)-2,3,5,6-tetrafluorophe-nylethynyl) phenylethynylbenzene (FFPB) (reprinted with permission from [53], copyright 2014 by the American Chemical Society); (**E**) Image correction from lateral force analysis for pentacene imaged with NC-AFM on Cu(111) and NaCl (2 ML)/Cu(111) (reprinted with permission from [57], copyright 2014 by the American Physical Society).

As mentioned earlier, the flexibility of the CO probe was found to be essential for revealing the bond order of aromatic hydrocarbons. Detailed three-dimensional (3D) force maps collected above the hexagonal face of a surface-adsorbed C_{60} molecule [29] showed that at close tip-sample separations, the tilting of the CO is responsible for significantly amplifying the differences in apparent length of the p and h bonds, making the difference between them discernible within the lateral accuracy afforded by NC-AFM. Moreover, as shown in Figure 2B, tip-sample flexure also leads to a striking sharpening effect of the bonds, particularly at small tip-sample separations. Interestingly, similar to the noted difference between Cl- and CO-terminated tips, it was found that on the same dibenzo(cd,n)naphtho(3,2,1,8-pqra)perylene (DBNP) molecule, whilst a CO tip achieved exceptionally sharp resolution, once again enhancing the length of the aromatic bonds, a single-atom Xe tip showed no such distortion, providing an image much closer to the true atomic positions of the molecule [38]. This was attributed to the CO molecule's greater ability to bend at the tip apex.

An elegant example illustrating the effect of flexure in the tip-sample junction was provided by Boneschanscher et al., who investigated graphene on Ir(111) with CO-mediated NC-AFM [56]. Due to the lattice mismatch between graphene and the underlying Ir(111), the graphene buckles (following a moiré pattern), creating a smoothly-corrugated surface with a vertical distance ideally sized, such that internal bond resolution can be achieved simultaneously on both the top and hollow sites of the sheet. A single large NC-AFM image, taken at constant height, can therefore resolve a large number of equivalent carbon rings with a smoothly-varying change in tip-sample separation (and therefore, the degree of tip-sample repulsion) in an environment where edge effects and the asymmetry of the molecule are no longer a problem. Following this measurement, analysis of the area for each carbon ring revealed that hexagons at the moiré top sites were significantly larger than those in the hollow positions (see Figure 2C), showing deviations up to ±5% from the average value (after taking into account background forces), much larger than the accepted values of bond length variation. Once again, this was explained as due to variations in the degree of CO flexibility at different tip-sample separations. Moreover, the experimental data were supported by molecular mechanics simulations (modelled with a Lennard–Jones-type potential, see the next section) that modelled a flexible CO tip, fully reproducing the variation in apparent bond length, generating complete simulated images at relatively little computational cost.

An interesting result arising from the Lennard–Jones model was the observation of distorted carbon rings in simulations of pentacene [56], where the carbon rings appear elongated across the molecule's short axis, taking on a similar appearance to the initial experimental observations on NaCl [24]. This effect was particularly noticeable for tips modelled with a smaller lateral spring constant describing the bending of the tip-terminating CO molecule (0.3 N·m^{-1}), i.e., for a given lateral force, the distortions of the flexible tip were larger compared to those using a greater value of stiffness. (This model will be revisited in the next section when discussing apparent bonds). A similar observation was also made by Moll et al. who examined image distortions in a partially-fluorinated hydrocarbon molecule FFPB [53], as shown in Figure 2D. In this case, using a density functional theory (DFT) description, it was found that the variation in distortion between the fluorinated and non-fluorinated carbon rings depended on both variations in the electronic density across the molecule and the flexibility of the tip.

In an elegant experiment showcasing the precision of 3D force-field acquisition with NC-AFM, Neu et al. [57] demonstrated that a "scaling constant", unique for each CO tip, can be determined by assuming that the in-plane distortion scales linearly with the lateral forces acting on the CO. The lateral forces are determined by first integrating the measured frequency shift ($\Delta f(x,y,z)$) twice over z, obtaining $U(x,y,z)$, before then taking the lateral gradient in x or y obtaining the lateral forces, Fx and Fy [13]. The NC-AFM images were then distortion corrected via the linear relation, producing the de-skewed images shown in Figure 2E. What is particularly striking is the correction observed for pentacene adsorbed on NaCl, where the lateral forces were found to be considerably larger. Importantly, based on repeated measurements, Neu et al. noted that there is no universal value for the optimal scaling constant, such that its value is unique for each individual tip apex, necessitating that it must be determined for each individual experiment. Although 3D measurements are certainly challenging, this procedure could in principle be applied to much larger molecules, where more complex distortions and lateral forces are present, therefore obtaining images with significantly reduced distortion (similar to Xe-terminated tips), whilst retaining the increased clarity and sharpness of NC-AFM images obtained with CO-terminated tips.

A major downside of tip flexibility, as discussed in detail in the next section, is the appearance of spurious features in the bond structure of molecules. In the earlier mentioned study by Gross et al. [27], investigating the unknown structure of cephalandole A, shown in Figure 1C, an apparent bond, similar in appearance to the C–C bonds, was observed between the deprotonated nitrogen and one of the nearby C–H units. The feature was very tentatively assigned to potential hydrogen bonding; however, it was noted that such a feature was not reproduced in the DFT simulations. Similarly, the conformational determination of DBTH [28,49] (shown in Figure 1D) shows a pronounced sharp feature connecting the two sulphur atoms of the molecule, despite no such bond being present. This is in addition to more subtle observations, such as the images of non-bonded Au-PTCDA clusters [48]. Despite NC-AFM images clearly resolving the adsorbed Au atom separated ~3 Å from the C–H units of a PTCDA molecule, faint connecting features can be seen between the two that might mistakenly imply bond formation.

It is therefore abundantly clear that relaxations in the tip-sample junction, particularly for CO-terminated tips, are critical to understand the appearance of molecules imaged with submolecular resolution, even in the most simple of cases. Whilst tip-induced distortions can be cleverly exploited to image molecules with exceptional clarity and reveal information even down to the

Pauling bond order, they can equally pose significant challenges with respect to image interpretation and in some cases make identifying the "true" structure particularly problematic, as we will see in the following section.

4. RESOLVING INTER-MOLECULAR BONDS—FACT OR FICTION?

With the difficulties associated with interpreting whether features arise from tip flexibility or real molecular bonds, it is clear that one of the greatest current challenges of CO-mediated NC-AFM is to attain exceptional resolution, down to the bond order of individual C–C bonds, whilst still maintaining correct information on the real molecular geometry. This raises an important question: to what extent can we trust the features present in submolecular NC-AFM images? That the observed length of a C–C bond can appear almost double the size of its real value, far beyond typical covalent bond lengths, suggests that in cases of complex molecules and molecular assemblies, significant care must be taken in image interpretation. Investigations with flexible tips are therefore ongoing, with the limitations of the technique far from clearly understood.

The first SPM images of apparent intermolecular bonding were observed by Weiss et al. using the STHM technique on the herringbone phase of PTCDA on Au(111) [58] where both the internal bond structure of the molecule and sharp features extending across regions of intermolecular bonding were observed, as shown in Figure 3A. Later, it was shown that a variety of molecular terminations could achieve the same resolution [17], each resolving exceptionally-sharp features between the individual PTCDA molecules. Although for many years, the true origin of the observed contrast remained unclear, the molecular tip was assumed to mediate the interaction, somehow acting as a transducer, modulating the observed tunnel current signal via either Pauli repulsion or longer range electrostatic forces [16,17,58]. Similar to the results described below, through an elegant model incorporating flexible molecularly-terminated tips and a numerical model for the tunnelling process through the mediating molecule, it has been recently shown that the observed intermolecular features are a direct consequence of the flexible tip geometry [59,60].

CO-mediated NC-AFM images of molecules stabilised via intermolecular bonds were initially reported by Zhang and Chen et al. [61] who investigated assemblies of 8-hydroxyquinoline (8-hq) molecules stabilised via hydrogen bonding. In their report, a CO-terminated tip was argued to not only resolve the internal structure of 8-hq, but also to visualise features appearing between the molecules located in many of the expected locations for hydrogen bonding (see Figure 3B). The experimental measurements were presented alongside DFT calculations of the different observed structures, which, via an analysis of the charge density difference (CDD), confirmed the presence of hydrogen bonding, although its contribution to the total electron density (TED) was orders of magnitude smaller than that of the internal C–C bonds. (The TED has been used as a crude approximation for CO-mediated NC-AFM images, as to a first estimation, the regions of highest TED should correlate with the regions of highest Pauli repulsion sensed by the NC-AFM probe). Whether the intermolecular features were real, however, or simply an imaging artefact, remained unclear, as no account was taken of the tip-sample interactions in the molecule-only calculations presented (which, as described throughout the previous section, can have a profound effect on imaging).

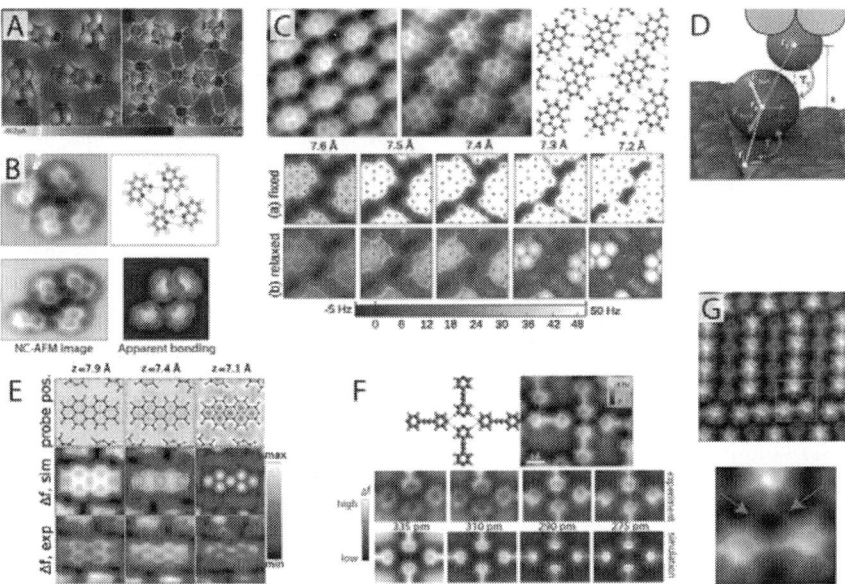

Figure 3. Intermolecular artefacts in hydrogen bonded assemblies. (**A**) First ever observation of apparent intermolecular bonds using scanning probe microscopy (SPM) observed via the STHM technique (reprinted with permission from [58], copyright 2010 by the American Chemical Society); (**B**) CO-mediated NC-AFM image of 8-hydroxyquinoline (8-hq) assembly exhibiting features in the locations of hydrogen bonding (from [61], reprinted with permission from AAAS) shown with simulated data (bottom right) [59]; (**C**) NC-AFM image revealing apparent intermolecular features in hydrogen bonded assemblies of NTCDI [40] also shown with simulated data confirming the apparent nature of the intermolecular features [59]; (**D**) Schematic of the flexible tip model used to simulate images of artificial intermolecular bonding [59]; (**E**) Experimental and simualted NC-AFM images of a PTCDA island shown with the simulated CO probe position [59]; (**F**) Experimental and simulated NC-AFM images of hydrogen bonded bis(para-pyridyl)acetylene (BPPA) molecules revealing apparent intermolecular bonds even where none are present (see red arrow) (reprinted with permission from [62], copyright 2014 by the American Physical Society); (**G**) NC-AFM image of apparent halogen bonding between the Fluorine atoms (see pink arrows in zoom of area marked with a yellow box) of three BPEPE-F18 (fluoro-substituted phenyleneethynylene) molecules (reprinted with permission from [63], copyright 2015 by the American Chemical Society). All figures from [59] reprinted with permission, copyright 2014 by the American Physical Society

Shortly after, we reported [40] very similar features observed for hydrogen-bonded assemblies of naphthalene tetracarboxylic diimide (NTCDI) molecules on the Ag:Si(111)-(3√×3√) surface, as shown in Figure 3C, where once again, exceptionally clear features were observed connecting the molecules together. In this report, the force field above the hydrogen bonded molecule was measured via collection of a high-density 3D grid of $\Delta f(z)$ force-spectroscopy data. These results were compared with dispersion-corrected DFT calculations modelling the interaction of a number of tip terminations with the NTCDI island, which determined that an NTCDI-terminated tip, with its C=O group pointing towards the surface, was most likely responsible for observing the submolecular resolution. As an attempt to model the influence of the tip on the observed images, two-dimensional line profiles of F(z) curves were simulated across both the C–C and hydrogen bonds of the NTCDI island. Although this reproduced well the apparent height of the C–C and hydrogen bond features observed in the experimental images, the DFT calculations were unable to capture the striking sharp appearance of the apparent intermolecular features.

An important development in the interpretation of intermolecular features in CO-mediated NC-AFM came from Hapala et al. [59,60] and Hämäläinen et al. [62] who showed that the flexibility of the molecular tip, rather than the presence (or lack thereof) of an intermolecular bond, determines whether such interconnecting features are observed between hydrogen-bonded molecules, strongly suggesting that such features cannot be directly assigned to direct visualisation of intermolecular bonding. Using a similar model (available at the following link: [64]) to that described by Boneschanscher and Hämäläinen et al. [56], Hapala et al. modelled the molecularly-terminated NC-AFM probe as the outermost atom of a metal tip with a single probe particle at its apex. The probe particle was subject to three primary forces: (i) the tip-surface force, modelled as the sum of all Lennard–Jones forces acting between the probe and the molecular layer; (ii) a radial force connecting the probe particle to the tip base at a distance tunable to the particular molecular termination modelled; and (iii) a harmonic restoring force modelled with a lateral stiffness typically between 0.3 and 1.5 N·m^{-1} (see Figure 3D for a schematic). By tuning two primary parameters, namely the probe particle radius (to radii representing, for example, oxygen or xenon) and the lateral stiffness, Hapala et al. were able to reproduce almost all of the primary features observed in images of intermolecular and internal bond structure resolution images with NC-AFM and STHM [59], as well as IETS measurements [60]. The IETS simulations also suggested that submolecular resolution imaging may strongly depend on the electrostatic force, raising the intriguing possibility that such images could be used to obtain information on the surface electrostatic potential.

The most important aspect of the model is the complete absence of any electronic structure information, such as the electronic density associated with chemical bonding. Despite this, just a simple summation of Lennard–Jones forces, centred at the location of each constituent atom of the molecule, is enough to almost fully reproduce the experimental measurements. Figure 3E shows one such simulation for an island of PTCDA molecules where increased submolecular resolution is observed to directly correlate with the localisation of the probe particle position to regions of energy minima located off the molecular bonds, primarily inside the carbon rings and away from locations where adjacent molecules sit close to one another, such as regions of intermolecular bonding. This effect not only leads to a dramatic sharpening of the bonds observed in NC-AFM images, but due to the bending of the probe molecule, acts to normalise internal and intermolecular features, such that both are equally visible (compared to an inflexible tip structure, where internal features appear much brighter). The success of the model is demonstrated in numerous examples in the same paper, including the 8-hq and NTCDI systems shown in Figure 3B,C respectively, where experimental NC-AFM images were fully reproduced, despite the complete absence of intermolecular bonding between the molecules.

Despite such strong evidence, it is tempting to ask whether such distinctions between the origin for the features matter. After all, in the end, are not the features located at the hydrogen bonds anyway, whether they are directly responsible for the image contrast or not? This exceptionally important question was addressed by Hämäläinen et al. [62] in an elegant experiment examining bis(para-pyridyl)acetylene (BPPA) molecules, organised as tetramers in islands stabilised via hydrogen bonding. Due to clever experimental design, the molecules arrange end-on as tetramers, each forming a single hydrogen bond with its neighbour, but never the molecule directly opposite, as shown in Figure 3F. Importantly, not only do the opposing molecules face each other with two nitrogen atoms, unable to form a bond, but they do so at a separation comparable to the distance between the C–H and N atoms that can form a bond; therefore if a feature is observed in the N–N junction it can only be the result of apparent bonding due to probe relaxations, confirming the artificial nature of the observed features. This is confirmed in Figure 3F throughout a sequence of NC-AFM images taken at decreasing tip-sample separation, where not only are interconnecting features observed in the hydrogen bonding locations, but also in the non-bonded N–N junction. Moreover, using their own flexible tip model, the connecting features were fully reproduced in simulated images, with the apparent bond appearing with the same brightness as over the regions of hydrogen bonding.

It is then clear that in the case of predominantly electrostatic interactions, such as hydrogen bonding, the observation of apparent intermolecular features cannot be interpreted as the identification of real intermolecular bonds. Although, in exceptionally simple systems, a majority of features may well indeed correlate with real hydrogen bonds, with no prior knowledge with which to compare, CO-mediated NC-AFM simply cannot be used to identify the existence or not of hydrogen bonding. Indeed, very recent results on alternative systems now almost completely rule out that such features can be ascribed to intermolecular bonding. For example, measurements performed on fluoro-substituted phenyleneethynylene (bis(2,3,5,6- tetrafluoro-4-(2,3,4,5,6-pentafluorophenylethynyl)phenyl)-ethyne (BPEPE-F18) molecule [63] show that apparent bonds can be observed between C–F units of the molecule, as shown in Figure 3G. BPEPE-F18 interacts via halogen bonding, a purely electrostatic interaction with no accumulation of electron density between the molecules, therefore ruling out the possibility that NC-AFM directly images the bond itself. Additionally, unpublished results from our group [65] involving an examination of islands of close-packed C_{60} fullerene molecules also exhibit sharp, directional, interconnecting features between molecules, despite the purely van der Waals nature of the bonding interaction, the large separation between molecules, and the highly non-planar geometry. It therefore appears that despite a huge amount of excitement surrounding early results on hydrogen-bonded molecules, it is unfortunately the case that hydrogen bonds are neither directly imaged, nor can they be indirectly implied from the observation of apparent interconnecting features.

5. PROSPECTS OF NC-AFM IN SUPRAMOLECULAR STUDIES

There are a wide variety of intermolecular interactions available to stabilise 2D supramolecular networks. Even, for the moment, remaining within the confines of hydrogen bonding interactions, many complex structures can be formed exploiting a variety of molecules and hydrogen bond donors and acceptors. In the pioneering paper by Theobald et al. [66], for instance, a mixed phase of PTCDI and melamine was found to produce large porous networks of well-defined size, capable of templating the subsequent growth of C_{60} fullerenes, as shown in Figure 4A. In this case, the molecules arranged in a well-defined manner, maximising the number of hydrogen bonds in a way that can be easily understood. In addition, a recent AFM investigation has shown that such structures are stable across a range of insulating materials, even under ambient conditions [67]. In other systems with more complex arrangements [68,69,70] and varying molecular species [71,72,73,74], however, the location and number of hydrogen bonds formed is not always trivial to answer, often requiring simulation input. The prospect of a technique capable of single bond resolution is therefore extremely attractive, provided, of course, that the necessary care is taken in the image analysis so as not to mistakenly assign artificial interconnecting features as real bonds.

Prior to the advent of CO-mediated submolecular imaging, supramolecular systems were, and continue to be, studied in great detail with NC-AFM, primarily on bulk insulating substrates otherwise inaccessible by other techniques. These include, as reviewed in detail elsewhere [75,76], investigations spanning prototypical molecules, such as C_{60} [77,78] and PTCDA [79,80,81] to a variety of small molecules, many of which are capable of forming hydrogen bonded networks [82,83,84,85,86,87]. Central to these studies is the use of cantilever NC-AFM operated in feedback at room temperature with single molecule resolution. There is therefore significant potential for submolecular NC-AFM to complement these investigations and reveal even greater detail on their interactions.

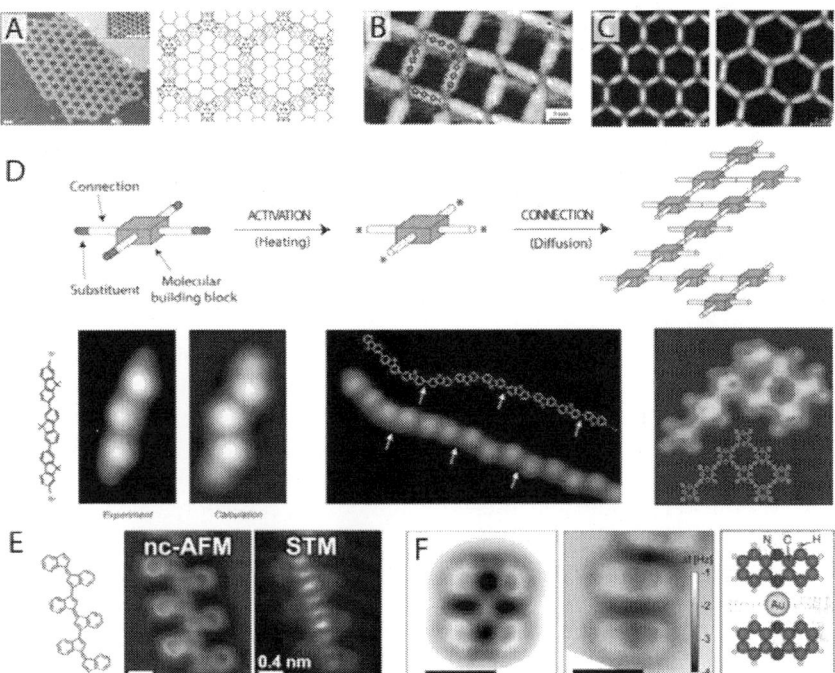

Figure 4. Potential for submolecular imaging of supramolecular systems. (**A**) STM image of a complex porous hydrogen bonded network comprising a mixed phase of perylene tetra-carboxylic di-imide (PTCDI) and melamine (reprinted by permission from Macmillan Publishers Ltd.: Nature [66], copyright 2003); (**B**) Nanocavities formed on Cu(100) by 4,1′,4′,1″-terphenyl-1,4″-dicarboxylic acid (TDA) molecules metal-coordinated with Fe atoms imaged in STM (reprinted by permission from Macmillan Publishers Ltd.: Nature Materials [88], copyright 2004); (**C**) Porous metal-coordinated networks of tunable size formed by NC-Ph$_n$-CN molecules coordinated with Co atoms on Ag(111) imaged with STM (reprinted with permission from [89], copyright 2007 by the American Chemical Society); (**D**) Schematic and STM images of covalently-polymerised molecular assemblies of tetraphenyl porphyrin 2D networks (**left**) (reprinted by permission from Macmillan Publishers Ltd.: Nature Nanotechnology [90], copyright 2007) and dibromoterfluorene (DBTF) molecular wires (**right**) (from [91], reprinted with permission from AAAS); (**E**) CO-mediated NC-AFM (**left**) and STM (**right**) images of a covalently-linked oligomer following on-surface cyclisation on Au(111) (reprinted with permission from [92], copyright 2014 by the American Chemical Society); (**F**) CO-mediated NC-AFM image of a bonded (**left**) and non-bonded (**right**) phenazine-gold complex (schematic shows bonded geometry) (reprinted with permission from [93], copyright 2013 by the American Chemical Society).

There is of course much more to 2D supramolecular chemistry than hydrogen bonding, as discussed in many excellent review articles [94,95,96,97,98]. Molecular networks stabilised through metal coordination are a particularly interesting case as the strong coordination bonding interaction allows a variety of 2D molecular networks with tunable properties to be grown, such as networks with functional cavities [88] and tunable pore size [89] as shown in Figure 4B,C. The benefit of investigating such networks using CO-mediated NC-AFM lies in the fact that a sizeable coordinating metal atom must be present to mediate the bonding, providing a clear feature that could potentially be observed in the NC-AFM image. Moreover, metal coordination bonds tend to be longer than other forms of bonding, as there are typically two bonds formed with the coordinating atom, requiring that the molecules should be reasonably well separated. Apparent bonding due to tip flexibility and the close proximity of molecules should therefore be reduced, if not completely avoided. Although very few studies are currently available, Albrecht et al. [93] have examined co-adsorption of a phenazine and single gold adatoms on a thin film of NaCl on Cu(111) before and after formation of a metal-coordinated complex, as shown in Figure 4F. The phenazine-gold-phenazine complex is shown in the left panel of Figure 4F, with a non-bonded pair in the centre for comparison. Clearly, in the case of the metal complex, a distinct additional feature joins the two molecules together at their centre, as shown in the cartoon in the right panel, inducing a slight distortion to the molecule, causing the centre rings to appear darker (closer to the surface) than for the unbonded molecule.

As noted above, metal coordination networks benefit NC-AFM imaging, as the organometallic bond lengths are reasonably well defined and usually longer than that of the close proximity of features required to produce the apparent hydrogen bonds observed in the previous section. In some systems, it may even be possible to identify the position of the coordinating ion itself, which may appear much larger than a typical hydrocarbon atom (similar to a $C\equiv C$ bond appearing significantly brighter and wider than a C–C bond, as seen in Figure 1G) offering much greater precision in characterising metal-coordinated structures than currently available via STM. On the other hand, metal-coordinating ions could be rather difficult to detect, due to their preference to sit closer towards the metal substrate than the molecules they bind to (and therefore, potentially obscured from the scanning probe, which typically operates at a constant height to obtain submolecular resolution). In these cases, novel protocols, such as the multi-pass lift-off technique proposed by Moreno et al. [22] or high-density 3D force field measurements, may become invaluable.

A final class of supramolecular structure—although, perhaps, use of the term supramolecular is no longer quite appropriate at this point—is the formation of large covalently-bound molecular networks. As originally shown by Grill et al. [90,91], single molecule precursor units, such as tetraphenyl porphyrin molecules, can be partially substituted with halogen atoms at their periphery. Following thermal activation on a catalytically-active surface such as gold or copper, the carbon-halogen bond is broken and the free radical molecules diffuse and bind together into 2D networks or 1D wires, depending on their design, stabilised via covalent C–C bonds (in effect, resulting in a single large molecule), as shown in Figure 4D. Examination of such covalently-bound structures with NC-AFM is now a particularly active area of research, with just a few examples currently published. In Figure 4E one such example is shown for Oligo-acetylene derivatives obtained following on-surface radical cyclisation [92]. Compared to hydrogen-bonded and metal-coordinated structures, the linking carbon bonds are not only exceptionally clear, but also appear with the same brightness as the linked molecules, which themselves exhibit no distortion from their isolated form. To an extent this should of course be expected, as the tip is once again imaging single C–C bonds, rather than intermolecular features. Although it is perhaps too early to tell, it may therefore be possible for NC-AFM to clearly distinguish between completely covalently-linked molecules and those simply still linked by precursor metalorganic bonds or exhibiting apparent bonding due to their close proximity, particularly in the case of less flexible tip terminations, such as Xe.

6. CONCLUSIONS AND PERSPECTIVES

In the few short years since its first demonstration by Gross et al. in 2009, submolecular imaging with CO-mediated NC-AFM has experienced a huge amount of growth with broad application across a wide range of molecular systems. As the number of methods, substrate materials and molecular tip terminations available to achieve submolecular resolution increases, and ever more research groups overcome the technical difficulties associated with its implementation, adoption of the technique is reaching a tipping point, where we are now seeing the beginnings of a huge expansion of submolecular NC-AFM that will provide unprecedented insights into a vast array of molecular and supramolecular systems.

Many of the results reviewed here highlight the important interplay between the choice of molecular tip termination and its flexibility in determining the real atomic structure of molecules and their self-assembled structures. Whilst in many cases, the flexibility of the molecular tip can be cleverly exploited in order to reveal exceptionally-sharp resolution corresponding to internal bond structure, in some cases, even revealing the Pauling order of C–C bonds, it can equally induce distortions in the NC-AFM image causing significant deviations from the true atomic structure of the molecule. This is an important ongoing area of investigation, with techniques now being developed to apply image corrections based on lateral force measurement and background subtraction, but also alternative tip terminations that significantly reduce image distortions whilst still providing submolecular resolution (e.g., Xe and Br tip terminations). One route towards reducing tip flexibility may be not only to investigate more rigid or strongly tip-adsorbing probe molecule species, but also the preparation of more reactive metal tips [99], via controlled pick up of single metal atoms, capable of more stable, and potentially more directional bonding with a tip-adsorbed probe molecule. Additionally, recent observations of submolecular imaging on semiconducting and ionic substrates, such as silicon and TiO_2, suggest that such surfaces may prove to be conducive for preparing rigid crystalline tip terminations. Often terminated with single hydrogen and oxygen atoms [22,100,101,102] such tip structures may dramatically reduce the effect of tip flexibility.

Understanding the nature of the tip-sample interaction and the best methods to reduce tip-induced distortions will be exceptionally important for future studies of supramolecular assemblies and on-surface chemistry [103]. Although in many respects, it is still early days, CO-mediated NC-AFM, operated and interpreted with care, may have the capability to offer unparalleled insight into supramolecular structures stabilised through a variety of intermolecular forces. As described in Section 4, there are already many studies reporting the exquisite detail achievable on hydrogen bonded molecular systems. Although NC-AFM cannot be said to directly image the hydrogen bonds themselves, or indeed accurately assign either their presence or absence, interpretation of the internal bond structure makes assignment of the orientation, separation and exact location of each individual molecule possible with single atom precision. This in itself may become invaluable in interpreting more complex supramolecular structures in the future. Over the coming years, the potential for NC-AFM to investigate various other supramolecular systems, such as those stabilised through metal-coordination and covalent bonds, will also be realised, opening the way for characterisation of molecular structures with unprecedented detail, pointing the way to an exciting future for NC-AFM in supramolecular chemistry.

ACKNOWLEDGMENTS

I would like to thank Philip Moriarty and Adam Sweetman for their comments and many useful discussions, Pavel Jelinek and Leo Gross for helpful comments on the manuscript, and the Engineering and Physical Sciences Research Council (EPSRC), as well as the Leverhulme Trust for the award of fellowship's EP/J500483/1 and ECF-2015-005, respectively.

REFERENCES

1.   Müller, E.W.; Bahadur, K. Field ionization of gases at a metal surface and the resolution of the field Ion microscope. Phys. Rev. **1956**, 102, 624–631.

2.   Müller, E.W. Field ion microscopy. Science **1965**, 149, 591.

3.   Binnig, G.; Rohrer, H.; Gerber, C.; Weibel, E. Surface studies by scanning tunneling microscopy. Phys. Rev. Lett. **1982**, 49, 57–61.

4.   Stroscio, J.A.; Kaiser, W.J. Scanning Tunneling Microscopy; Academic Press: San Diego, CA, USA, 1993.

5.   Wiesendanger, R. Scanning Probe Microscopy and Spectroscopy: Methods And Applications; Cambridge University Press: Cambridge, UK, 1994.

6. Giessibl, F.J. Advances in atomic force microscopy. Rev. Mod. Phys. **2003**, 75, 949–983.

7. Morita, S.; Wiesendanger, R.; Meyer, E. Noncontact Atomic Force Microscopy, Vol. 1; Springer: Heidelberg, Germany, 2002.

8. Morita, S.; Giessibl, F.J.; Wiesendanger, R. Noncontact Atomic Force Microscopy, Vol. 2; Springer: Heidelberg, Germany, 2009.

9. Morita, S.; Giessibl, F.J.; Meyer, E.; Wiesendanger, R. Noncontact Atomic Force Microscopy, Vol. 3; Springer: Heidelberg, Germany, 2015.

10. Sugimoto, Y.; Abe, M.; Hirayama, S.; Oyabu, N.; Custance, O.; Morita, S. Atom inlays performed at room temperature using atomic force microscopy. Nat. Mater. **2005**, 4, 156–159.

11. Sugimoto, Y.; Pou, P.; Custance, O.; Jelinek, P.; Abe, M.; Perez, R.; Morita, S. Complex patterning by vertical interchange atom manipulation using atomic force microscopy. Science **2008**, 322, 413–417.

12. Sugimoto, Y.; Pou, P.; Abe, M.; Jelinek, P.; Perez, R.; Morita, S.; Custance, O. Chemical identification of individual surface atoms by atomic force microscopy. Nature **2007**, 446, 64–67.

13. Ternes, M.; Lutz, C.P.; Hirjibehedin, C.F.; Giessibl, F.J.; Heinrich, A.J. The force needed to move an atom on a surface. Science **2008**, 319, 1066–1069.

14. Temirov, R.; Soubatch, S.; Neucheva, O.; Lassise, A.C.; Tautz, F.S. A novel method achieving ultra-high geometrical resolution in scanning tunnelling microscopy. New J. Phys. **2008**, 10, 053012.

15. Gross, L. Recent advances in submolecular resolution with scanning probe microscopy. Nat. Chem. **2011**, 3, 273–278.

16. Weiss, C.; Wagner, C.; Kleimann, C.; Rohlfing, M.; Tautz, F.S.; Temirov, R. Imaging Pauli repulsion in scanning tunneling microscopy. Phys. Rev. Lett. **2010**, 105, 086103.

17. Kichin, G.; Weiss, C.; Wagner, C.; Tautz, F.S.; Temirov, R. Single molecule and single atom sensors for atomic resolution imaging of chemically complex surfaces. J. Am. Chem. Soc. **2011**, 133, 16847–16851.

18. Giessibl, F.J. High-speed force sensor for force microscopy and profilometry utilizing a quartz tuning fork. Appl. Phys. Lett. **1998**, 73, 3956.

19. Albrecht, T.R.; Grütter, P.; Horne, D.; Rugar, D. Frequency modulation detection using high-Q cantilevers for enhanced force microscope sensitivity. J. Appl. Phys. **1991**, 69, 668.

20. Sweetman, A.; Jarvis, S.; Danza, R.; Bamidele, J.; Gangopadhyay, S.; Shaw, G.A.; Kantorovich, L.; Moriarty, P. Toggling bistable atoms via mechanical switching of bond angle. Phys. Rev. Lett. **2011**, 106, 136101.

21. Shaw, G.A.; Pratt, J.; Jabbour, Z. Small mass measurements for tuning fork-based atomic force microscope cantilever spring constant calibration. Conf. Proc. Soc. Exp. Mech. Ser. **2011**, 2, 49–56.

22. Moreno, C.; Stetsovych, O.; Shimizu, T.K.; Custance, O. Imaging three-dimensional surface objects with submolecular resolution by atomic force microscopy. Nano Lett. **2015**, 15, 2257–2262.

23. Iwata, K.; Yamazaki, S.; Mutombo, P.; Hapala, P.; Ondráček, M.; Jelínek, P.; Sugimoto, Y. Chemical structure imaging of a single molecule by atomic force microscopy at room temperature. Nat. Commun. **2015**, 6, 7766.

24. Gross, L.; Mohn, F.; Moll, N.; Liljeroth, P.; Meyer, G. The chemical structure of a molecule resolved by atomic force microscopy. Science **2009**, 325, 1110–1114.

25. Moll, N.; Gross, L.; Mohn, F.; Curioni, A.; Meyer, G. The mechanisms underlying the enhanced resolution of atomic force microscopy with functionalized tips. New J. Phys. **2010**, 12, 125020.

26. Moll, N.; Gross, L.; Mohn, F.; Curioni, A.; Meyer, G. A simple model of molecular imaging with noncontact atomic force microscopy. New J. Phys. **2012**, 14, 083023.

27. Gross, L.; Mohn, F.; Moll, N.; Meyer, G.; Ebel, R.; Abdel-Mageed, W.M.; Jaspars, M. Organic structure determination using atomic-resolution scanning probe microscopy. Nat. Chem. **2010**, 2, 821–825.

28. Pavliček, N.; Fleury, B.; Neu, M.; Niedenführ, J.; Herranz-Lancho, C.; Ruben, M.; Repp, J. Atomic force microscopy reveals bistable configurations of dibenzo[a,h]thianthrene and their interconversion pathway. Phys. Rev. Lett. **2012**, 108, 086101.

29. Gross, L.; Mohn, F.; Moll, N.; Schuler, B.; Criado, A.; Guitián, E.; Peña, D.; Gourdon, A.; Meyer, G. Bond-order discrimination by atomic force microscopy. Science **2012**, 337, 1326–1329.

30. Chiang, C.I.; Xu, C.; Han, Z.; Ho, W. Real-space imaging of molecular structure and chemical bonding by single-molecule inelastic tunneling probe. Science **2014**, 344, 885–888.

31. De Oteyza, D.G.; Gorman, P.; Chen, Y.C.; Wickenburg, S.; Riss, A.; Mowbray, D.J.; Etkin, G.; Pedramrazi, Z.; Tsai, H.Z.; Rubio, A.; et al. Direct imaging of covalent bond structure in single-molecule chemical reactions. Science **2013**, 340, 1434–1437.

32. Sweetman, A.M.; Jarvis, S.P.; Rahe, P.; Champness, N.R.; Kantorovich, L.N.; Moriarty, P.J. Intramolecular bonds resolved on a semiconductor surface. Phys. Rev. B **2014**, 90, 165425.

33. Jarvis, S.P.; Sweetman, A.M.; Lekkas, I.; Champness, N.R.; Kantorovich, L.; Moriarty, P. Simulated structure and imaging of NTCDI on Si(111)-7×7 : A combined STM, NC-AFM and DFT study. J. Phys. Condens. Matter **2014**, 27, 054004.

34. Boneschanscher, M.P.; van der Lit, J.; Sun, Z.; Swart, I.; Liljeroth, P.; Vanmaekelbergh, D. Quantitative atomic resolution force imaging on epitaxial graphene with reactive and nonreactive AFM probes. ACS Nano **2012**, 6, 10216–10221.

35. Eigler, D.M.; Lutz, C.P.; Rudge, W.E. An atomic switch realized with the scanning tunnelling microscope. Nature **1991**, 352, 600–603.

36. Bartels, L.; Meyer, G.; Rieder, K.H. Controlled vertical manipulation of single CO molecules with the scanning tunneling microscope: A route to chemical contrast. Appl. Phys. Lett. **1997**, 71, 213–215.

37. Bartels, L.; Meyer, G.; Rieder, K.H.; Velic, D.; Knoesel, E.; Hotzel, A.; Wolf, M.; Ertl, G. Dynamics of electron-induced manipulation of individual CO molecules on Cu(111). Phys. Rev. Lett. **1998**, 80, 2004–2007.

38. Mohn, F.; Schuler, B.; Gross, L.; Meyer, G. Different tips for high-resolution atomic force microscopy and scanning tunneling microscopy of single molecules. Appl. Phys. Lett. **2013**, 102, 073109.

39. Huber, F.; Matencio, S.; Weymouth, A.J.; Ocal, C.; Barrena, E.; J, G.F. Intramolecular force contrast and dynamic current-distance measurements at room temperature. Phys. Rev. Lett. **2015**, 115, 066101.

40. Sweetman, A.M.; Jarvis, S.P.; Sang, H.; Lekkas, I.; Rahe, P.; Wang, Y.; Wang, J.; Champness, N.R.; Kantorovich, L.; Moriarty, P. Mapping the force field of a hydrogen-bonded assembly. Nat. Commun. **2014**, 5, 3931.

41. Sang, H.; Jarvis, S.P.; Zhou, Z.; Sharp, P.; Moriarty, P.; Wang, J.; Wang, Y.; Kantorovich, L. Identifying tips for intramolecular NC-AFM imaging via in situ fingerprinting. Sci. Rep. **2014**, 4, 6678.

42. Guillermet, O.; Gauthier, S.; Joachim, C.; de Mendoza, P.; Lauterbach, T.; Echavarren, A. STM and AFM high resolution intramolecular imaging of a single decastarphene molecule. Chem. Phys. Lett. **2011**, 511, 482–485.

43. Kawai, S.; Sadeghi, A.; Feng, X.; Lifen, P.; Pawlak, R.; Glatzel, T.; Willand, A.; Orita, A.; Otera, J.; Goedecker, S.; et al. Obtaining detailed structural information about supramolecular systems on surfaces by combining high-resolution force microscopy with ab initio calculations. ACS Nano **2013**, 7, 9098–9105.

44. Pawlak, R.; Kawai, S.; Fremy, S.; Glatzel, T.; Meyer, E. Atomic-scale mechanical properties of orientated C60 molecules revealed by noncontact atomic force microscopy. ACS Nano **2011**, 5, 6349–6354.

45. Pawlak, R.; Kawai, S.; Fremy, S.; Glatzel, T.; Meyer, E. High-resolution imaging of C_{60} molecules using tuning-fork-based non-contact atomic force microscopy. J. Phys. Condens. Matter **2012**, 24, 084005.

46. Chiutu, C.; Sweetman, A.M.; Lakin, A.J.; Stannard, A.; Jarvis, S.; Kantorovich, L.; Dunn, J.L.; Moriarty, P. Precise orientation of a single C_{60} molecule on the tip of a scanning probe microscope. Phys. Rev. Lett. **2012**, 108, 268302.

47. Hanssen, K.O.; Schuler, B.; Williams, A.J.; Demissie, T.B.; Hansen, E.; Andersen, J.H.; Svenson, J.; Blinov, K.; Repisky, M.; Mohn, F.; et al. A combined atomic force microscopy and computational approach for the structural elucidation of breitfussin A and B: Highly modified halogenated dipeptides from Thuiaria breitfussi. Angew. Chem. Int. Ed. Engl. **2012**, 51, 12238–12241.

48. Mohn, F.; Repp, J.; Gross, L.; Meyer, G.; Dyer, M.S.; Persson, M. Reversible bond formation in a gold-atom-organic-molecule complex as a molecular switch. Phys. Rev. Lett. **2010**, 105, 266102.

49. Pavliček, N.; Herranz-Lancho, C.; Fleury, B.; Neu, M.; Niedenführ, J.; Ruben, M.; Repp, J. High-resolution scanning tunneling and atomic force microscopy of stereochemically resolved dibenzo[a,h]thianthrene molecules. Phys. Status Solidi B **2013**, 250, 2424–2430.

50. Schuler, B.; Liu, W.; Tkatchenko, A.; Moll, N.; Meyer, G.; Mistry, A.; Fox, D.; Gross, L. Adsorption geometry determination of single molecules by atomic force microscopy. Phys. Rev. Lett. **2013**, 111, 106103.

51. Mistry, A.; Moreton, B.; Schuler, B.; Mohn, F.; Meyer, G.; Gross, L.; Williams, A.; Scott, P.; Costantini, G.; Fox, D.J. The synthesis and STM/AFM imaging of "olympicene" benzo[cd]pyrenes. Chemistry **2015**, 21, 2011–2018.

52. Mohn, F.; Gross, L.; Moll, N.; Meyer, G. Imaging the charge distribution within a single molecule. Nat. Nanotechnol. **2012**, 7, 227–231.

53. Moll, N.; Schuler, B.; Kawai, S.; Xu, F.; Peng, L.; Orita, A.; Otera, J.; Curioni, A.; Neu, M.; Repp, J.; et al. Image distortions of a partially fluorinated hydrocarbon molecule in atomic force microscopy with carbon monoxide terminated tips. Nano Lett. **2014**, 14, 6127.

54. Schuler, B.; Collazos, S.; Gross, L.; Meyer, G.; Pérez, D.; Guitián, E.; Peña, D. From perylene to a 22-ring aromatic hydrocarbon in one-pot. Angew. Chem. **2014**, 126, 9150–9152.

55. Sun, Z.; Boneschanscher, M.P.; Swart, I.; Vanmaekelbergh, D.; Liljeroth, P. Quantitative atomic force microscopy with carbon monoxide terminated tips. Phys. Rev. Lett. **2011**, 106, 046104.

56. Boneschanscher, M.P.; Ha, S.K.; Liljeroth, P.; Swart, I. Sample corrugation affects the apparent bond lengths in atomic force microscopy. ACS Nano **2014**, 8, 3006–3014.

57. Neu, M.; Moll, N.; Gross, L.; Meyer, G.; Giessibl, F.J.; Repp, J. Image correction for atomic force microscopy images with functionalized tips. Phys. Rev. B **2014**, 89, 205407.

58. Weiss, C.; Wagner, C.; Temirov, R.; Tautz, F.S. Direct imaging of intermolecular bonds in scanning tunneling microscopy. J. Am. Chem. Soc. **2010**, 132, 11864–11865.

59. Hapala, P.; Kichin, G.; Wagner, C.; Tautz, F.S.; Temirov, R.; Jelínek, P. Mechanism of high-resolution STM/AFM imaging with functionalized tips. Phys. Rev. B **2014**, 90, 085421.

60. Hapala, P.; Temirov, R.; Tautz, F.S. Origin of high-resolution IETS-STM images of organic molecules with functionalized tips. Phys. Rev. Lett. **2014**, 226101, 1–5.

61. Zhang, J.; Chen, P.; Yuan, B.; Ji, W.; Cheng, Z.; Qiu, X. Real-space identification of intermolecular bonding with atomic force microscopy. Science **2013**, 342, 611–614.

62. Hämäläinen, S.K.; van der Heijden, N.; van der Lit, J.; den Hartog, S.; Liljeroth, P.; Swart, I. Intermolecular contrast in atomic force microscopy images without intermolecular bonds. Phys. Rev. Lett. **2014**, 113, 186102.

63. Kawai, S.; Sadeghi, A.; Xu, F.; Peng, L.; Orita, A.; Otera, J.; Goedecker, S.; Meyer, E. Extended halogen bonding between fully fluorinated aromatic molecules. ACS Nano **2015**, 9, 2574–2583.

64. PyProbe Web Interface. Available online: http://nanosurf.fzu.cz/ppr/ (accessed on 25 June 2015).

65. Jarvis, S.P.; Rashid, M.A.; Sweetman, A.M.; Leaf, J.; Taylor, S.; Dunn, J.L.; Moriarty, P.J. Intermolecular artefacts observed in 2D C_{60} assemblies with NC-AFM. 2015. submitted.

66. Theobald, J.A.; Oxtoby, N.S.; Phillips, M.a.; Champness, N.R.; Beton, P.H. Controlling molecular deposition and layer structure with supramolecular surface assemblies. Nature **2003**, 424, 1029–1031.

67. Korolkov, V.V.; Svatek, S.A.; Allen, S.; Roberts, C.J.; Tendler, S.J.B.; Taniguchi, T.; Watanabe, K.; Champness, N.R.; Beton, P.H. Bimolecular porous supramolecular networks deposited from solution on layered materials: Graphite, boron nitride and molybdenum disulphide. Chem. Commun. **2014**, 50, 8882–8885.

68. Staniec, P.A.; Perdigão, L.M.A.; Saywell, A.; Champness, N.R.; Beton, P.H. Hierarchical organisation on a two-dimensional supramolecular network. ChemPhysChem **2007**, 8, 2177–2181.

69. Saywell, A.; Magnano, G.; Satterley, C.J.; Perdigão, L.M.A.; Champness, N.R.; Beton, P.H.; O'Shea, J.N. Electrospray deposition of C_{60} on a hydrogen-bonded supramolecular network. J. Phys. Chem. C **2008**, 112, 7706–7709.

70. Perdigão, L.M.A.; Perkins, E.W.; Ma, J.; Staniec, P.A.; Rogers, B.L.; Champness, N.R.; Beton, P.H. Bimolecular networks and supramolecular traps on Au(111). J. Phys. Chem. B **2006**, 110, 12539–12542.

71. Pawin, G.; Wong, K.L.; Kwon, K.Y.; Bartels, L. A homomolecular porous network at a Cu(111) surface. Science **2006**, 313, 961–962.

72. Mura, M.; Sun, X.; Silly, F.; Jonkman, H.T.; Briggs, G.A.D.; Castell, M.R.; Kantorovich, L.N. Experimental and theoretical analysis of H-bonded supramolecular assemblies of PTCDA molecules. Phys. Rev. B **2010**, 81, 195412.

73. Silly, F.; Shaw, A.Q.; Castell, M.R.; Briggs, G.A.D.; Mura, M.; Martsinovich, N.; Kantorovich, L. Melamine structures on the Au(111) surface. J. Phys. Chem. C **2008**, 112, 11476–11480.

74. Staniec, P.A.; Perdigão, L.M.A.; Rogers, B.L.; Champness, N.R.; Beton, P.H. Honeycomb networks and chiral superstructures formed by cyanuric acid and melamine on Au(111). J. Phys. Chem. C **2007**, 111, 886–893.

75. Rahe, P.; Kittelmann, M.; Neff, J.L.; Nimmrich, M.; Reichling, M.; Maass, P.; Kühnle, A. Tuning molecular self-assembly on bulk insulator surfaces by anchoring of the organic building blocks. Adv. Mater. **2013**, 25, 3948–3956.

76. Burke, S.A.; Topple, J.M.; Grütter, P. Molecular dewetting on insulators. J. Phys. Condens. Matter **2009**, 21, 423101.

77. Burke, S.A.; Mativetsky, J.M.; Hoffmann, R.; Grütter, P. Nucleation and submonolayer growth of C_{60} on KBr. Phys. Rev. Lett. **2005**, 94, 1–4.

78. Mativetsky, J.M.; Burke, S.a.; Hoffmann, R.; Sun, Y.; Grutter, P. Molecular resolution imaging of C_{60} on Au(111) by non-contact atomic force microscopy. Nanotechnology **2004**, 15, S40–S43.

79. Mativetsky, J.M.; Burke, S.A.; Fostner, S.; Grutter, P. Templated growth of 3,4,9,10-perylenetetracarboxylic dianhydride molecules on a nanostructured insulator. Nanotechnology **2007**, 18, 105303.

80. Pakarinen, O.H.; Mativetsky, J.M.; Gulans, A.; Puska, M.J.; Foster, A.S.; Grutter, P. Role of van der Waals forces in the adsorption and diffusion of organic molecules on an insulating surface. Phys. Rev. B **2009**, 80, 1–5.

81. Burke, S.A.; Ji, W.; Mativetsky, J.M.; Topple, J.M.; Fostner, S.; Gao, H.J.; Guo, H.; Grütter, P. Strain induced dewetting of a molecular system: Bimodal growth of PTCDA on NaCl. Phys. Rev. Lett. **2008**, 100, 7–10.

82. Amrous, A.; Bocquet, F.; Nony, L.; Para, F.; Loppacher, C.; Lamare, S.; Palmino, F.; Cherioux, F.; Gao, D.Z.; Canova, F.F.; et al. Molecular design

and control over the morphology of self-assembled films on ionic substrates. Adv. Mater. Interfaces **2014**, 1, 1400414.

83. Bocquet, F.; Nony, L.; Mannsfeld, S.C.B.; Oison, V.; Pawlak, R.; Porte, L.; Loppacher, C. Inhomogeneous relaxation of a molecular layer on an insulator due to compressive stress. Phys. Rev. Lett. **2012**, 108, 1–5.

84. Pawlak, R.; Nony, L.; Bocquet, F.; Oison, V.; Sassi, M.; Debierre, J.m.; Loppacher, C.; Porte, L. Supramolecular assemblies of 1,4-benzene diboronic acid on KCl(001). J. Phys. Chem. **2010**, 114, 9290.

85. Rahe, P.; Nimmrich, M.; Kühnle, A. Substrate templating upon self-assembly of hydrogen-bonded molecular networks on an insulating surface. Small **2012**, 8, 2969–2977.

86. Kittelmann, M.; Rahe, P.; Nimmrich, M.; Hauke, C.M.; Gourdon, A.; Kühnle, A. On-surface covalent linking of organic building blocks on a bulk insulator. ACS Nano **2011**, 5, 8420–8425.

87. Rahe, P.; Nimmrich, M.; Greuling, A.; Schütte, J.; Stará, I.G.; Rybáček, J.; Huerta-Angeles, G.; Starý, I.; Rohlfing, M.; Kühnle, A. Toward molecular nanowires self-assembled on an insulating substrate: Heptahelicene-2-carboxylic acid on Calcite (1014). J. Phys. Chem. C **2010**, 114, 1547–1552.

88. Stepanow, S.; Lingenfelder, M.; Dmitriev, A.; Spillmann, H.; Delvigne, E.; Lin, N.; Deng, X.; Cai, C.; Barth, J.V.; Kern, K. Steering molecular organization and host-guest interactions using two-dimensional nanoporous coordination systems. Nat. Mater. **2004**, 3, 229–233.

89. Schlickum, U.; Decker, R.; Klappenberger, F.; Zoppellaro, G.; Klyatskaya, S.; Ruben, M.; Silanes, I.; Arnau, A.; Kern, K.; Brune, H.; et al. Metal-organic honeycomb nanomeshes with tunable cavity size. Nano Lett. **2007**, 7, 3813–3817.

90. Grill, L.; Dyer, M.; Lafferentz, L.; Persson, M.; Peters, M.V.; Hecht, S. Nano-architectures by covalent assembly of molecular building blocks. Nat. Nanotechnol. **2007**, 2, 687–691.

91. Lafferentz, L.; Ample, F.; Yu, H.; Hecht, S.; Joachim, C.; Grill, L. Conductance of a single conjugated polymer as a continuous function of its length. Science **2009**, 323, 1193–1198.

92. Riss, A.; Wickenburg, S.; Gorman, P.; Tan, L.Z.; Tsai, H.Z.; de Oteyza, D.G.; Chen, Y.C.; Bradley, A.J.; Ugeda, M.M.; Etkin, G.; et al. Local electronic and chemical structure of oligo-acetylene derivatives formed through radical cyclizations at a surface. Nano Lett. **2014**, 14, 2251–2255.

93. Albrecht, F.; Neu, M.; Quest, C.; Swart, I.; Repp, J. Formation and characterization of a molecule-metal-molecule bridge in real space. J. Am. Chem. Soc. **2013**, 135, 9200.

94. Elemans, J.A.A.W.; Lei, S.; de Feyter, S. Molecular and supramolecular networks on surfaces: From two-dimensional crystal engineering to reactivity. Angew. Chem. Int. Ed. Engl. **2009**, 48, 7298–7333.

95. Slater (née Phillips), A.G.; Beton, P.H.; Champness, N.R. Two-dimensional supramolecular chemistry on surfaces. Chem. Sci. **2011**, 2, 1440.

96. Kudernac, T.; Lei, S.; Elemans, J.A.A.W.; de Feyter, S. Two-dimensional supramolecular self-assembly: Nanoporous networks on surfaces. Chem. Soc. Rev. **2009**, 38, 402–421.

97. Barth, J.V.; Costantini, G.; Kern, K. Engineering atomic and molecular nanostructures at surfaces. Nature **2005**, 437, 671–679.

98. Barth, J.V. Molecular architectonic on metal surfaces. Annu. Rev. Phys. Chem. **2007**, 58, 375–407. [Google Scholar] [CrossRef] [PubMed]

99. Frenking, G.; Frohlich, N. The nature of the bonding in transition-metal compounds. Chem. Rev. **2000**, 100, 717–774.

100. Yurtsever, A.; Sugimoto, Y.; Tanaka, H.; Abe, M.; Morita, S.; Ondráček, M.; Pou, P.; Pérez, R.; Jelínek, P. Force mapping on a partially H-covered Si(111)-(7×7) surface: Influence of tip and surface reactivity. Phys. Rev. B **2013**, 87, 155403.

101. Sharp, P.; Jarvis, S.; Woolley, R.; Sweetman, A.; Kantorovich, L.; Pakes, C.; Moriarty, P. Identifying passivated dynamic force microscopy tips on H:Si(100). Appl. Phys. Lett. **2012**, 100, 233120.

102. Jarvis, S.; Sweetman, A.; Bamidele, J.; Kantorovich, L.; Moriarty, P. Role of orbital overlap in atomic manipulation. Phys. Rev. B **2012**, 85, 235305.

103. Lindner, R.; Kühnle, A. On-surface reactions. ChemPhysChem **2015**, 16, 1582–1592.

CHAPTER 5

Improving Atomic Force Microscopy Imaging by a Direct Inverse Asymmetric PI Hysteresis Model

Dong Wang [1,2], Peng Yu[1], Feifei Wang[1,2], Ho-Yin Chan[3], Lei Zhou [1], Zaili Dong [1], Lianqing Liu [1],* and Wen Jung Li [1,3],*

[1] State Key Laboratory of Robotics, Shenyang Institute of Automation, Chinese Academy of Sciences, Shenyang 110016, China

[2] University of Chinese Academy of Sciences, Beijing 100049, China

[3] Department of Mechanical and Biomedical Engineering, City University of Hong Kong, Kowloon, Hong Kong, China

ABSTRACT

A modified Prandtl–Ishlinskii (PI) model, referred to as a direct inverse asymmetric PI (DIAPI) model in this paper, was implemented to reduce the displacement error between a predicted model and the actual trajectory of a piezoelectric actuator which is commonly found in AFM systems. Due to the nonlinearity of the piezoelectric actuator, the standard symmetric PI model cannot precisely describe the asymmetric motion of the actuator. In order to improve the accuracy of AFM scans, two series of slope parameters were introduced in the PI model to describe both the voltage-increase-loop (trace) and voltage-decrease-loop (retrace). A feedforward controller based on the DIAPI model was implemented to compensate hysteresis. Performance of the DIAPI model and the feedforward controller were validated by scanning micro-lenses and standard silicon grating using a custom-built AFM.

Keywords: atomic force microscope; hysteresis; piezoelectric actuator; direct inverse asymmetric PI model; feedforward control

1. INTRODUCTION

The atomic force microscope (AFM) is a powerful tool for nanoscale imaging and manipulation [1,2]. In an AFM, piezoelectric materials are often used as nanopositioning actuators to drive a scanning probe because of their high resolution capabilities and fast response time [3]. However, the inherent nonlinearities in piezoelectric actuators, especially the hysteresis [4], will lead to displacement errors in horizontal direction in AFM scanning images [5].

Several methods have been proposed to compensate the hysteresis effect. They can be generally classified into two categories: (1) feedback control; and (2) model-based feedforward control. Feedback control method is known to have modeling errors that are simpler to handle, and they have errors caused by parameter variations that can be minimized [6]. However, the hysteresis effect can make a feedback control system unstable [7]. As for model-based feedforward control, it improves performance without incurring the stability problems associated with feedback design. However, the challenges associated with model-based feedforward control are model accuracy and computational complexity [8].

The keys to successfully develop a feedforward control include: (1) develop a hysteresis model which is close to the real hysteresis curve; and (2) realize a feedforward controller based on an inverse model to linearize the response of the actuators, as illustrated in Figure 1. The most commonly used feedforward models include Maxwell's slip model [9], Duhem model [10], Krasnosel'skii-Pokrovskii operator [11], Preisach model [12] and Prandtl-Ishlinskii (PI) model [13]. Among all these hysteresis models, the PI model is the most suitable for real-time applications, such as AFM real-time scanning. This is because it has a much simpler implementation procedure and it also has a unique analytical inverse model [14,15]. However, feedforward controller based on conventional symmetric PI (SPI) model will lead to inevitable error when compensating asymmetric hysteresis of piezoelectric actuators which have asymmetric voltage-displacement response (see Figure 2).

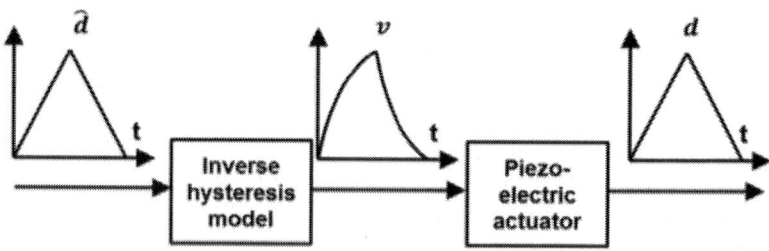

Figure 1. Illustration of the operating principle of a model-based feedforward controller. Through the inverse PI model, a desired linear displacement $\hat{d}(i)$ is transformed into an non-linear voltage $v(i)$. The piezoelectric actuator travels on a linear trajectory $d(i)$ when driven by $v(i)$.

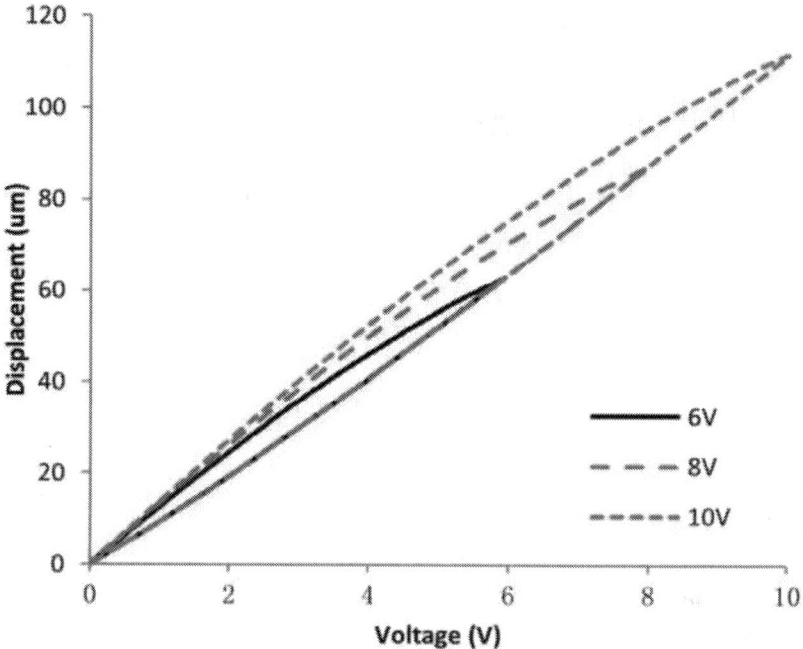

Figure 2. Asymmetric hysteresis loops of a piezoelectric actuator under different driving peak voltages. The hysteresis effect becomes more obvious with increasing driving peak voltages.

In order to compensate asymmetric hysteresis more accurately, some feedforward controllers based on modified PI (MPI) hysteresis models have been investigated, such as adding new components to the conventional SPI model [3,16], using a generalized play operator for characterizing asymmetric hysteresis nonlinearity [17], and using different width parameters in trace and retrace branches [18]. Among these MPI models, the one proposed in [18] is easier to implement because the model does not involve operator modifications and extra components. However, it is central-symmetric and cannot describe the asymmetric hysteresis as shown in Figure 2.

To characterize and compensate asymmetric hysteresis without increasing model complexity, we describe in this paper our work on developing an inverse asymmetric PI (API) hysteresis model with different slope parameters in trace and retrace branches. In order to reduce computation, instead of establishing an inverse model using conventional three-step method (see Figure 3a), a two-step method proposed in [19] by Qin, et al., which can model inverse hysteresis directly, was used. We will present in this paper the experimental results from implementing this DIAPI hysteresis model and the related feedforward controller in a custom-built AFM.

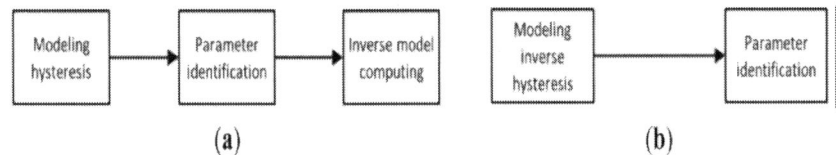

Figure 3. Procedure of establishing inverse hysteresis model: **(a)** the conventional method; **(b)** the modified method.

Our team has previously demonstrated in 2011 that an asymmetric PI model gives more accurate AFM images than a symmetric model [5]. In the current work, we made the following revisions and extensions to the prior work: (1) based on the work of Qin, et al. [19], the direct inverse modeling process is used instead of computing an inverse model based on a forward model; (2) in order to illustrate the performance of API model in AFM imaging, a normalized square L_2 norm (NSL2) that describes dissimilarity was used instead of the normalized product correlation that describes similarity; (3) a micro-lens AFM-scan experiment was added to further evaluate the performance of the direct inverse modeling process, i.e., see Section 3.3; (4) in order to illustrate the asymmetric hysteresis of the piezoelectric actuator, the difference between the slope parameters is shown in this paper.

The rest of the paper is organized as follows: Section 2 presents the principle of the DIAPI model and characterization of parameters; Section 3 discusses the experimental results to interpret the advantages of using the DIAPI model, which include less error and less computational requirement; finally, our conclusions are presented in Section 4.

2. SYMMETRIC & ASYMMETRIC PI INVERSE MODELS

In this section, both SPI and DIAPI models used in this study are presented.

2.1. Definition of Backlash Operator

In this study, forward hysteresis model means the mapping of driving voltage to actuator displacement and inverse hysteresis model means the mapping of actuator displacement to driving voltage. Both of these PI models have same form which is a phenomenological model composed of many elementary rate-independent symmetric backlash operators. Backlash operator can be expressed in two forms: (1) play operator [3] (see Figure 4a); (2) one side play (OSP) operator [16] (see Figure 4b). Since the input voltage of the piezoelectric actuator in our custom-built AFM is positive, the OSP operator is preferred.

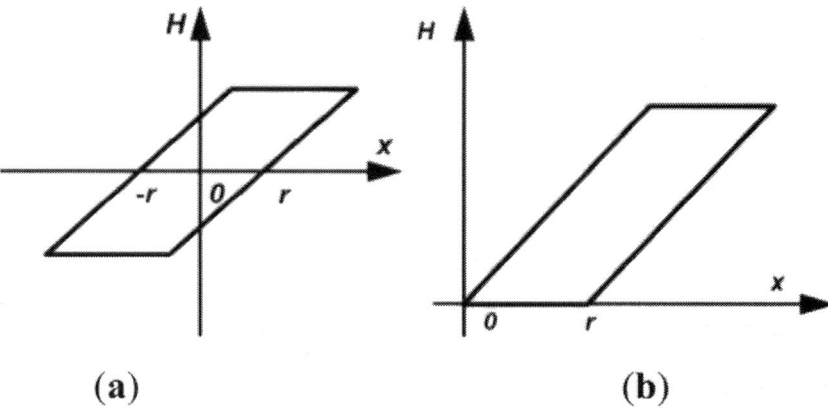

Figure 4. Backlash operators: (**a**) play operator; (**b**) OSP operator.

2.2. Modeling of Hysteresis by Conventional PI Inverse Model

According to the definition in [16], the conventional SPI inverse model can be expressed by:

$$v(i) = \vec{\omega}\vec{H} = \sum_{j=1}^{N} \omega(j)H(i,j)$$

(1)

$$H(i,j) = \max\{d(i) - r(j), \min[d(i), H(i-1)]\}\)$$

(2)

where v is the output of the PI model (and represents the input voltage of the piezoelectric actuator), H is an OSP operator, N is the number of backlash operators, ω is the slope of the backlash (ω = 1 means a 45° slope), d is the input of the SPI inverse model (and represents the displacement of the piezoelectric actuator), and r is width of the backlash. This type of SPI inverse model uses a single slope to describe both the trace and retrace section, which will lead to intrinsic errors when describing an asymmetric hysteresis induced by a piezoelectric actuator as shown in Figure 2.

2.3. Modeling of Hysteresis by DIAPI Model

In order to solve the problem mentioned, we established a DIAPI inverse model that use separate slope parameters to describe trace and retrace branches respectively. The DIAPI inverse model is defined by:

$$v_a(i) = \vec{m}\vec{\omega}_a\vec{H}(i) = \vec{m}[\vec{\omega}_{af} \quad \vec{\omega}_{ab}]^T \vec{H}(i)$$

(3)

where $\vec{m} = [1\ 0]$ or $\vec{m} = [0\ 1]$ for trace and retrace section, respectively. Two slope parameters $\vec{\omega}_{af}$ and $\vec{\omega}_{ab}$ are used to describe trace and retrace branches, respectively. The initial condition of the backlash operator is defined by:

$$H(0) = \max\{d(0) - r, \min[d(0) + r, h_0]\}$$

(4)

where h_0 is set to 0 if the piezoelectric actuator starts from its de-energized state.

In this DIAPI inverse model, three parameters have to be determined: (1) number of backlash operators N; (2) width parameters $\vec{r}$; and (3) slope parameters $\vec{\omega}_{af}$ and $\vec{\omega}_{ab}$. The number of backlash operator N is set to be 10 (as discussed in [3]). The width parameters $\vec{r}$ are given by:

$$r_i = \frac{i-1}{N}[\max(d) - \min(d)], \qquad i = 1 \cdots N$$

(5)

The slope values $\vec{\omega}_{af}$ and $\vec{\omega}_{ab}$ can be determined by minimizing the following error function of a least-squares fit method:

$$E(\omega) = \sum_i^m [v_a(i) - v_{actual}(i)]^2$$

(6)

where $v_{actual}(i)$ is the control voltage of piezoelectric actuator.

3. EXPERIMENTAL RESULTS

A custom-built AFM using piezoelectric actuators was built and experiments were performed using this system to validate the DIAPI hysteresis model and the corresponding model-based feedforward controller.

3.1. Experimental Setup

A schematic view of the custom-built AFM is shown in Figure 5a. The AFM system mainly consists of two parts: a controller and a scan head module (see Figure 5b). The controller contains a PC, two DAQ cards and a three-channel high voltage amplifier. The scan head module (shown in Figure 5b) consists of two nano-positioning stages, a probe holder, an optical microscope, a two-dimension position sensitive device (PSD, an optical position sensor that can measure the position of a light spot in one or two-dimension on a sensor surface) and mirrors. A nanopositioning stage with a 12 μm travel range and a sensitivity of 0.6 μm/V is used as the actuator in vertical (Z-axis) direction. A nanopositioning stage, with a travel range of 100 μm × 100 μm and a sensitivity of 10 μm/V, is used as the actuator in horizontal (XY) plane. The displacement of the nano-positioning stage in the XY direction is measured by a capacitive sensor (CS05, Micro-Epsilon Messtechnik GmbH&Co. KG, Ortenburg, Germany), which has a measurement range of 0~500 μm and a dynamic resolution of 10 nm at an 8.5 kHz sampling frequency. The surrounding temperature of the experimental setup is maintained at 20 °C.

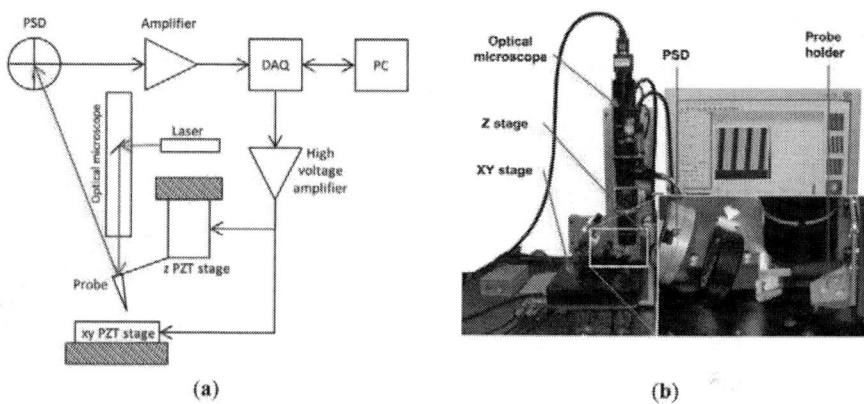

(a)   (b)

Figure 5. A custom-built AFM system. **(a)** Schematic view of the custom-built AFM; **(b)** AFM scan head.

3.2. Model Prediction Experiment

The DIAPI model is used to model the inverse hysteresis of the nanopositioning stage in the X and Y directions, e.g., horizontal direction. As the modeling steps and results are the same in both directions, only those in X direction are shown. Considering AFM commonly works in raster scan mode in the horizontal direction with low scan speed (commonly less than 10 Hz), the hysteresis loop can be approximated as rate-independent. Therefore, the nanopositioning stage is excited using a 1 Hz triangular waveform input with various amplitudes and the displacement was measured by the CS05 capacitive sensor with 1 KSPS sampling rate. The hysteresis loop with 100 μm displacement that was excited by 9.1 V amplitude input voltage was used for parameter identification.

According to the algorithm presented in Section 2, the parameters of the DIAPI model have been calculated and are listed in Table 1. To illustrate asymmetry of hysteresis, difference between slope parameters in trace and retrace branches is calculated by:

$$e_\omega = \left| \frac{\omega_{af} - \omega_{ab}}{\omega_{ab}} \right|$$

(7)

and shown in Table 1. It can be seen that the difference between two slope parameters can be large (at most 90%). Therefore, modeling error of SPI model will be inevitably large if the same slope parameters used in trace and retrace branches.

Table 1. Recognized parameters of the DIAPI model.

i	r(i)	ω_{af} (i)	ω_{ab} (i)	e_ω
1	0	0.1068	0.1274	16%
2	11.1	−0.0080	−0.0204	61%
3	22.2	−0.0046	−0.0094	51%
4	33.3	−0.0034	−0.0066	48%
5	44.4	−0.0034	−0.0049	31%
6	55.5	−0.0008	−0.0040	80%
7	66.6	−0.0027	−0.0036	25%
8	77.7	−0.0003	−0.0030	90%
9	88.8	−0.0011	−0.0010	10%
10	99.9	−0.0012	−0.0025	52%

Experimental results are shown in Figures 6 and 7. In Figure 6, errors between the SPI prediction and the actual hysteresis loop are larger than predicted by the DIAPI model. Ten experiments were performed and showed consistent results. Three representative data sets of the modeling errors in trace and retrace branches from the two models are shown in Figure 7a and Figure 7b, respectively. A smaller error means the corresponding model is better in predicting the hysteresis. From Figure 7a,b, it can be seen that errors of the DIAPI model are obviously smaller than that of SPI model especially in range of 20 μm~60 μm, 90 μm~100 μm during trace and 0 μm~20 μm, 50 μm~80 μm, 90 μm~100 μm during retrace. The RMS and maximum errors of the SPI and DIAPI model in 10 experiments are listed in Table 2. The mean RMS errors of SPI model and DIAPI model are 0.08 V (with s.d. of 0.003) and 0.02 V (with s.d. of 0.002), respectively. The mean RMS error of DIAPI model is 75% smaller than that of the SPI model.

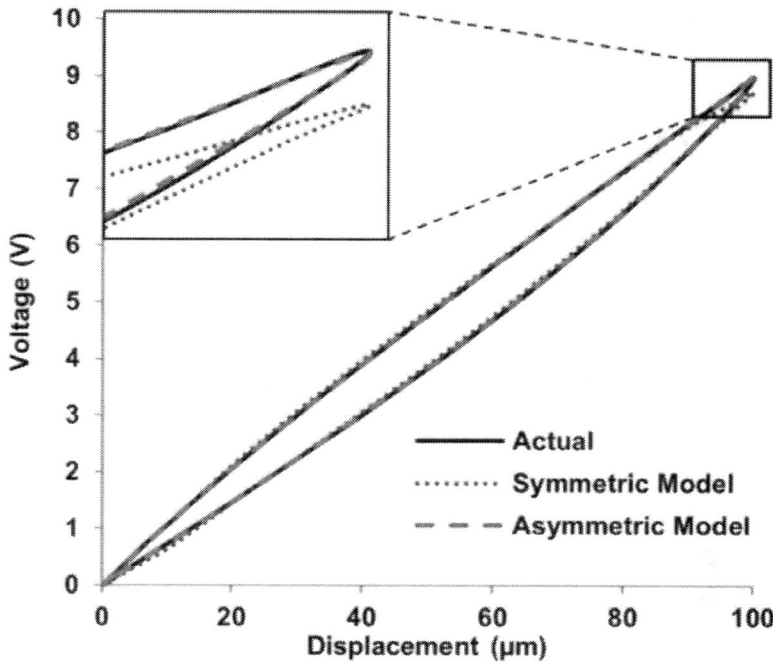

Figure 6. Actual hysteresis loop and PI model loops.

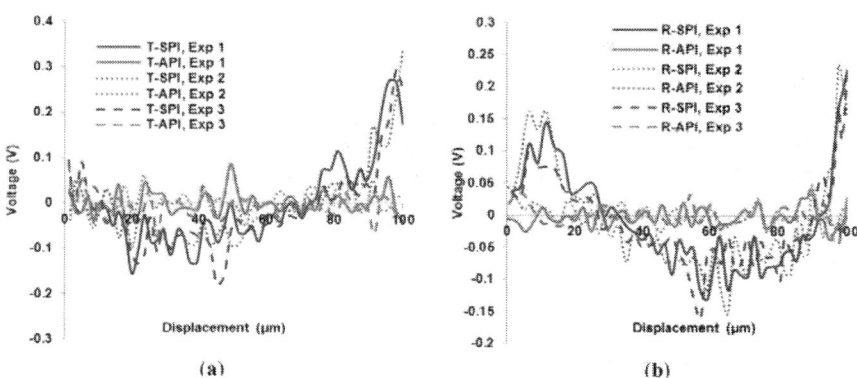

Figure 7. (a) Trace and (b) retrace errors of the SPI and DIAPI models are labeled as T-SPI and T-API, respectively. Ten experiments were performed to show the consistency of experimental results. Three representative data sets are shown in this figure.

Table 2. Errors of SPI and DIAPI model.

Date Set	RMS Errors (V)		Maximum Errors (V)	
	SPI Model	**DIAPI Model**	**SPI Model**	**DIAPI Model**
1	0.078	0.021	0.260	0.068
2	0.083	0.019	0.277	0.062
3	0.082	0.021	0.297	0.054
4	0.079	0.023	0.293	0.061
5	0.073	0.021	0.283	0.064
6	0.080	0.024	0.287	0.068
7	0.085	0.022	0.284	0.067
8	0.081	0.023	0.271	0.056
9	0.081	0.021	0.263	0.064
10	0.076	0.020	0.259	0.065
Mean	0.080	0.021	0.277	0.063
σ	0.003	0.002	0.013	0.005

Compared to conventional three-step methods that are used in indirect inverse API (IIAPI) models, the DIAPI model uses less computational time because the inverse model computing step is eliminated. In order to quantitatively compare the computational time consumption of the two models, IIAPI and DIAPI models were tested with various hysteresis loop sampling rate (M) and operator quantity (N). The IIAPI model from [5] is used for this comparison. The identification and computation algorithms were coded in Matlab, which was installed in a PC with an i5 CPU and 8 G RAM. To reduce the effect of contingency, the algorithms were run 10,000 times for each model and the mean run-time was used for comparison. The time-saving percentages of DIAPI model over IIAPI model are shown in Figure 8.

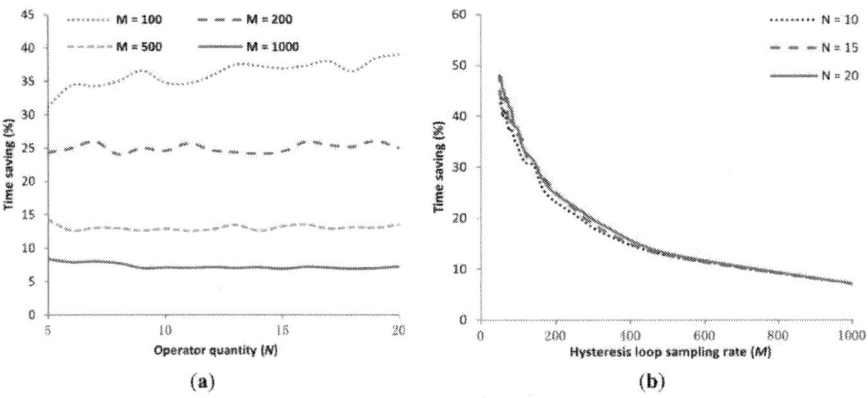

Figure 8. Time saving percentage of DIAPI model over IIAPI model with **(a)** various hysteresis loop sampling rate and **(b)** operator quantity.

As shown in Figure 8a, the time-saving percentage fluctuates slightly when the quantity of N increases. Also, as shown in Figure 8b, the time-saving percentage decreases when M increases, i.e., when more samples are used in a hysteresis loop. According to experiments, both models are more accurate when M increases, but the accuracy does not further improve much when M exceeds 200. Hence, referring to Figure 8b, we can conclude that DIAPI model can save ~25% (M = 200) more run-time than the IIAPI model.

3.3. AFM Imaging of Micro-Lenses

To demonstrate the improved performance of the DIAPI model in AFM imaging, spherical micro-lenses which are made of Polydimethylsiloxane (PDMS) were used and scanned in contact mode. The type of AFM probe used is MLCT which is manufactured by Bruker (Santa Barbara, CA, USA). The micro-lenses were first scanned using SEM and a commercial AFM (Dimension 3100, Bruker) using feedforward calibration (closed loop off) as a reference. The SEM image and section curves are shown in Figure 9a,b. Next, in order to illustrate image distortion caused by hysteresis of the piezoelectric stage, the micro-lens was scanned by the custom-built AFM without calibration. As shown in Figure 10a,b, the location of the micro-lenses in trace and retrace image are shifted because of the hysteresis of the piezoelectric stage. Detailed differences can be observed in cross-section curves (see Figure 10c), e.g., trace curve and retrace curves do not overlap because of hysteresis.

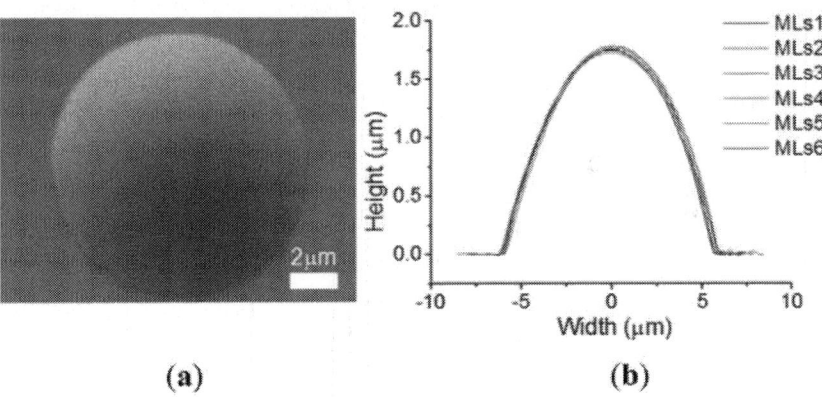

(a)   (b)

Figure 9. (a) An SEM image of a PDMS micro-lens; (b) The section curves of PDMS micro-lenses scanned by Bruker AFM. MLs 1-6 represent section curves of six micro-lenses.

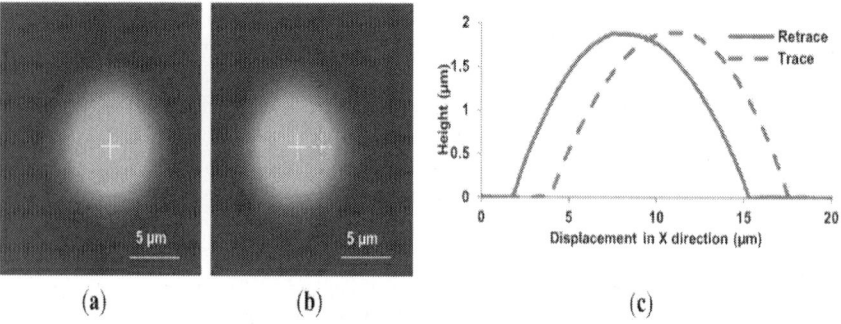

(a)   (b)   (c)

Figure 10. Height images of a PDMS micro-lens scanned using the custom-built AFM without calibration. (**a–c**) are trace image, retrace image and cross-section curves, respectively. The white cross represents location of the micro-lens center. In the retraced image, the dashed white cross represents the location of the micro-lens center in the trace image. The center of the micro-lens is shifted about 2.5 μm because of hysteresis.

The feedforward controller based on the DIAPI model was implemented on custom-built AFM for hysteresis calibration. The parameters of inverse model were calculated using Equations (5) and (6). For comparison, the feedforward controller based on SPI model was also implemented. Figures 10, 11 and 12 show height images and cross-section curves of micro-lenses scanned using the custom-built AFM based on feedforward controller using SPI and DIAPI models and Bruker AFM, respectively. As can be seen from height image (Figures 11a,b and 12a,b) of micro-lenses, distortion of the image was corrected. The trace and retrace images become visually identical. Furthermore, the cross-section curves based on DIAPI model (see Figure 12c) coincide better than that of Bruker AFM (see Figure 13c), while the one based on SPI model (see Figure 11c) coincide the worst.

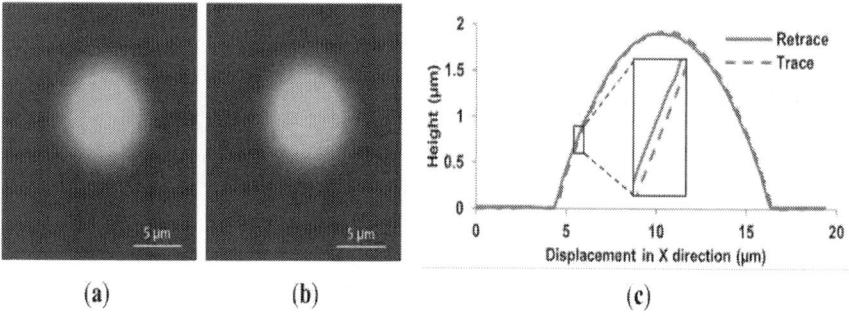

(a)   (b)   (c)

Figure 11. Height images of the PDMS micro-lens scanned using the custom-built AFM with the SPI model calibration. (**a–c**) are trace image, retrace image and cross-section curves, respectively. Inset shows a 5× magnified view of cross-section between 5.4 μm and 6 μm.

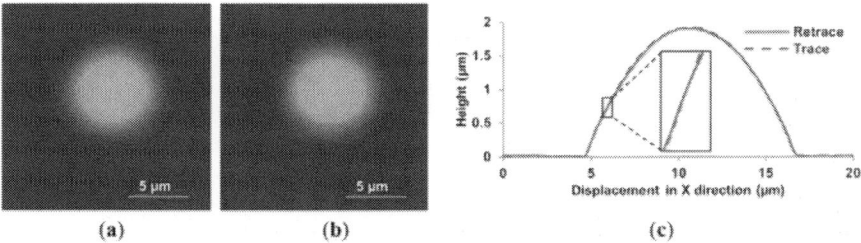

Figure 12. Height images of PDMS micro-lens scanned using the custom-built AFM with DIAPI model calibration. (**a–c**) are trace image, retrace image and cross-section curves, respectively. Inset shows a 5× magnified view of the cross-section between 5.4 μm and 6 μm.

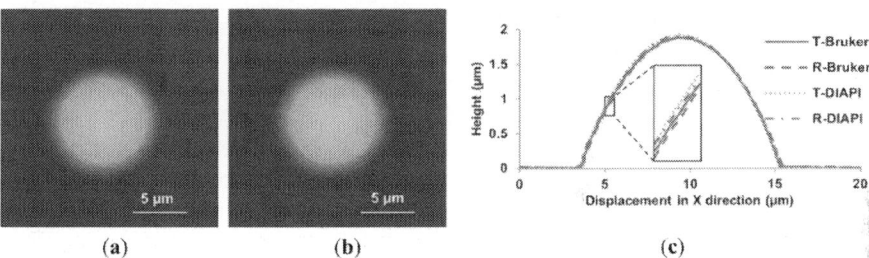

Figure 13. Height images of PDMS micro-lens scanned using the Bruker AFM. (**a,b**) are trace and retrace images scanned by the Bruker AFM, respectively; (**c**) shows the cross-section curves scanned by the Bruker AFM and the custom-built AFM based on DIAPI model. T-Bruker and R-Bruker represent trace and retrace cross-section curves scanned using the Bruker AFM, respectively; T-DIAPI and R-DIAPI represent trace and retrace cross-section curves using the custom-built AFM based on DIAPI model, respectively. Inset shows a 5× magnified view of the cross-section between 5.1 μm and 5.7 μm.

In order to quantitatively compare the difference between AFM images based on DIAPI and SPI models, normalized square L_2 norm (NSL2) describing dissimilarity between trace and retrace images can be given by [20]:

$$D = \sum_{i=1}^{n} \left(\frac{x_i - \bar{x}}{\sigma_x} - \frac{y_i - \bar{y}}{\sigma_y} \right)^2$$

$$\sigma_x = \sqrt{\frac{1}{n}\sum_{i=1}^{n}(x_i - \bar{x})^2}, \ \sigma_y = \sqrt{\frac{1}{n}\sum_{i=1}^{n}(y_i - \bar{y})^2}, \ \bar{x} = \frac{1}{n}\sum_{i=1}^{n}x_i, \ \bar{y} = \frac{1}{n}\sum_{i=1}^{n}y_i$$

(8)

where D represents NSL2, i.e., dissimilarity between two images, the smaller the better; x_i and y_i are values of each pixel in trace and retrace images after binarization, respectively. The threshold of binarization is the mean gray value of image. To illustrate the asymmetry of hysteresis, the difference between the NSL2 parameter in trace and retrace images is calculated by:

$$e_D = \frac{D_S - D_A}{D_S}$$

(9)

Table 3 summarizes the NSL2 between the trace and retrace images of the micro-lenses scanned by the custom-built AFM and Bruker AFM, respectively. D_N, D_S, D_A and D_B represent the NSL2 without calibration, SPI model calibration, DIAPI model calibration, and Bruker AFM feedforward calibration, respectively. Without calibration, it is obvious that the location of the micro-lens in the trace image is different from that in the retrace image (see Figure 10a,b). The mean NSL2 of 10 experiments is 45,814. After the model-based feedforward controllers were implemented, the trace and retrace images had no major image shift and the means NSL2 of 10 experiments dropped to 2322 (SPI model) and 1292 (DIAPI model). As reference, the mean NSL2 of 10 experiments scanned using the Bruker AFM was 1545. Hence, we can conclude that the PI models will definitely improve AFM imaging consistency during the trace and retrace steps. The NSL2 between the trace and retrace images based on DIAPI model is 44.3% smaller than that of SPI model and 16.3% ((i.e., $(D_B - D_A)/D_B$) smaller than that of Bruker AFM with feedforward calibration.

Table 3. NSL2 between trace and retrace image of micro-lenses.

Number	D_N (Without Calibration)	D_S (SPI Model)	D_A (DIAPI Model)	E_D (%)	D_B (Bruker AFM)
1	45,887	2011	1136	43.5	1404
2	46,104	2674	1042	61.0	1302
3	46,091	2481	1096	55.8	1898
4	47,938	2570	1336	48.0	1778
5	45,711	2563	1539	40.0	1883
6	43,865	1908	1399	26.7	1400
7	45,833	1924	1308	32.0	1298
8	43,024	2543	1106	56.5	1390
9	47,135	2679	1318	50.8	1453
10	46,556	1868	1649	11.7	1648
Mean	45,814	2322	1293	44.3	1545
σ	1436.3	346.0	200.7		235.3

3.4. AFM Imaging of Silicon Grating

Besides scanning the micro-lens, a silicon grating (HS-500MG, Innovative Solutions, Sofia, Bulgaria) with pitches (500 nm high and 5 μm period) was also scanned using the custom-built AFM under same machine settings of that used in micro-lens scan. The profiles of the pitches are shown in SEM image (Figure 14). The trace and retrace images of the pitches without calibration are shown in

Figure 15. We can see that pitches of trace image and retrace image were shifted. In trace image, the pitch (3.6 μm) on the left is wider than that on the right (3.2 μm). In retrace image, the pitch (3.2 μm) on the left is narrower than that on the right (3.6 μm). Figures 16 and 17 are images of the pitches after implementing the SPI and DIAPI models, respectively. We can see from the figures that trace and retrace image become visually equal after calibration. In order to further confirm the improvement of AFM imaging based on DIAPI model, silicon grating images scanned using the Bruker AFM are shown in Figure 18.

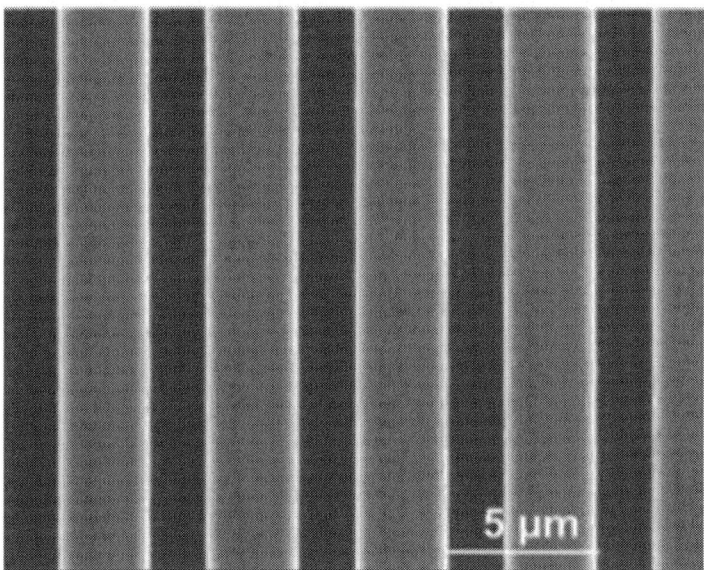

Figure 14. A SEM image of a silicon grating.

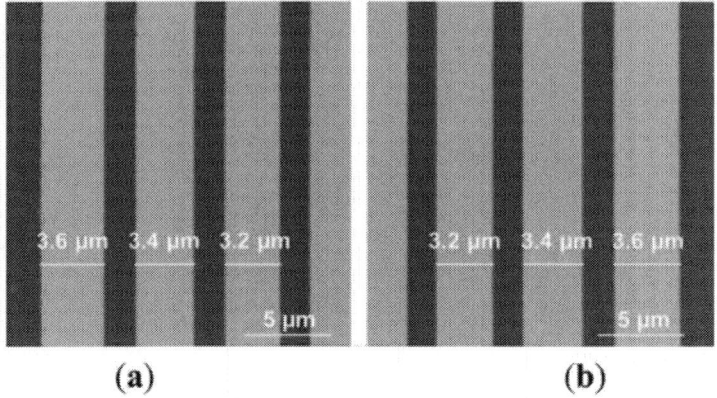

Figure 15. Height images of pitches in silicon grating scanned using custom-built AFM without calibration. (**a**) Trace image; (**b**) retrace image.

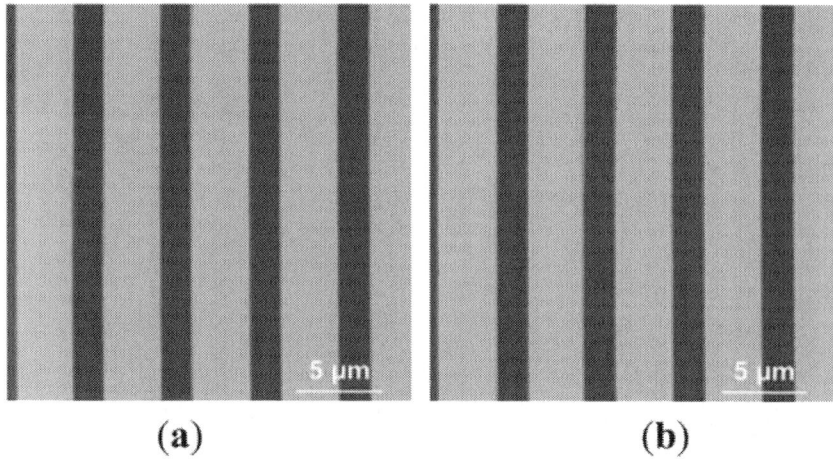

Figure 16. Height images of pitches in silicon grating scanned using the custom-built AFM with SPI model calibration. (**a**) Trace image; (**b**) retrace image.

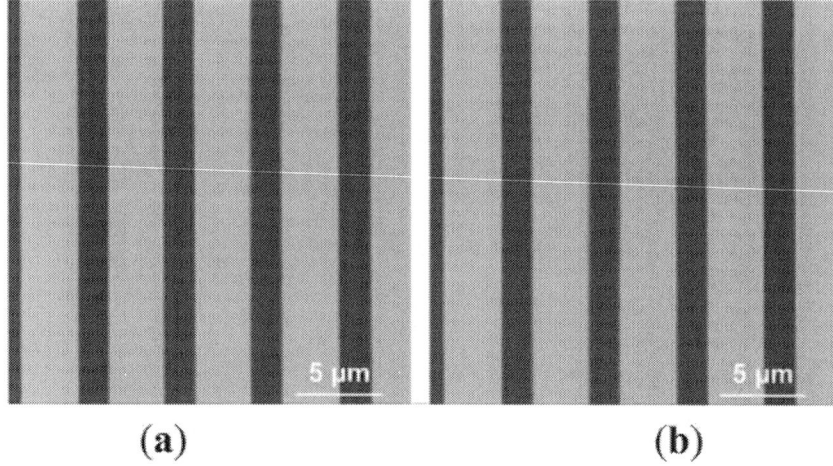

Figure 17. Height images of pitches in silicon grating scanned using the custom-built AFM with DIAPI model calibration. (**a**) Trace image; (**b**) retrace image.

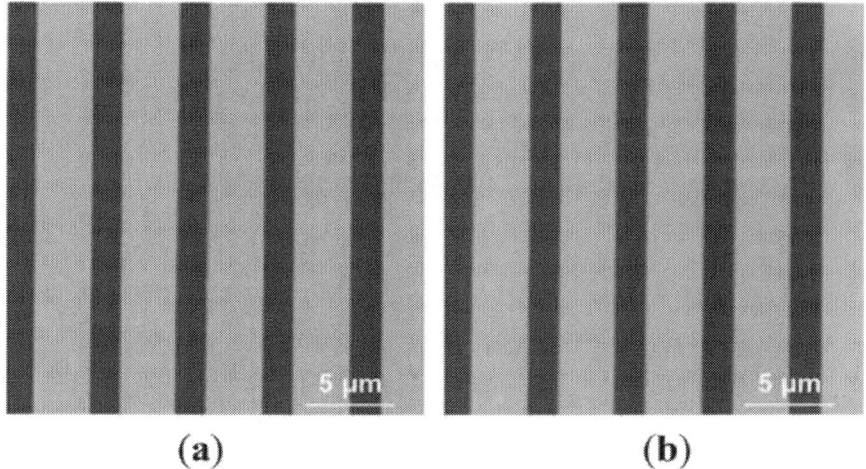

(a)   **(b)**

Figure 18. Height images of pitches in silicon grating scanned using the Bruker AFM. (**a**) Trace image; (**b**) retrace image.

Table 4 summarizes the NSL2 between trace and retrace images of the pitches scanned by the custom-built AFM based on the SPI and DIAPI models. Without calibration (i.e., PI model implementation), mean NSL2 of 10 experiments is ~178,272. After model-based feedforward controllers were implemented, the mean NSL2 dropped to 6467 (SPI model) and 4023 (DIAPI model). As reference, the mean NSL2 of 10 experiments scanned using the Bruker AFM was 4717. Hence, there is a 37.8% improvement of the DIAPI model over the SPI model, and a 14.7% (i.e., $(D_B-D_A)/D_B$) improvement over the Bruker AFM.

Table 4. NSL2 between trace and retrace image of pitches.

Number	D_N (Without Calibration)	D_S (SPI Model)	D_A (DIAPI Model)	E_D (%)	D_B (Bruker AFM)
1	177,690	6474	4066	37.2	4890
2	180,457	6804	3772	44.6	4359
3	169,440	6184	3781	38.9	4451
4	185,401	6707	3869	42.3	4413
5	180,865	6675	4343	34.9	5176
6	182,269	6037	3866	36.0	4876
7	175,774	5981	4321	27.8	4771
8	180,452	6472	3857	40.4	4971
9	171,847	7069	4415	37.5	5007
10	178,529	6267	3944	37.1	4251
Mean	178,272	6467	4023	37.8	4717
σ	4818.7	352.5	247.3		320.9

It should be noted that the hysteresis of piezoelectric actuators is less significant in a small travel range, i.e., asymmetric hysteresis is less significant for small scan range of an AFM. Therefore, when the scan range decreases, the difference between the SPI and DIAPI models becomes smaller, and the NSL2 difference between the trace and retrace images based on SPI and DIAPI models also become smaller. Hence, the advantage of the DIAPI model over the SPI model is less obvious when the AFM scan range is smaller.

4. CONCLUSIONS

In this paper, an asymmetric PI model using two series slope parameters in trace and retrace branches is proposed to model the inverse asymmetric hysteresis of a piezoelectric actuator. The parameters of the inverse model were identified by a least-square fit instead of computation based on the loading curve. A feedforward controller based on the inverse DIAPI model was implemented on a custom-built AFM for validation. Model prediction experiment shows that mean RMS error of DIAPI model is 75% smaller than that of the SPI model, and the establishment of the DIAPI model is 25% faster than that of the IIAPI model. We have shown experimentally that both the SPI and DIAPI models are effective in improving the precision of the resulting AFM scanned images. Furthermore, the DIAPI model yields significantly lower NSL2 values between the trace and retrace scanned images, e.g., 44.3% and 37.8% lower in micro-lenses and silicon grating scanning experiments, respectively.

ACKNOWLEDGMENTS

This work is supported by the National Natural Science Foundation of China (Grant No. 61304251), the National High-tech R&D Program of China (863 Program, Project No. 2012AA041204), and the Instrument Developing Project of the Chinese Academy of Sciences (Project No. YZ201245).

REFERENCES

1. Krohs, F.; Onal, C.; Sitti, M.; Fatikow, S. Towards automated nanoassembly with the atomic force microscope: A versatile drift compensation procedure. J. Dyn. Syst. Meas. Control **2009**, 131.
2. Tian, X.J.; Wang, Y.C.; Xi, N.; Dong, Z.L. A pilot study on nano forces in AFM-based robotic nanomanipulation. Robot **2007**, 29, 363–367.
3. Ang, W.T.; Khosla, P.K.; Riviere, C.N. Feedforward controller with inverse rate-dependent model for piezoelectric actuators in trajectory-tracking applications. IEEE-ASME T. Mech. **2007**, 12, 134–142.

4. Liu, L.Q.; Xi, N.; Zhang, J.B.; Li, G.Y.; Wang, Y.C.; Dong, Z.L. System positioning error compensated by local scan in atomic force microscope based nanomanipulation. Proceedings of the IEEE International Conference on Nano/Micro Engineered and Molecular Systems (NEMS), Sanya, China, 6–9 January 2008; pp. 1113–1118.

5. Wang, D.; Dong, Z.L.; Jiao, N.D.; Yuan, S.; Zhou, L.; Li, W.J. An asymmetric PI hysteresis model for piezoceramics in nanoscale AFM imaging. Proceedings of the IEEE International Conference on Nano/Micro Engineered and Molecular Systems (NEMS), Kaohsiung, Taiwan, 20–23 February 2011; pp. 1075–1079.

6. Leang, K.K.; Devasia, S. Feedback-linearized inverse feedforward for creep, hysteresis, and vibration compensation in AFM piezoactuators. IEEE Trans. Control Syst. Technol. **2007**, 15, 927–935. [Google Scholar]

7. Tao, G.; Kokotovic, P.V. Adaptive control of plants with unknown hysteresis. IEEE Trans. Autom. Control **1995**, 40, 200–212.

8. Devasia, S.; Eleftheriou, E.; Moheimani, S.O.R. A survey of control issues in nanopositioning. IEEE Trans. Control Syst. Technol. **2007**, 15, 802–823.

9. Goldfarb, M.; Celanovic, N. Modeling piezoelectric stack actuators for control of micromanipulation. IEEE Control Syst. **1997**, 17, 69–79.

10. Ouyang, R.Y.; Andrieu, V.; Jayawardhana, B. On the characterization of the Duhem hysteresis operator with clockwise input–output dynamics. Syst. Control Lett. **2013**, 62, 286–293.

11. Zhou, M.L.; Zhang, Q.; Wang, J.Y. Feedforward-feedback hybrid control for magnetic shape memory alloy actuators based on the Krasnosel'skii-Pokrovskii model. PLoS One **2014**, 9.

12. Li, Z.; Su, C.Y.; Chai, T.Y. Compensation of hysteresis nonlinearity in magnetostrictive actuators with inverse multiplicative structure for Preisach model. IEEE Trans. Autom. Sci. Eng. **2013**, 11, 613–619.

13. Aljanaideh, O.; Aljanaideh, M.; Rakheja, S.; Su, C.Y. Compensation of rate-dependent hysteresis nonlinearities in a magnetostrictive actuator using an inverse Prandtl–Ishlinskii model. Smart Mater. Struct. **2013**, 22.

14. Kuhnen, K.; Janocha, H. Inverse feedforward controller for complex hysteretic nonlinearities in smart-material systems. Control Intell. Syst. **2001**, 29, 74–83.

15. Mokaberi, B.; Requicha, A.A.G. Compensation of scanner creep and hysteresis for AFM nanomanipulation. IEEE Trans. Autom. Sci. Eng. **2008**, 5, 197–206.

16. Gu, G.Y.; Zhu, L.M.; Su, C.Y. Modeling and compensation of asymmetric hysteresis nonlinearity for piezoceramic actuators with a modified Prandtl-Ishlinskii model. IEEE Trans. Ind. Electron. **2014**, 61, 1583–1595.

17. Aljanaideh, M.; Rakheja, S.; Su, C.Y. An analytical generalized Prandtl–Ishlinskii model inversion for hysteresis compensation in micropositioning control. IEEE-ASME Trans. Mech. **2011**, 16, 734–744.

18. Bashash, S.; Jalili, N. Robust multiple frequency trajectory tracking control of piezoelectrically driven micro/nanopositioning systems. IEEE Trans. Control Syst. Technol. **2007**, 15, 867–878.

19. Qin, Y.D.; Tian, Y.L.; Zhang, D.W.; Shirinzadeh, B.; Fatikow, S. A novel direct inverse modeling approach for hysteresis compensation of piezoelectric actuator in feedforward applications. IEEE-ASME Trans. Mech. **2013**, 18, 981–989.

20. Evangelidis, G.D.; Psarakis, E.Z. Parametric image alignment using enhanced correlation coefficient maximization. IEEE Trans. Pattern Anal. **2008**, 30, 1858–1865.

CHAPTER 6

Molecular Processes Studied at a Single-Molecule Level Using DNA Origami Nanostructures and Atomic Force Microscopy

Ilko Bald [1,2,]* *and Adrian Keller* [3,]*

[1] Institute of Chemistry—Physical Chemistry, Universität Potsdam, Karl-Liebknecht-Straße 24-25, D-14476 Potsdam, Germany
[2] BAM Federal Institute of Materials Research and Testing, Richard-Willstätter Str. 11, D-12489 Berlin, Germany
[3] Technical and Macromolecular Chemistry, University of Paderborn,

ABSTRACT

DNA origami nanostructures allow for the arrangement of different functionalities such as proteins, specific DNA structures, nanoparticles, and various chemical modifications with unprecedented precision. The arranged functional entities can be visualized by atomic force microscopy (AFM) which enables the study of molecular processes at a single-molecular level. Examples comprise the investigation of chemical reactions, electron-induced bond breaking, enzymatic binding and cleavage events, and conformational transitions in DNA. In this paper, we provide an overview of the advances achieved in the field of single-molecule investigations by applying atomic force microscopy to functionalized DNA origami substrates.

Keywords: DNA origami; atomic force microscopy; single-molecule analysis; DNA radiation damage; protein binding; enzyme reactions; G quadruplexes

1. INTRODUCTION

During the last three decades, the field of structural DNA nanotechnology has developed a variety of techniques to assemble DNA into increasingly complex nanostructures [1]. The unique self-assembly capabilities of DNA result from the strong specificity of Watson-Crick base pairing. By controlling the nucleobase sequence of DNA strands, different segments along a given strand can be programmed to pair with different partners, thus enabling the formation of branched junctions. These junctions can be used as building blocks and further assembled into larger arrangements, albeit only with rather moderate assembly yields [2]. However, with the introduction of the DNA origami technique by Rothemund in 2006 [3], the rapid, high-yield assembly of complex, well-defined DNA nanostructures suddenly became feasible.

In the DNA origami technique, a long, single-stranded (ss) DNA scaffold (typically a viral genome) is folded into a nanoscale shape by hybridization with a number of short synthetic oligonucleotides, so-called staple strands. Each staple strand is partially complementary to different separated segments of the scaffold strand which causes the scaffold to fold upon hybridization. The resulting DNA origami then consists completely of double-stranded (ds) DNA which is held together by periodic crossovers of the staple strands. The shape of the DNA origami is "programmed" by the sequences of the individual staple strands. Figure 1a shows schematically how a circular DNA strand assembles into a triangular shape upon addition of specific staple strands.

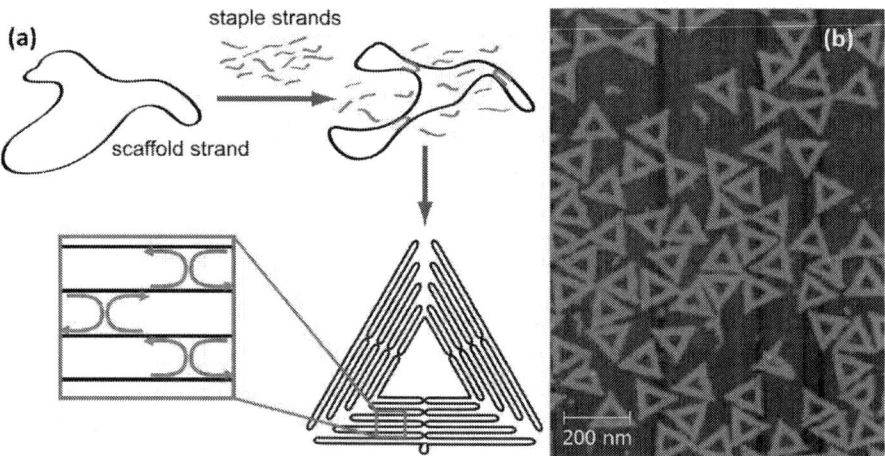

Figure 1. (a) Scheme for synthesizing a triangular DNA origami nanostructure: A circular single-stranded scaffold is folded into a triangular shape by addition of ~200 specifically binding oligonucleotides, so-called staple strands; (b) Atomic force microscopy (AFM) image of the resulting triangular DNA origami immobilized on a mica surface.

DNA origami assembly is typically performed with a high excess of staple strands in Mg^{2+}-containing buffer which screens the electrostatic repulsion between the negatively charged DNA strands. The solution is rapidly heated above the melting temperature of the DNA, i.e., to 60–90 °C, and slowly cooled down to room temperature. During cooling down, the individual staple strands have enough time to find their complementary sequences on the scaffold and fold it into the desired shape.

Most 2D DNA origami consist of the M13mp18 viral DNA scaffold and around 200 staple strands [3]. Each staple strand can be extended to protrude from the DNA origami surface, which results in more than 200 unique sites that can be modified systematically with respect to length, nucleobase sequence, and hybridization state, and synthesized to carry various chemical modifications. Therefore, DNA origami templates are frequently used as locally addressable supports (so-called "molecular breadboards", see Figure 2 left) for the precise arrangement of functional entities such as plasmonic nanoparticles [4,5], quantum dots [6,7], fluorophores [8,9], and proteins [10,11], which enables their use as templates for the study of chemical reactions at a single-molecule level. For this purpose, also DNA origami frames can be employed (see Figure 2, right) which facilitate the incorporation of single DNA strands with high structural control as substrates for biochemical reactions.

The shapes of DNA origami nanostructures are typically analyzed using single-molecule microscopy techniques [12], the most important being atomic force microscopy (AFM) and transmission electron microscopy (TEM). TEM can yield high-resolution images of single nanostructures, but requires sophisticated sample preparation and image analysis. AFM has the advantage that it can be operated under atmospheric conditions and even in liquids, and the requirements for sample preparation are minimal.

In AFM a cantilever with a sharp tip (radius of apex curvature 1–10 nm) is scanned over the surface and the interaction forces (long range attractive van-der-Waals forces, electrostatic interactions and the shorter range Pauli repulsion) between tip and sample lead to a deflection of the cantilever (in contact mode) or to a reduced vibrational amplitude (in dynamic mode). As a result the topography of a sample's surface is recorded with lateral resolution of 1–10 nm and height resolution down to 0.1 nm. Under ultra-high vacuum single molecules can be imaged with atomic resolution and even chemical reactions between individual molecules can be observed [13]. Under atmospheric conditions AFM is typically used to image (bio)macromolecules such as DNA and proteins, and due to the accurate height information, the volume of single macromolecules can also be determined. Apart from the topographical information other materials properties can be probed by AFM that are based on the interaction forces, in particular mechanical properties and surface charges.

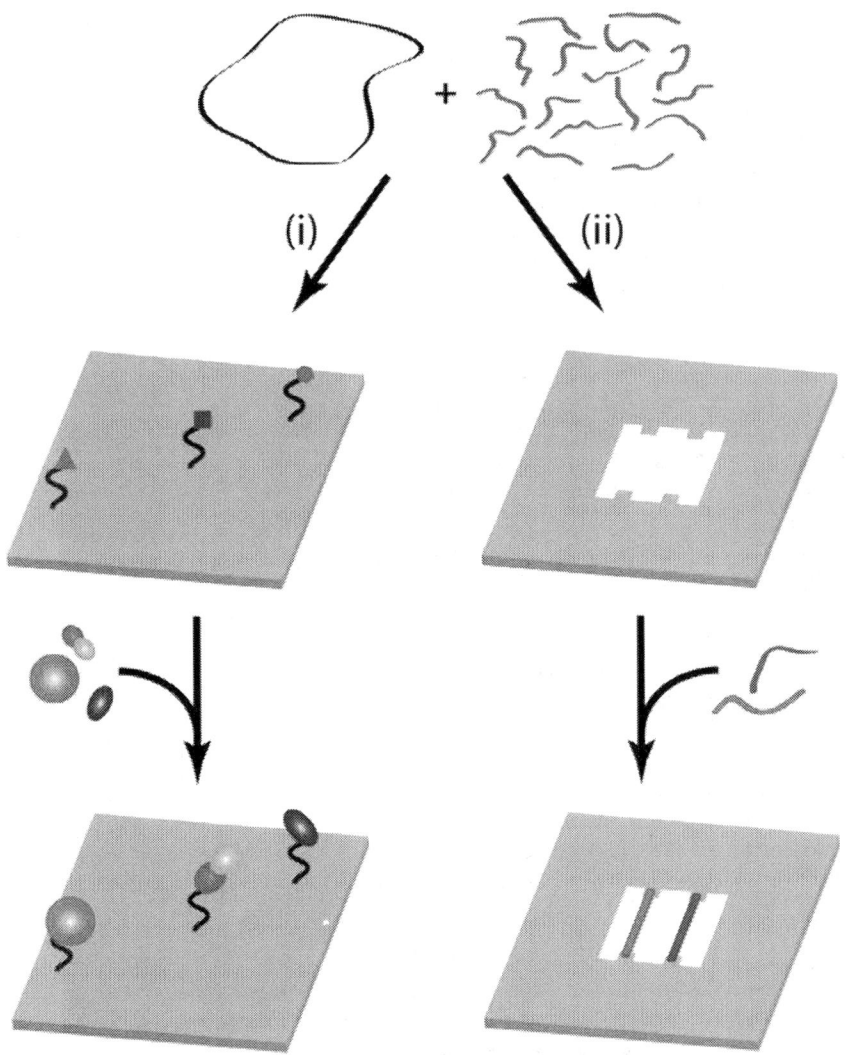

Figure 2. DNA origami nanostructures as substrates for the study of chemical reactions. In strategy (i), a DNA origami is used as a molecular breadboard to arrange molecular entities with high lateral precision while in strategy (ii), a DNA origami frame is constructed for the incorporation of DNA strands of well-defined structure and topology which can then serve as substrates for biochemical reactions.

The typical scan speed of AFM is in the range of several minutes per frame. However, recent developments in high-speed AFM (HSAFM) enable the study of dynamic processes at the sub-second time-scale with scan speeds up to 33 frames per second using small cantilevers with high resonance frequency ($\approx$MHz) and stiff and compact piezoscanners [14].

Other analytical tools that have been used to analyze functionalized DNA origami structures are based on optical methods and can also be operated down to a single-molecule level. Fluorescence spectroscopy is used to detect single molecules and to study, for instance, energy transfer pathways [15,16]. Other optical spectroscopy techniques are based on light scattering, most importantly surface-enhanced Raman scattering (SERS), which was recently demonstrated on a few-molecule level with gold-nanoparticle functionalized DNA origami structures [17,18,19,20]. Single-molecule binding events have also been detected by measuring mechanical transitions in DNA origami using optical tweezers [21]. This review focuses on the application of AFM to study molecular processes on DNA origami nanostructures.

Soon after the invention of the DNA origami technique by Paul Rothemund in 2006 it was demonstrated that DNA origami tiles can be used as the molecular analogues of macroscopic DNA chips [22]. Rectangular DNA origami structures have been decorated with V shaped pairs of 20 nt long ssDNA and the hybridization with the complementary 40 nt long RNA target sequences was detected by AFM (Figure 3). Twelve binding sites for one target sequence have been placed on each probe tile and the detection was based on the increased stiffness of dsDNA compared to ssDNA. It was estimated that a detection limit of 1000 RNA molecules could be reached if a sample volume of 1 nL was used [22]. To detect RNA with a DNA origami and AFM based approach a similar strategy was recently demonstrated using strand displacement reactions and streptavidin (SAv)-quantum dot reporters [23].

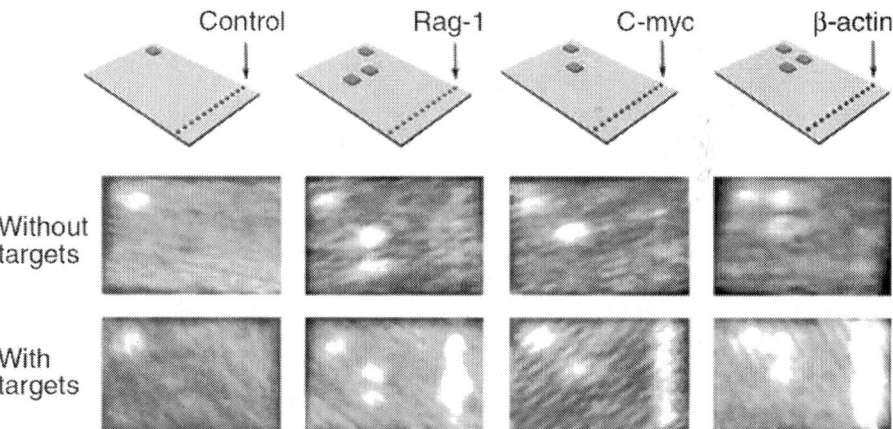

Figure 3. RNA detection using bar-coded DNA origami tiles and AFM analysis. The probe molecules are aligned in a row on the right side of the tiles, and different probes are distinguished from each other with the bar-code located on the top left corner of the tiles. Hybridization is visualized with AFM due to the higher stiffness of double-stranded DNA compared to single-stranded DNA. From Ref. [22]. Reprinted with permission from AAAS.

This example shows the potential of DNA origami nanostructures to place molecular binding sites with nanometer precision and to study binding events and molecular processes on a nanometer level by AFM.

2. CHEMICAL REACTIONS

The presence of dsDNA of a certain length (e.g., 40 nt as in the example discussed above) and hairpin loops of a certain size attached to a DNA origami platform can directly be visualized by AFM. Smaller molecules and soft materials are more difficult to observe directly with AFM. Hence, chemical reactions between single molecules can be followed by AFM by visualizing the presence of certain functional groups using SAv as a marker. SAv binds strongly to biotin (Bt) and with its diameter of 5 nm can be easily identified in AFM images. For a detailed discussion of the Bt-SAv interaction, see Section 3.2.

2.1. Single-Molecule Chemical Reactions

By using SAv-Bt binding the cleavage of single chemical bonds could be visualized on a single-molecule level. Voigt et al. placed three different types of linkers on a rectangular DNA origami platform by extending selected staple strands [24]. It was demonstrated that subsequently the disulfide containing linker can be cleaved with a reducing agent, and a linker containing an electron-rich double bond could be cleaved by using photo generated singlet oxygen. Each cleavage step was followed by observing the disappearing SAv markers at the respective positions of the linkers on the DNA origami rectangle. In the same study the formation of chemical bonds was demonstrated using the same strategy. Subsequently, a click reaction (formation of a triazole from an alkyne and an azide), formation of an amide from an NHS-ester and an amine followed by another click reaction was realized on different positions of the DNA origami and each step was observed by AFM. The yield of the reactions could be determined by analyzing about 200 DNA origami structures.

DNA origami unfolds its real power when the local control of the chemical modifications, i.e., their relative positions, is used to study chemical reactions as a function of distance between the reactants. This was demonstrated by Helmig et al. with a photosensitizer placed in the center of a DNA origami rectangle and four singlet oxygen cleavable (SOC) linkers placed at different distances (18 and 36 nm) from the photosensitizer (Figure 4) [25]. Upon UV irradiation singlet oxygen is produced in the center of the DNA origami, which diffuses along the DNA origami. The SOC linkers are cleaved with different yields depending on the distance to the photosensitizer. The reaction was observed on a single-molecule level using AFM and the diffusion controlled linker cleavage induced by singlet oxygen could be analyzed [25].

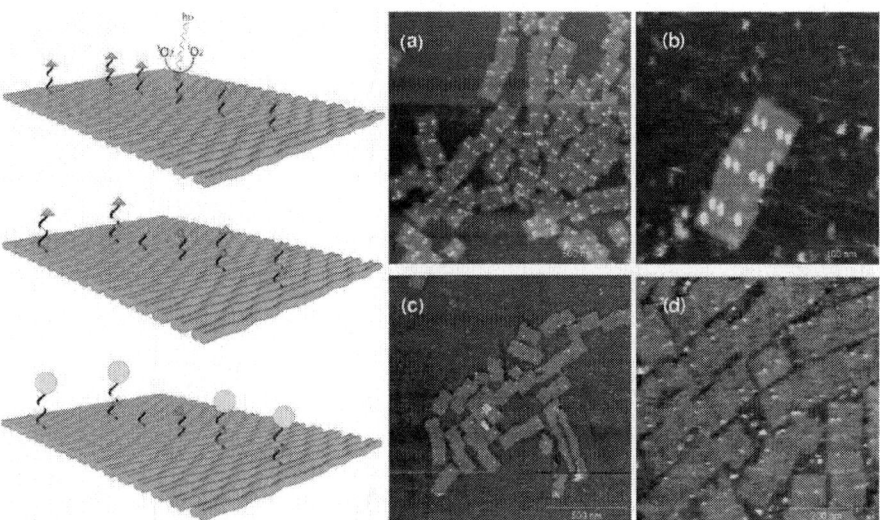

Figure 4. Left: Schematic presentation of the rectangular DNA origami platform used to study singlet oxygen production, diffusion and bond cleavage. Singlet oxygen is produced by an IPS photosensitizer in the center and induces bond cleavage in the SOC linkers placed at different distance from the IPS. After UV-irradiation and thus reaction with singlet oxygen SAv is added, which binds to the remaining strands to visualize them for AFM analyses. On the right examples of AFM images are shown. (**a**) Non-irradiated DNA origami platforms with the SOC linkers and one marker in the corner of the rectangle. A zoom-in is shown in (**b**). UV irradiation results in singlet oxygen production and cleavage of the SOC linkers, which is shown in (**c,d**). Adapted with permission from Ref. [25]. Copyright 2010 American Chemical Society.

2.2. Electron-Induced Processes

DNA origami platforms have also been used to study reactions of low-energy electrons with DNA oligonucleotides of defined nucleotide sequence [26,27,28]. Electron-induced DNA strand cleavage represents a central elementary reaction in the damage of DNA by high-energy radiation. A detailed knowledge of these processes is beneficial for accurate risk estimates for living organisms exposed to high-energy radiation such as γ- and X-rays and for the improvement of tumor radiation therapy [29,30].

For biological systems DNA strand breaks represent the most important type of radiation damage, since a double strand break can basically not be repaired by enzymes and thus results in cell death. Low-energy electrons (E < 20 eV) are generated in copious amounts along the radiation track of the primary high-energy radiation [31]. It was demonstrated in several studies that low-energy electrons are able to directly induce DNA strand breaks with high cross sections even at energies as low as 0.8 eV by the dissociative electron attachment (DEA) mechanism [32,33]. The determination of electron-induced DNA strand break yields of oligonucleotides of well-defined sequence is a very challenging task, and due to a limited sensitivity of traditional chemical analysis tools such as HPLC only very short oligonucleotides up to a maximum of 4 nt can be analyzed within a reasonable time scale [34]. To study the influence of nucleotide sequence on the electron-induced strand breakage, novel experimental approaches have to be developed and an important step was done by using DNA origami structures as platforms for DNA nanoarrays, whose radiation damage can be studied by AFM [26,27].

The experiments are based on a similar procedure as the single-molecule studies of chemical reactions by Voigt et al. [24] and Helmig et al. [25] described above. SAv binding to biotinylated oligonucleotides protruding from the DNA origami platform is used to visualize intact target strands. The system consisting of target oligonucleotides placed at specific positions on the DNA origami platform is in the following referred to as DNA nanoarray. After irradiation of the DNA nanoarrays in ultrahigh vacuum (UHV) with low-energy electrons DNA strand breaks might occur in the target sequences. The remaining intact strands can be visualized by AFM after addition of SAv. An analysis of a sufficient number of DNA origami structures yields the relative number of strand breaks (N_{SB}), and from the slope of the linear regime of the fluence dependence of N_{SB} the yields of DNA strand cleavage at a specific electron energy can be determined. The experimental procedure is schematically shown in Figure 5. The DNA origami technique has several advantages compared to other experimental approaches: (i) Since the detection of DNA strand breaks at a single-molecule level allows for maximum sensitivity, only miniscule amounts of material, corresponding to sub-monolayer coverage, are required; (ii) Two or more different oligonucleotide sequences can be directly compared within one irradiation experiment, which allows for efficient parallel investigation of a number of different DNA structures; (iii) Absolute strand break cross sections are readily accessible due to the single-molecule technique; (iv) The DNA nanoarray technique can be extended to quantify double strand breaks and to investigate higher-order DNA structures. Furthermore, the technique can be applied to other radiation sources such as UV radiation.

In a proof-of-principle experiment a triangular DNA origami platform was decorated with nine protruding strands containing a disulfide bond, two T nucleotides and the Bt modification: 5'-TTSS [26]. The disulfide bond is very sensitive to electron-induced bond cleavage and was used as a test system. Figure 6 shows examples of AFM images from control samples (non-irradiated) and samples irradiated with different fluences of 18 eV electrons. Figure 6e shows that the number of strand breaks increases linearly with the fluence until a saturation level is reached. From the slope of the linear fit in the low-fluence regime a strand break yield of 1.0×10^{-3} was determined [26].

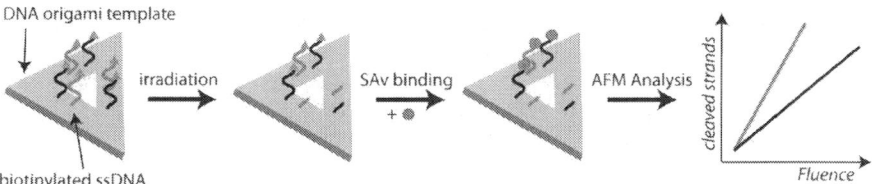

Figure 5. Scheme of the experimental sequence used to quantify electron-induced strand breakages [26]. Triangular DNA origami nanostructures with six protruding strands with two different target structures (indicated in green and black) are irradiated with low-energy electrons.

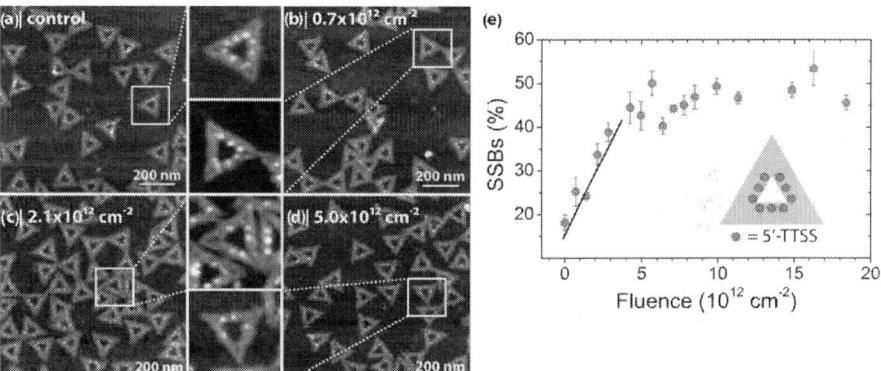

Figure 6. Irradiation of triangular DNA origami structures carrying nine disulfide-containing strands with 18 eV electrons [26]. (**a**) AFM image of a non-irradiated control sample. The sequence of the protruding strands is 5'-TTSS. The DNA origami structures were immobilized on silicon and exposed to SAv, which had a binding efficiency of 80%–90%; (**b–d**) AFM images of samples irradiated with different fluences of 18 eV electrons. Due to electron-induced bond cleavage in the protruding strands the number of specifically bound SAv decreases with electron fluence; (**e**) Plot of the relative number of strand breaks (in %) vs. electron fluence. The DNA origami design is shown in the inset of (e).

The strand break yields of two different target sequences can be determined in the same irradiation experiment by placing two different sequences on the DNA origami target and arranging them in an asymmetric pattern to make them distinguishable in the AFM images. In this way, the strand break yield for the 5'-TTSS sequence could be directly compared with a 5'-TT sequence without disulfide bond [26]. The damage of the Bt label has to be taken into account as well and in summary the damage yields for 18 eV electron induced damage to the different sequences were found to be: 7.1×10^{-14} cm^2 (5'-TTSS), 1.7×10^{-14} cm^2 (5'-TT) and 1.1×10^{-14} cm^2 (Bt) [27]. Very recently, it was demonstrated that the strand break yields determined with the DNA nanoarray method correspond directly to the absolute cross sections for DNA strand breakage [28].

3. PROTEIN BINDING REACTIONS

A variety of methods for the site-specific immobilization of different protein species on DNA origami nanostructures have been reported. Approaches used so far include aptamer-protein binding [35,36], chelate complex formation with His-tagged proteins [37], the sequence-specific binding of zinc-finger proteins to dsDNA [11], and most prominently SAv-Bt binding [10,38,39,40,41]. Recently, also the growth and arrangement of amyloid fibrils on DNA origami templates has been demonstrated [42]. Due to the comparatively large size of most proteins, AFM provides a powerful tool for the identification of individual proteins arranged in complex arrays on DNA origami substrates. Consequently, AFM of functionalized DNA origami has been employed at several instances to monitor and quantify protein binding dynamics and selectivity.

3.1. Aptamer-Protein Binding

Aptamers are short ssDNA sequences which recognize and bind specific protein sites via their 3D structure by wrapping around the target molecule [43]. They can be easily arranged on DNA origami nanostructures by extending selected staple strands with the desired aptamer sequence [35]. In this way, Rinker et al. exploited the spatial addressability of DNA origami to arrange two different aptamers into parallel lines with controlled distance [36]. The different aptamer sequences recognized two opposite sites of the blood coagulation protein thrombin which has a diameter of about 4 nm. By varying the distance between the aptamers, AFM revealed that binding occurred selectively at an aptamer-aptamer distance of 5.3 nm while almost no binding was observed for a distance of 20.7 nm. In these experiments, the rigidity of the DNA origami substrate plays a key role as it minimizes spatial fluctuations in the aptamer-aptamer distance and thus enhances binding selectivity.

A similar DNA origami design has recently been used to detect the DNA repair activity of human O^6-alkylguanine-DNA alkyltransferase (hAGT) [44]. To this end, DNA origami were again decorated with parallel lines of the two thrombin-binding aptamers. Introduction of one methylated guanine into one of the aptamers disrupted the 3D structure of the aptamer and thus suppressed thrombin binding. The methylated guanine, however, could be repaired by incubation with hAGT which fully restored the thrombin-binding activity of the aptamer.

3.2. Streptavidin-Biotin Binding

The binding of SAv to Bt is among the strongest non-covalent interactions, exhibiting a dissociation constant K_d of 10^{-14} M–10^{-15} M [45]. SAv has a spherical shape and consists of four identical subunits, each of which can bind one Bt molecule. Due to its size of ~5 nm in diameter and the well-established modification of oligonucleotides with Bt, SAv is widely used as a marker to visualize individual sites on DNA origami by AFM (see Section 2). Furthermore, it has been demonstrated that biotinylated ssDNA strands of sufficient length can "thread" through the holes in the 2D DNA origami sheet, enabling efficient SAv binding also for DNA origami substrates that are adsorbed in a face-down configuration, i.e., with the Bt modifications facing toward the solid surface and away from solution [26,46].

The dynamics of SAv-Bt binding on DNA origami substrates was studied at a single-molecule level by Wu et al. [40]. After immobilization of DNA origami with four Bt modifications on mica, the authors exposed the substrates to a constant concentration of SAv and monitored the amount of specifically bound SAv on the DNA origami surfaces over time by time-lapse AFM. Over a course of 30 min, the ratio of bound SAv per available Bt site slowly increased until saturation was reached at close to 100%. By repeatedly scanning the same area, it was further shown that the SAv-Bt binding is strong enough not to be affected by the forces exerted by the AFM tip.

Reversible SAv immobilization on DNA origami was demonstrated by Wong et al. [41]. To this end, 2D DNA origami nanostructures carrying both Bt and desthiobiotin (dBt) modifications were employed. Similar in its structure to Bt, dBt is also able to bind SAv but with a much higher dissociation constant ($K_d \sim 10^{-11}$ M). Therefore, upon exposure to the DNA origami, SAv was binding to both modifications which resulted in the appearance of a pattern associated with the Morse code "OOOO" as observed by AFM (see Figure 7a). The binding yield of SAv to Bt was found to be ~95% while SAv-dBt binding resulted in a yield of only ~84%, reflecting the higher dissociation constant. The authors then added free Bt to the SAv-decorated DNA origami. In the case of the Bt-immobilized SAv, free Bt was captured by the remaining three unoccupied SAv subunits but the protein remained at its predefined binding site on the origami surface. For SAv bound to dBt, however, free Bt displaced dBt due to its higher binding affinity, thus releasing the immobilized SAv from the DNA origami. In the AFM images, the selective release could be observed by the appearance of the Morse code for "NANO" (see Figure 7b). The authors furthermore demonstrated that the original message can be restored after washing away the free Bt by a second exposure to SAv, which then binds to the newly unmasked dBt sites.

SAv can be immobilized on a variety of inorganic species such as gold nanoparticles [47] or quantum dots (QDs) [48] which enables their coupling to biotinylated molecules. By investigating the binding of SAv-coated QDs to different DNA origami designs, Ko et al. demonstrated the great potential of AFM in combination with DNA origami substrates not only to visualize but to gain quantitative information about binding reactions [7]. The QDs used in their study had a diameter of ~20 nm and were coated with 5–10 SAv molecules per QD. The authors first studied the effect of monovalent, i.e., one Bt per binding site, vs. trivalent binding, i.e., three Bt molecules per binding site. Trivalent binding was found to drastically enhance the binding yield from ~22% to ~90%. By varying the distance between the individual binding sites on the origami surface between 22 and 50 nm, steric hindrance effects in QD binding were assessed. Steric hindrance led both to a reduced binding yield and to poor spatial precision in the QD arrangement when the spacing between binding sites was less than twice the hydrodynamic radius of the QDs. Using AFM, the authors could also quantitatively assess the QD binding kinetics and found that the SAv-QD DNA origami hybrids have a K_d that exceeds that of pure SAv-Bt by up to seven orders of magnitude. Increasing the linker length, i.e., the length of the biotinylated protruding strands, led to a considerable increase in reaction rate while K_d was barely affected.

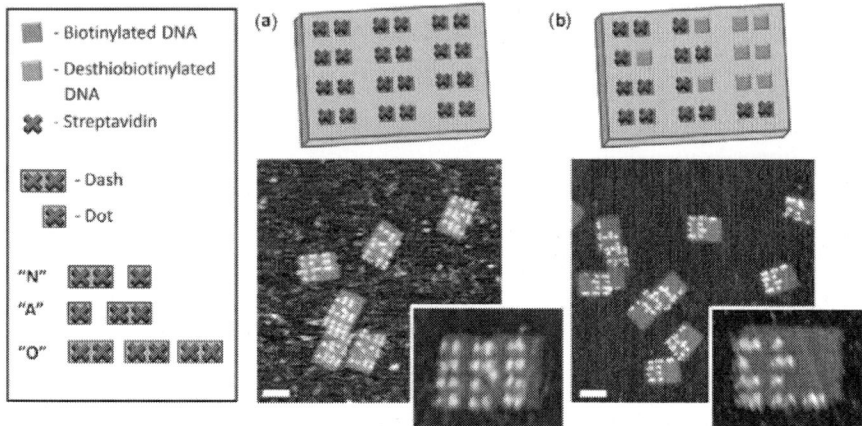

Figure 7. Design of the SAv Morse code arrangement on DNA origami substrates with corresponding code translation. After SAv binding, the DNA origami display the Morse code for "OOOO" (**a**). After addition of free Bt, the selective release of SAv from the origami results in the message "NANO" (**b**). Reprinted with permission from Ref. [41]. Copyright 2013 American Chemical Society.

3.3. DNA-Binding Proteins

Many DNA-binding proteins induce distortions in DNA upon binding which requires a certain flexibility of the DNA strand. DNA origami substrates provide the unique opportunity to immobilize target DNA strands for protein binding and control their structural properties in a well-defined manner.

Yamamoto et al. recently used this approach to study the cooperative binding of the transcription factors Sox2 and Pax6 to the DC5 element of the δ1-crystallin gene in dependence of DNA tension [49]. To this end, the authors employed a DNA origami frame in which two DNA strands exhibiting the DC5 element were immobilized (see Figure 8a). By controlling the lengths of the strands, the degree of tension could be varied. An orientation marker in the DNA origami frame enabled the unambiguous identification of the individual DNA strands in AFM images while the bound proteins appeared as protrusions on the strands. For the more flexible strand, the authors observed an almost twofold increase in Sox2 binding compared to the shorter, tensed strand. Furthermore, the Sox2 binding to the flexible strand was often accompanied by a kink in the substrate strand (see Figure 8b) which indicates that although Sox2 can bind to stiff DNA, the binding affinity is reduced as bending of the substrate is prohibited. In retinal tissues, complexes of Sox2 and Pax6 are involved in the expression of the lens protein δ-crystallin. Although Pax6 alone did not bind to either the tensed or the relaxed strand, cooperative binding was observed in the presence of both proteins with the Sox2-Pax6 complex being characterized by an increased volume compared to Sox2 alone. In addition, it was observed that the presence of Pax6 almost doubled the binding yield to the relaxed DNA strand while it resulted only in a marginally increased binding to the tensed strand. These results show that the cooperativity depends on the initial binding and bending of the DNA strand by Sox2 which makes it accessible for Pax6 and leads to a further increase in Sox2-Pax6 complex formation on the DNA strand.

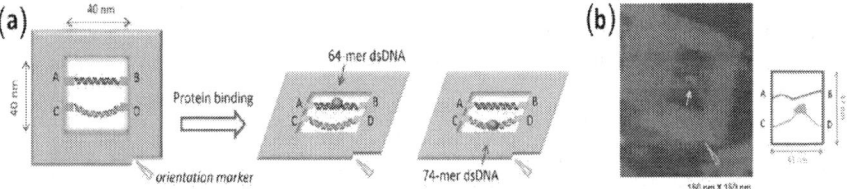

Figure 8. (a) Scheme of the DNA origami frame with a tensed 64-mer (red) and a relaxed 74-mer DNA substrate (orange) for Sox2 binding; **(b)** AFM image of Sox2 bound to the relaxed substrate strand **(left)** and representation of the kinked geometry of the substrate strand obtained from the AFM image **(right)**. Adapted with permission from Ref. [49]. Copyright 2014 American Chemical Society.

4. ENZYMATIC REACTIONS

Enzymes are proteins that are able of catalyzing biochemical reactions. By decorating DNA nanostructures with enzyme cascades or multi-enzyme complexes in different configurations, such reactions can be studied in dependence of various parameters including enzyme distance, arrangement, and stoichiometry [50,51]. Many biological processes such as DNA replication, transcription, and repair, however, involve the direct and highly specific interaction of enzymes with target DNA sequences, with enzyme activity often being influenced by the target DNA's structural properties. Immobilizing target DNA sequences on DNA origami substrates and controlling their structural properties therefore enables the investigation of DNA-binding enzymes under well-defined conditions at a single-molecule level. In particular, HSAFM of such DNA origami substrates allows for the real-time study of enzyme binding, diffusion, and activity. For instance, by attaching a long dsDNA featuring a T7 promoter region to a rigid DNA origami substrate, the movement of a T7 RNA polymerase along the dsDNA and the resulting transcription product could be directly observed by HSAFM [52].

Endo et al. have used the DNA origami frame discussed in Section 3.3. to study the activity of various enzymes in dependence of DNA tension. Their first experiments focused for instance on the DNA methylation enzyme M.EcoRI which specifically recognizes the sequence GAATTC [53]. They found, that the yields of M.EcoRI binding varied significantly for the two different strands: a binding yield of 87% was observed for the relaxed strand while the binding yield for the tensed strand was only 13%. This demonstrates the importance of structural alterations induced by M.EcoRI upon binding. In line with this observation, also diffusion of the enzyme along the DNA strand was inhibited on the tensed strands. Finally, the activity of M.EcoRI was investigated in dependence of DNA tension by using a second enzyme, R.EcoRI, which was able to cleave both the tensed and relaxed non-methylated strands. Methylation, however, inhibits cleavage by R.EcoRI. Therefore, the authors investigated the yield of R.EcoRI cleavage after treatment with M.EcoRI for the two strands and found that 87% of the tensed strands were cleaved but only 57% of the relaxed strands. This demonstrates that DNA tension hinders DNA methylation by M.EcoRI.

The same approach was also used to study the activity of the base excision repair enzymes hOgg1 and PDG which selectively remove 8-oxoguanine and pyrimidine dimers, respectively, by introducing these sites into the target strands opposite to a nick [54]. Base excision thus results in a double strand break which can be identified in the AFM images. For both enzymes, base excision was suppressed for the tensed DNA strands, i.e., 11% vs. 32% for hOgg1 and 7% vs. 29% for PDG. The authors were furthermore able to directly visualize the individual reaction steps of PDG base excision, i.e., binding to the target strand, diffusion along the strand, strand cleavage, and PDG dissociation, in real time by HSAFM.

Using a modified DNA origami frame with a crossed geometry of attachment sites for the target strands, Suzuki et al. investigated the site-specific recombination between two loxP sites by the E.coli phage P1 Cre recombinase with HSAFM [55]. In particular, the authors immobilized two DNA strands containing loxP sites across the edges of the DNA frame (Figure 9a, left) either in parallel or antiparallel configuration. Upon exposure to Cre, four Cre monomers were binding to the two strands and forming a synaptic tetramer complex. The first strand exchange then resulted in an intermediate Holliday junction (Figure 9a, center), while the second strand exchange yielded the recombinant products that had different connection patterns and were therefore distinguishable from the substrates (Figure 9a, right). The time course of the reaction as monitored by HSAFM is shown in Figure 9b for the antiparallel configuration. The synaptic complex, its disintegration into four monomers, and the recombinant products can be clearly seen. For the parallel configuration of loxP sites, a synaptic complex was observed but no recombination. From these experiments, the authors concluded that antiparallel loxP sites can be converted from the substrates to recombinant products through the formation of synaptic complexes while parallel sites do not recombine efficiently even though they can be synapsed. In further experiments using Holliday junctions as starting substrates tethered to two different DNA origami frames, the authors could also establish a relation between the preferential cleavage site of Cre in the loxP sequence and the topological state of the Holliday junction which again demonstrated the importance of substrate bending by the enzyme for the reaction. Using a similar approach, also the influence of Holliday junction flexibility on the activity of the RecU Holliday junction resolvase was investigated [56]. It was found, that although the resolvase could bind both to flexible and inflexible Holliday junctions, flexible junctions were resolved more efficiently.

5. CONFORMATIONAL TRANSITIONS OF DNA

5.1. Guanine Quadruplexes

G quadruplexes can be formed from telomeric DNA, which is a single stranded G rich sequence present at the ends of eukaryotic chromosomes. Telomeric DNA protects and stabilizes the genome, and its length determines the lifetime of a cell [57]. In mammals the telomeric DNA consists of the repeat unit 5'-(TTAGGG)$_n$ [58]. In the presence of monovalent cations four G bases form a tetrad through intermolecular Hoogsteen-type hydrogen bonding, and two or three G tetrads can be stacked on top of each other stabilized by π-π interactions and coordination with central cations. The formation of single G quadruplexes upon addition of K^+ ions was directly observed by AFM using the DNA origami frame introduced above [59]. From two parallel DNA duplexes spanning two sides of the frame single stranded extensions were introduced containing the G tracts 5'-GGGTTAGGGTTAGGGTTT and 5'-TTTGGGT, respectively. After addition of K^+ an intermolecular G quadruplex is formed resulting in a transition from the parallel DNA duplex to an X shape, which can be directly observed by AFM.

In a very recent study it was demonstrated that the formation of the G quadruplex might proceed via two transition states, a G hairpin (i.e., a G-G mismatched Hoogsteen duplex), and a G triplex (i.e., a partly folded structure with three tandem G repeats bound by Hoogsteen hydrogen bonding, see Figure 10a) [60]. The formation of the transition states was demonstrated with two types of structures, a tetramolecular antiparallel and a (3+1)-type G quadruplex structure. It was found that the G-hairpin structure forms also in presence of Mg^{2+} ions (with a yield of 64%) and the yield did not increase notably when adding K^+ ions indicating that the G hairpin selectively prefers Mg^{2+} over K^+ ions. The G triplex structure was formed with considerable yield in the presence of both Mg^{2+} and K^+ ions (43% and 54%, respectively), whereas the highest yield of the G quadruplex structure was obtained only in presence of K^+ ions (76%). This study suggests that the intermediate states are preferentially formed when Mg^{2+} is present, and the transition to the G quadruplex is most efficient with K^+.

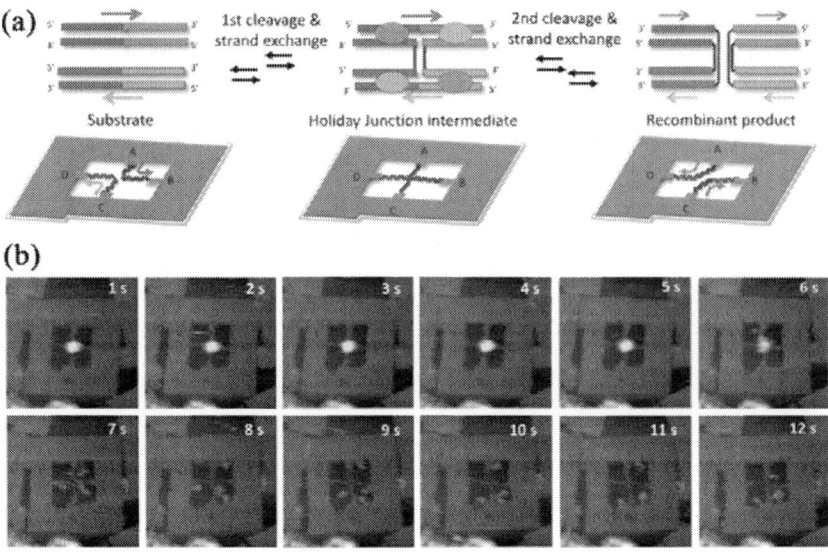

Figure 9. (a) Scheme of the DNA origami frame with a crossed geometry of attachment sites for the immobilization of loxP-containing DNA strands in antiparallel configuration and pathway of Cre-loxP recombination via an intermediate Holliday junction; (b) HSAFM images of the reaction, starting with the synaptic complex. Adapted with permission from Ref. [55]. Copyright 2013 American Chemical Society.

G quadruplexes are also of considerable medical relevance. In most cancer cells the enzyme telomerase is overexpressed by maintaining and extending the telomere sequences thereby extending the life span of cells. In the context of cancer therapy numerous quadruplex-binding ligands have been identified [61]. The DNA origami based single-molecule analysis of G quadruplex-ligand binding has very recently been demonstrated to be very attractive, since the G quadruplex structure (based on number of involved strands, strand polarity and sequence) can be effectively controlled on the single-molecule level, which is not the case for bulk solutions that usually consist of a complex mixture of different G quadruplex structures [62]. In this way the binding of a pyrido-dicarboxamide (PDC) ligand modified with Bt to four G tracts was studied. The G tracts were incorporated into two G-G mismatched duplexes attached to the DNA frame. Binding of the PDC ligand induced G quadruplex formation, which was again visualized by AFM via the X shape of the interlinked duplexes in the center of the DNA frame. The presence of the PDC linker was further confirmed by the addition of SAv, which bound to the Bt modification of PDC thereby visualizing the position of PDC. A reversible PDC-induced G quadruplex formation and unbinding was observed by HSAFM most likely representing formation of intermediate states before a stable final configuration was adopted. The HSAFM measurements also revealed that PDC-induced G quadruplex formation is considerably slower than K^+ induced binding [62].

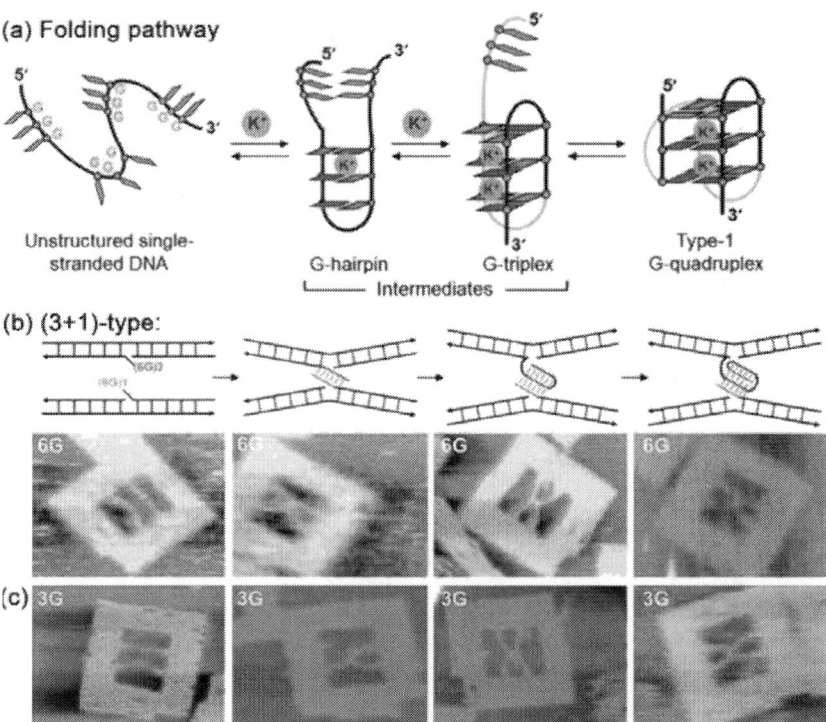

Figure 10. (a) Proposed folding pathway of the human telomeric type-1 G-quadruplex including the intermediates G hairpin and G triplex; **(b)** Scheme of different strand designs within the DNA origami frame that are able to form the G hairpin, G triplex, or a G quadruplex in presence of K^+. Corresponding AFM images of the different designs show that without K^+, but a (3+1)-type (6G in (b) and 3G in (c)) design no G-quadruplexes are formed (parallel arrangement). In the presence of K^+ and using DNA strands capable of forming G hairpins, G triplexes, or G quadruplexes, respectively, the X shapes are observed. Adapted from Ref. [60]. Copyright Wiley VCH Verlag GmbH & Co KGaA. Reproduced with permission.

With the same technique also the nucleocapsid protein (NCp) of the human deficiency virus type 1 (HIV-1) was studied [63]. G-rich sequences were shown to inhibit HIV-1 replication [64]. HIV-1 NCp is a multifunctional protein that promotes and stabilizes G quadruplex structures. By using the DNA origami frame and HSAFM the real-time NCp induced G quadruplex formation was imaged and translocation events (1-D sliding, 3-D hopping, loop formation) were captured as part of the NCp-telomer searching mechanism [63].

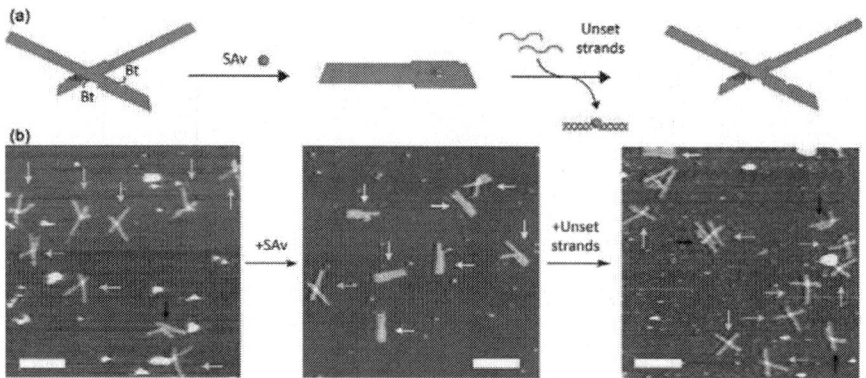

Figure 11. (**a**) Shape transition of the nanomechanical DNA origami device upon SAv binding and release; (**b**) Corresponding AFM images. Adapted by permission from Macmillan Publishers Ltd: Ref. [38], copyright 2011.

5.2. Nanomechanical DNA Origami Devices for Molecular Detection

Kuzuya et al. designed a nanomechanical DNA origami device which undergoes a shape transition upon target binding and release [38]. The device consists of two symmetric levers attached to each other via an immobile Holliday junction to resemble a plier geometry (see Figure 11a). Each of the levers further featured a single Bt modification. When adsorbed to mica surfaces, AFM revealed that the adsorbed devices predominantly had a cross geometry (see Figure 11b, left). In the presence of SAv, the Bt modifications of each lever could bind to two different subunits of one SAv molecule which led to a shape transition of the device. In the AFM images, the SAv-bound devices predominantly displayed a parallel geometry (see Figure 11b, center). The Bt modifications, however, were not attached directly to staple strands. Instead, biotinylated oligonucleotides were hybridized to protruding strands so that the Bt modifications and therefore the captured SAv molecules could be released again via a strand displacement reaction which reversed the shape transition (see Figure 11b, right). The versatility of the nanomechanical DNA origami devices was further demonstrated by employing a number of different capture and release mechanisms to detect a variety of molecular species using the same shape transition. In particular, anti-fluorescein IgG was detected via FAM modifications, Na^+ and K^+ via G quadruplex formation, Ag^+ via C-C mismatch stabilization, miRNA via strand displacement reactions, and ATP via aptamer modifications.

6. CONCLUSIONS

With the DNA origami technique, well-defined nanostructures of arbitrary shape can be synthesized. Due to the high structural control provided by the technique and the possibility to arrange molecular entities with highest precision, DNA origami have therefore become a promising substrate for investigating molecular processes at a single-molecule level. In combination with AFM as a high-resolution imaging technique, not only the visualization but also the quantitative study of chemical and biochemical reactions becomes possible.

The works discussed in this review have demonstrated the great potential of this approach and incredible advances have been made in past few years. The applications range from the kinetics of protein binding reactions, to molecular detection with highest sensitivity, to enzymatic reactions, to radiation-induced bond cleavage. The progress made in HSAFM imaging has even enabled the real-time study of enzyme binding, diffusion, and activity in dependence of a number of parameters, most importantly DNA tension and structure.

Further advances in the field can be expected in the years to come. The already large number of studies focusing on G quadruplexes may be further extended to other higher order structures such as i-motifs while the quantitative investigation of radiation-induced reactions by AFM on DNA origami substrates is only in its early infancy. The previous studies of electron-induced strand breakage in ssDNA may easily be extended to more complex DNA and even DNA-protein targets, and to other types of radiation such as photons and ions. AFM imaging of DNA origami nanostructures is thus likely to develop into a well-established technique for the single-molecule investigation of a large number of molecular processes and reactions.

ACKNOWLEDGMENTS

This work was supported by a Marie Curie FP7 Integration Grant within the 7th European Union Framework Programme and by the Deutsche Forschungsgemeinschaft (DFG).

REFERENCES

1. Seeman, N.C. Nanomaterials based on DNA. Annu. Rev. Biochem. **2010**, 79, 65–87.
2. Tørring, T.; Voigt, N.V.; Nangreave, J.; Yan, H.; Gothelf, K.V. DNA origami: A quantum leap for self-assembly of complex structures. Chem. Soc. Rev. **2011**, 40, 5636–5646.
3. Rothemund, P.W.K. Folding DNA to create nanoscale shapes and patterns. Nature **2006**, 440, 297–302.

4. Kuzyk, A.; Schreiber, R.; Fan, Z.Y.; Pardatscher, G.; Roller, E.M.; Hogele, A.; Simmel, F.C.; Govorov, A.O.; Liedl, T. DNA-based self-assembly of chiral plasmonic nanostructures with tailored optical response. Nature **2012**, 483, 311–314.

5. Ding, B.Q.; Deng, Z.T.; Yan, H.; Cabrini, S.; Zuckermann, R.N.; Bokor, J. Gold nanoparticle self-similar chain structure organized by DNA origami. J. Am. Chem. Soc. **2010**, 132, 3248–3249.

6. Bui, H.; Onodera, C.; Kidwell, C.; Tan, Y.; Graugnard, E.; Kuang, W.; Lee, J.; Knowlton, W.B.; Yurke, B.; Hughes, W.L. Programmable periodicity of quantum dot arrays with DNA origami nanotubes. Nano Lett. **2010**, 10, 3367–3372.

7. Ko, S.H.; Gallatin, G.M.; Liddle, J.A. Nanomanufacturing with DNA origami: Factors affecting the kinetics and yield of quantum dot binding. Adv. Funct. Mater. **2012**, 22, 1015–1023.

8. Steinhauer, C.; Jungmann, R.; Sobey, T.; Simmel, F.; Tinnefeld, P. DNA origami as a nanoscopic ruler for super-resolution microscopy. Angew. Chem. Int. Ed. **2009**, 48, 8870–8873.

9. Stein, I.H.; Steinhauer, C.; Tinnefeld, P. Single-molecule four-color FRET visualizes energy-transfer paths on DNA origami. J. Am. Chem. Soc. **2011**, 133, 4193–4195.

10. Kuzyk, A.; Laitinen, K.T.; Törmä, P. DNA origami as a nanoscale template for protein assembly. Nanotechnology **2009**, 20, 235305.

11. Nakata, E.; Liew, F.F.; Uwatoko, C.; Kiyonaka, S.; Mori, Y.; Katsuda, Y.; Endo, M.; Sugiyama, H.; Morii, T. Zinc-finger proteins for site-specific protein positioning on DNA-origami structures. Angew. Chem. Int. Ed. **2012**, 51, 2421–2424.

12. Birkedal, V.; Dong, M.; Golas, M.M.; Sander, B.; Andersen, E.S.; Gothelf, K.V.; Besenbacher, F.; Kjems, J. Single molecule microscopy methods for the study of DNA origami structures. Microsc. Res. Tech. **2011**, 74, 688–698.

13. De Oteyza, D.G.; Gorman, P.; Chen, Y.-C.; Wickenburg, S.; Riss, A.; Mowbray, D.J.; Etkin, G.; Pedramrazi, Z.; Tsai, H.-Z.; Rubio, A.; et al. Direct imaging of covalent bond structure in single-molecule chemical reactions. Science **2013**, 340, 1434–1437.

14. Rajendran, A.; Endo, M.; Sugiyama, H. State-of-the-art high-speed atomic force microscopy for investigation of single-molecular dynamics of proteins. Chem. Rev. **2014**, 1493–1520.

15. Acuna, G.P.; Moller, F.M.; Holzmeister, P.; Beater, S.; Lalkens, B.; Tinnefeld, P. Fluorescence enhancement at docking sites of DNA-directed self-assembled nanoantennas. Science **2012**, 338, 506–510.

16. Acuna, G.P.; Bucher, M.; Stein, I.H.; Steinhauer, C.; Kuzyk, A.; Holzmeister, P.; Schreiber, R.; Moroz, A.; Stefani, F.D.; Liedl, T.; et al. Distance dependence of single-fluorophore quenching by gold nanoparticles studied on DNA origami. ACS Nano **2012**, 6, 3189–3195.

17. Prinz, J.; Schreiber, B.; Olejko, L.; Oertel, J.; Rackwitz, J.; Keller, A.; Bald, I. DNA origami substrates for highly sensitive surface-enhanced raman scattering. J. Phys. Chem. Lett. **2013**, 4, 4140–4145.

18. Thacker, V.V.; Herrmann, L.O.; Sigle, D.O.; Zhang, T.; Liedl, T.; Baumberg, J.J.; Keyser, U.F. DNA origami based assembly of gold nanoparticle dimers for surface-enhanced Raman scattering. Nat. Commum. **2014**, 5.

19. Pilo-Pais, M.; Watson, A.; Demers, S.; LaBean, T.H.; Finkelstein, G. Surface-enhanced Raman scattering plasmonic enhancement using DNA origami-based complex metallic nanostructures. Nano Lett. **2014**, 14, 2099–2104.

20. Kühler, P.; Roller, E.-M.; Schreiber, R.; Liedl, T.; Lohmüller, T.; Feldmann, J. Plasmonic DNA-origami nanoantennas for surface-enhanced Raman spectroscopy. Nano Lett. **2014**, 14, 2914–2919.

21. Koirala, D.; Shrestha, P.; Emura, T.; Hidaka, K.; Mandal, S.; Endo, M.; Sugiyama, H.; Mao, H. Single-molecule mechanochemical sensing using DNA origami nanostructures. Angew. Chem. Int. Ed. **2014**, 53, 8137–8141.

22. Ke, Y.G.; Lindsay, S.; Chang, Y.; Liu, Y.; Yan, H. Self-assembled water-soluble nucleic acid probe tiles for label-free RNA hybridization assays. Science **2008**, 319, 180–183.

23. Zhu, J.; Feng, X.; Lou, J.; Li, W.; Li, S.; Zhu, H.; Yang, L.; Zhang, A.; He, L.; Li, C.; et al. Accurate quantification of microRNA via single strand displacement reaction on DNA origami motif. PLoS One **2013**, 8, e69856.

24. Voigt, N.V.; Torring, T.; Rotaru, A.; Jacobsen, M.F.; Ravnsbaek, J.B.; Subramani, R.; Mamdouh, W.; Kjems, J.; Mokhir, A.; Besenbacher, F.; et al. Single-molecule chemical reactions on DNA origami. Nat. Nanotechnol. **2010**, 5, 200–203.

25. Helmig, S.; Rotaru, A.; Arian, D.; Kovbasyuk, L.; Arnbjerg, J.; Ogilby, P.R.; Kjems, J.; Mokhir, A.; Besenbacher, F.; Gothelf, K.V. Single molecule atomic force microscopy studies of photosensitizedsinglet oxygen behavior on a DNA origami template. ACS Nano **2010**, 4, 7475–7480.

26. Keller, A.; Bald, I.; Rotaru, A.; Cauet, E.; Gothelf, K.V.; Besenbacher, F. Probing electron-induced bond cleavage at the single-molecule level using DNA origami templates. ACS Nano **2012**, 6, 4392–4399.

27. Keller, A.; Kopyra, J.; Gothelf, K.V.; Bald, I. Electron-induced damage of biotin studied in the gas phase and in the condensed phase at a single-molecule level. New J. Phys. **2013**, 15, 083045.

28. Keller, A.; Rackwitz, J.; Cauet, E.; Lievin, J.; Körzdörfer, T.; Rotaru, A.; Gothelf, K.V.; Besenbacher, F.; Bald, I. Sequence dependence of electron-induced DNA strand breakage revealed by DNA nanoarrays. Unpublished work. 2014.

29. Baccarelli, I.; Bald, I.; Gianturco, F.A.; Illenberger, E.; Kopyra, J. Electron-induced damage of DNA and its components: Experiments and theoretical models. Phys. Rep.-Rev. Sec. Phys. Lett. **2011**, 508, 1–44.

30. Alizadeh, E.; Sanche, L. Precursors of solvated electrons in radiobiological physics and chemistry. Chem. Rev. **2012**, 112, 5578–5602.

31. Pimblott, S.M.; LaVerne, J.A. Production of low-energy electrons by ionizing radiation. Radiat. Phys. Chem. **2007**, 76, 1244–1247.

32. Boudaiffa, B.; Cloutier, P.; Hunting, D.; Huels, M.A.; Sanche, L. Resonant formation of DNA strand breaks by low-energy (3 to 20 eV) electrons. Science **2000**, 287, 1658–1660.

33. Martin, F.; Burrow, P.D.; Cai, Z.L.; Cloutier, P.; Hunting, D.; Sanche, L. DNA strand breaks induced by 0–4 eV electrons: The role of shape resonances. Phys. Rev. Lett. **2004**, 93, 068101.

34. Li, Z.J.; Cloutier, P.; Sanche, L.; Wagner, J.R. Low-energy electron-induced DNA damage: Effect of base sequence in oligonucleotide trimers. J. Am. Chem. Soc. **2010**, 132, 5422–5427.

35. Chhabra, R.; Sharma, J.; Ke, Y.; Liu, Y.; Rinker, S.; Lindsay, S.; Yan, H. Spatially addressable multiprotein nanoarrays templated by aptamer-tagged DNA nanoarchitectures. J. Am. Chem. Soc. **2007**, 129, 10304–10305.

36. Rinker, S.; Ke, Y.; Liu, Y.; Chhabra, R.; Yan, H. Self-assembled DNA nanostructures for distance-dependent multivalent ligand-protein binding. Nat. Nanotechnol. **2008**, 3, 418–422.

37. Shen, W.; Zhong, H.; Neff, D.; Norton, M.L. NTA directed protein nanopatterning on DNA origami nanoconstructs. J. Am. Chem. Soc. **2009**, 131, 6660–6661.

38. Kuzuya, A.; Sakai, Y.; Yamazaki, T.; Xu, Y.; Komiyama, M. Nanomechanical DNA origami 'single-molecule beacons' directly imaged by atomic force microscopy. Nat. Commun. **2011**, 2.

39. Numajiri, K.; Kimura, M.; Kuzuya, A.; Komiyama, M. Stepwise and reversible nanopatterning of proteins on a DNA origami scaffold. Chem. Commun. **2010**, 46, 5127–5129.

40. Wu, N.; Zhou, X.; Czajkowsky, D.M.; Ye, M.; Zeng, D.; Fu, Y.; Fan, C.; Hu, J.; Li, B. In situ monitoring of single molecule binding reactions with time-lapse atomic force microscopy on functionalized DNA origami. Nanoscale **2011**, 3, 2481–2484.

41. Wong, N.Y.; Xing, H.; Tan, L.H.; Lu, Y. Nano-encrypted morse code: A versatile approach to programmable and reversible nanoscale assembly and disassembly. J. Am. Chem. Soc. **2013**, 135, 2931–2934.

42. Udomprasert, A.; Bongiovanni, M.N.; Sha, R.; Sherman, W.B.; Wang, T.; Arora, P.S.; Canary, J.W.; Gras, S.L.; Seeman, N.C. Amyloid fibrils nucleated and organized by DNA origami constructions. Nat. Nanotechnol. **2014**, 9, 537–541.

43. Hermann, T. Adaptive recognition by nucleic acid aptamers. Science **2000**, 287, 820–825.

44. Tintoré, M.; Gállego, I.; Manning, B.; Eritja, R.; Fàbrega, C. DNA origami as a DNA repair nanosensor at the single-molecule level. Angew. Chem. Int. Ed. **2013**, 52, 7747–7750.

45. Green, N.M. Avidin. Adv. Protein Chem. **1975**, 29, 85–133.

46. Wu, N.; Czajkowsky, D.M.; Zhang, J.; Qu, J.; Ye, M.; Zeng, D.; Zhou, X.; Hu, J.; Shao, Z.; Li, B.; et al. Molecular threading and tunable molecular recognition on DNA origami nanostructures. J. Am. Chem. Soc. **2013**, 135, 12172–12175.

47. Sun, X.J.; Tolbert, L.P.; Hildebrand, J.G. Using laser scanning confocal microscopy as a guide for electron microscopic study: A simple method for correlation of light and electron microscopy. J. Histochem. Cytochem. **1995**, 43, 329–335.

48. Leduc, C.; Ruhnow, F.; Howard, J.; Diez, S. Detection of fractional steps in cargo movement by the collective operation of kinesin-1 motors. Proc. Nat. Acad. Sci. USA **2007**, 104, 10847–10852.

49. Yamamoto, S.; De, D.; Hidaka, K.; Kim, K.K.; Endo, M.; Sugiyama, H. Single molecule visualization and characterization of Sox2–Pax6 complex formation on a regulatory DNA element using a DNA origami frame. Nano Lett. **2014**, 14, 2286–2292.

50. Fu, J.; Yang, Y.R.; Johnson-Buck, A.; Liu, M.; Liu, Y.; Walter, N.G.; Woodbury, N.W.; Yan, H. Multi-enzyme complexes on DNA scaffolds capable of substrate channelling with an artificial swinging arm. Nat. Nanotechnol. **2014**, 9, 531–536.

51. Fu, J.; Liu, M.; Liu, Y.; Woodbury, N.W.; Yan, H. Interenzyme substrate diffusion for an enzyme cascade organized on spatially addressable DNA nanostructures. J. Am. Chem. Soc. **2012**, 134, 5516–5519.

52. Endo, M.; Tatsumi, K.; Terushima, K.; Katsuda, Y.; Hidaka, K.; Harada, Y.; Sugiyama, H. Direct visualization of the movement of a single T7 RNA polymerase and transcription on a DNA nanostructure. Angew. Chem. Int. Ed. **2012**, 51, 8778–8782.

53. Endo, M.; Katsuda, Y.; Hidaka, K.; Sugiyama, H. Regulation of DNA methylation using different tensions of double strands constructed in a defined DNA nanostructure. J. Am. Chem. Soc. **2010**, 132, 1592–1597.

54. Endo, M.; Katsuda, Y.; Hidaka, K.; Sugiyama, H. A versatile DNA nanochip for direct analysis of DNA base-excision repair. Angew. Chem. Int. Ed. **2010**, 122, 9602–9606.

55. Suzuki, Y.; Endo, M.; Katsuda, Y.; Ou, K.; Hidaka, K.; Sugiyama, H. DNA origami based visualization system for studying site-specific recombination events. J. Am. Chem. Soc. **2014**, 136, 211–218.

56. Suzuki, Y.; Endo, M.; Cañas, C.; Ayora, S.; Alonso, J.C.; Sugiyama, H.; Takeyasu, K. Direct analysis of Holliday junction resolving enzyme in a DNA origami nanostructure. Nucleic Acids Res. **2014**, 42, 7421–7428.

57. Neidle, S.; Parkinson, G.N. The structure of telomeric DNA. Curr. Opin. Struct. Biol. **2003**, 13, 275–283.

58. Cech, T.R. Life at the end of the chromosome: Telomeres and telomerase. Angew. Chem. Int. Ed. **2000**, 39, 34–43.

59. Sannohe, Y.; Endo, M.; Katsuda, Y.; Hidaka, K.; Sugiyama, H. Visualization of dynamic conformational switching of the G-quadruplex in a DNA nanostructure. J. Am. Chem. Soc. **2010**, 132, 16311–16313.

60. Rajendran, A.; Endo, M.; Hidaka, K.; Sugiyama, H. Direct and single-molecule visualization of the solution-state Structures of G-hairpin and G-triplex intermediates. Angew. Chem. Int. Ed. **2014**, 126, 4191–4196.

61. Müller, S.; Sanders, D.A.; di Antonio, M.; Matsis, S.; Riou, J.-F.; Rodriguez, R.; Balasubramanian, S. Pyridostatin analogues promote telomere dysfunction and long-term growth inhibition in human cancer cells. Org. Biomol. Chem. **2012**, 10, 6537–6546.

62. Rajendran, A.; Endo, M.; Hidaka, K.; Thao Tran, P.L.; Teulade-Fichou, M.-P.; Mergny, J.-L.; Sugiyama, H. G-quadruplex-binding ligand-induced DNA synapsis inside a DNA origami frame. RSC Adv. **2014**, 4, 6346–6355.

63. Rajendran, A.; Endo, M.; Hidaka, K.; Tran, P.L.T.; Mergny, J.-L.; Gorelick, R.J.; Sugiyama, H. HIV-1 nucleocapsid proteins as molecular chaperones for tetramolecular antiparallel G-quadruplex formation. J. Am. Chem. Soc. **2013**, 135, 18575–18585.

64. Esté, J.A.; Cabrera, C.; Schols, D.; Cherepanov, P.; Gutierrez, A.; Witvrouw, M.; Pannecouque, C.; Debyser, Z.; Rando, R.F.; Clotet, B.; et al. Human immunodeficiency virus glycoprotein gp120 as the primary target for the antiviral action of AR177 (Zintevir). Mol. Pharmacol. **1998**, 53, 340–345.

CHAPTER 7

Frequency Function in Atomic Force Microscopy Applied to a Liquid Environment

Po-Jen Shih

Department of Civil and Environmental Engineering, National University of Kaohsiung, CEE NUK, No. 700, Kaohsiung University Rd., Nanzih District, 81148, Kaohsiung, Taiwan

ABSTRACT

Scanning specimens in liquids using commercial atomic force microscopy (AFM) is very time-consuming due to the necessary try-and-error iteration for determining appropriate triggering frequencies and probes. In addition, the iteration easily contaminates the AFM tip and damages the samples, which consumes probes. One reason for this could be inaccuracy in the resonant frequency in the feedback system setup. This paper proposes a frequency function which varies with the tip-sample separation, and it helps to improve the frequency shift in the current feedback system of commercial AFMs. The frequency function is a closed-form equation, which allows for easy calculation, as confirmed by experimental data. It comprises three physical effects: the quasi-static equilibrium condition, the atomic forces gradient effect, and hydrodynamic load effect. While each of these has previously been developed in separate studies, this is the first time their combination has been used to represent the complete frequency phenomenon. To avoid "jump to contact" issues, experiments often use probes with relatively stiffer cantilevers, which inevitably reduce the force sensitivity in sensing low atomic forces. The proposed frequency function can also predict jump to contact behavior and, thus, the probe sensitivity could be increased and soft probes could be widely used. Additionally, various tip height behaviors coupling with the atomic forces gradient and hydrodynamic effects are discussed in the context of carbon nanotube probes.

Keywords: frequency shift function; jump to contact; liquid environment; atomic force microscopy

1. INTRODUCTION

In the last ten years, the atomic-scale resolution of frequency-modulation atomic force microscopy (FM AFM) has been taken advantage of to image and conduct force spectroscopy measurements of biological samples in liquids. This has enabled quantitative measurements of conservative and dissipative forces, and subnanometer resolution imaging with piconewton-order loading forces at the solid and liquid interface. The thermal noise in the liquid environment increases the vertical vibration amplitude of the probe, hindering the FM AFM techniques from adopting probes with stiffness greater than 40 N/m [1]. The stiff probe is sensitive enough for strong interactions; however, weak interactions require a soft probe. Soft probes are highly susceptible to the problem of "jump to contact," which is minor reason why the soft probe has not been adopted in FM AFM. On the other hand, Sader et al. [2] successfully overcame the amplitude problem of soft probes and constrained the vibration amplitude due to thermal noise under 2 nm for a probe stiffness of 3 N/m. This achievement indicates that the thermal noise problem of soft probes is solvable. To overcome the instability of "jump to contact", this paper provides a function which can calculate frequency shifts, thus predicting "jump to contact" occurrences. The proposed frequency function could also be used in the tapping-modulation (TM) AFM and other forms of AFM. This frequency function is important for feedback control of AFM, and it can be used for the inverse calculation of tip-sample interaction forces in liquids.

FM AFM is a promising technique to measure biological samples. Recently, Bruker [3] developed a new TM AFM technique, referred to as Peak Force Quantitative Nanomechanical Mapping (PF QNM), applied to measure living and soft samples. PF QNM, succeeding from TM AFM, replaces detecting frequency (or phases) shifts with the peak force as the measurement feedback. Different from typical AFMs driven around resonance frequencies (around a hundred kilohertz), PF QNM is driven at a frequency far below (a few kilohertz). The probe amplitude of PF QNM is around hundred nanometers like general TM AFM, but that of FM AFM is around few nanometers. PF QNM features in measuring wide range of Young's moduli of biological samples with noise lower than general TM AFM. However, FM AFM, because of its small tip amplitudes of tip, has subnanometer resolution for measuring biological samples [1].

A probe selected in biological AFM depends on modulations of AFM and the resolution required for the specimen. Most atomic force microscopy surface imaging of bio-specimens is performed by the contact-modulated approach. To avoid destruction of the biological specimen [4], probe stiffness remains within the low range, 0.03–0.5 N/m [5,6]. However, low-stiffness probes easily jump to contact and suffer impaired resolution. For this reason, interest in the use of noncontact AFM has grown. FM AFM, one type of noncontact technique, uses low amplitude and measures the frequency shift of the probe, enabling higher resolution than other techniques. Within a liquid environment, atomic resolution can be achieved with biological applications of FM AFM, achieving a maximum signal by initial values of amplitudes at 0.6~1.0 nm and leading a resonance frequency shift of tens of hertz [1]. Thermal noise makes control of the amplitude in the liquid difficult; the peak value of the power spectrum density of the thermal vibration is given by $(z_{th})_{peak} = \sqrt{2k_B TQ/\pi f_0 k}$, where f_0 and Q are resonance frequency and the Q factor of the microcantilever, k_B and T are Boltzmann's constant and absolute temperature, and k is the probe stiffness. Typically, we increase stiffness (the resonance frequency) to decrease the thermal amplitude of probe, thus, avoiding "jump to contact" issues. However, increasing stiffness decreases the sensitivity when measuring the atomic force. Reversing the assumption on increasing stiffness, we try to reduce the value of Q (for example, by improving the microcantilever shape) and consider the benefit of "jump to contact." It may be possible to develop a highly sensitive probe in liquid. Thus, this paper adapts the use of soft probes and provides a frequency shift function which predicts "jump to contact" behavior.

The frequency function setup in commercial AFM devices can be improved easily. Scanning specimens in liquids using commercial atomic force microscopy (AFM) is time-consuming due the need for a try-and-error iteration to determine appropriate triggering frequencies. In addition, this iteration easily contaminates the AFM tip and damages samples, which consumes probes. One significant reason for this is that the frequency function is not accurate. The frequency function is not a function of the separation between tip and specimen. This results in the TM AFM incorrectly using resonant frequency to trigger the microcantilever and recording the wrong signals to judge height of specimen when the tip approaches the sample. The resonance frequency of AFM probes in liquid has been commonly approximated by $f = f_{vac} (1 + \pi \rho_f b^2/4m)^{-1/2}$, where f_{vac} is the resonance frequency in vacuum, ρ_f and m are the fluid density and mass per unit length, and b is the width of the probe. This approximation is suitable for the probe in far field, but it does not adequately represent the resonance frequency when the probe approaches close to the specimen. Unfortunately, this equation is used in commercial AFM devices, and the differences could be on the order of a few hundred hertz. Two causes underlie the shift in resonance frequency in liquids: (1) the interaction of atomic forces between the tip and the specimen, and (2) the disturbance of liquid in the space between the probe and the boundary surface on which the specimen is placed. In the first, atomic forces manifest over a few nanometers, and the effects to the shift can again be divided into two types: (a) the probe's eigenfunction varying with the tip–sample separation (the quasi-static equilibrium condition [7]), and (b) the force gradient effect around the amplitude region (Giessibl's theorem [8]). The effect of the disturbance of liquid is mainly based on Green and Sader's theorem [9] and arises from the liquid pressure on the probe varying with probe-surface separation. This effect manifests over tens of micrometers. Sader et al. suggested a frequency equation that considers the atomic force effect and the liquid pressure effect [1]. The aforementioned frequency function contains an opened-integration ($\int \infty z$), which is ultimately problematic for AFM engineers. To improve it and allow for the use of soft probes, this paper provides an equation for the frequency shift as a function of tip-sample separation. The frequency function is a closed-from equation and is quick to obtain results. Experimental data are also proven with the proposed frequency function. Additionally, various tip height behaviors coupling with the atomic forces gradient and hydrodynamic effects are discussed in the context of carbon nanotube probes.

2. MATHEMATIC MODEL AND DISCUSSION

2.1. Quasi-Static Equilibrium Condition

The equation of motion from the Euler-Bernoulli beam is:

$$\frac{\partial^2}{\partial x^2}[EI(x)\frac{\partial^2 w(x,t)}{\partial x^2}] + m(x)\frac{\partial^2 w(x,t)}{\partial t^2} + c(x)\frac{\partial w(x,t)}{\partial t} = p(x,t)$$

(1)

The following discussion is limited to beams with uniform properties along their lengths, i.e., Young's modulus, moment of inertia, mass per unit length, and the damping coefficient given by the constants: E, I, m, and c. The total displacement is w(x,t) as the sum of displacements that would be induced by static application of the support motion, i.e., the quasi-static displacement, plus the additional displacement due to the dynamic inertial and viscous force effects [10]. At the clamped-end, $w(0,t) = 0$ and $w'(0,t) = 0$. At the tip-end, affected by the atomic force, $w''(0,t) = 0$ and $EIw'''(L,t) = m_{tip}\ddot{w}(L,t) + k_{ts}(z)w(L,t)$, in which m_{tip} is the mass of the tip, $k_{ts}(z)$ is the gradient of the force curve, and L is beam length. Applying the above boundary conditions to Equation (1) gives the eigen function:

$$(3R_k - R_{tip}\beta_n^4)(\sin\beta_n \cosh\beta_n - \cos\beta_n \sinh\beta_n) + \beta_n^3(1 + \cos\beta_n \cosh\beta_n) = 0$$

(2)

where β_n is a coefficient in the separation of variable method, and the value is relative to the natural frequency of the microcantilever. The stiffness ratio and the tip-mass ratio are $R_k = -k_{ts}/k_0$ and $R_{tip} = m_{tip}/mL$. $k_0 = 3EI/L^3$ is the stiffness of the microcantilever. If $k_{ts}(z)$ is assumed to be specific form (for example, Lennard-Jones form), β_n can be solved by numerical method. The eigenfrequency is defined by $f_n(R_k, R_{up}) = \sqrt{\frac{EI}{m}}(\beta_n/L)^2/2\pi.$ The first three modes of the eigenvalues β_1, β_2, and β_3 versus the stiffness ratio for certain tip-mass ratios are plotted in Figure 1. All eigenvalues are normalized by β_0, the eigenvalue when $R_k = 0$. Thus β_0 leads to f_0, which represents the resonance frequency in the far field. From above, the resonance frequency is not constant, as it varies with tip-sample separation. Experimental data obtained from Gotsmann et al. is also plotted [11]. The minimum stiffness ratios in the associated experiments are shown to be a little bit less than zero ($R_k \rightarrow 0^-$).

However, the first eigenvalue may sometimes have a trivial solution when $R_k \leq -1$. A critical point, $(R_k, \beta_1) = (-1,0)$ is utilized to determine where static equilibrium is broken and a jump occurs. This can be proven by applying an asymptotic series in β_1. Here, substituting $sin\beta_1 = \beta_1 - \beta_1^3/3! + \cdots$, $cos\beta_1 = \beta_1 - \beta_1^2/2! + \cdots$, $sinh\beta_1 = \beta_1 + \beta_1^3/3! + \cdots$, and $cosh\beta_1 = \beta_1 + \beta_1^2/2! + \cdots$ into Equation (2) gives $(3R_k - R_{tip}\beta_1^4)(2\beta_1^3/3) + 2\beta_1^3 = 0$, in which the higher order terms are neglected. Setting $\beta_1 \rightarrow 0$ leads to $R_k = -1$. This implies that the absolute value of the gradient of atomic force is the same as the stiffness of the beam. This corresponds to the case where the tip approaches the sample from afar, and the stiffness ratio decreases from 0 to -1. The minimum stiffness ratio of a stiff microcantilever may not reach -1, but a soft microcantilever may reach -1. Then a small disturbance brings the soft cantilever into the region $R_k < -1$, where the attractive force exceeds the restoring force, accelerating the tip toward the sample. In Figure 1, this nonlinear result breaks the assumption that the frequency shift is a linear function of the tip–sample gradient in the range $[-1, 0]$, $\Delta f = f_0/2(k_{rs}/k_0)$ as provided [12]. Figure 1 also shows that the second and third eigenvalues remain constant with the stiffness ratio. It is also clear that the tip-mass ratio influences the eigenvalues, but is independent of the tip-sample separation. From an experimental data point of view, typical probes are conservatively designed according to $(R_k \rightarrow 0^-)$. Since the performance of the probe can be predicted in our proposal, the stiffness ratio range for an effective probe can be expanded to $-1 < R_k < 0$.

2.2. Atomic Force Gradient Effect

Moreover, the frequency is a function of vibration amplitude, especially oscillation in the near field featuring rapid variation of the atomic force. Giessibl modified the results of Albrecht et al. [12] and utilized canonical perturbation theory to solve the frequency shift for large amplitudes [8]. His solution assumes that the gradient of the restoring force is larger than that of the atomic force, and that the cantilever motion is well described as approximately harmonic. Then the frequency shift is strictly proportional to f_n and $1/k_0$, i.e., $\Delta f = (f_n/2k_0)<k^*(A_0)>$, in which the bracket, $< >$, indicates averaging across one oscillation cycle and is the weighted average of the gradient of the force:

$$\left\langle k^*(A_0) \right\rangle = \frac{2}{\pi A_0^2} \int_{-A_0}^{A_0} k_{ts}(x)\sqrt{A_0^2 - x^2}\, dx$$

(3)

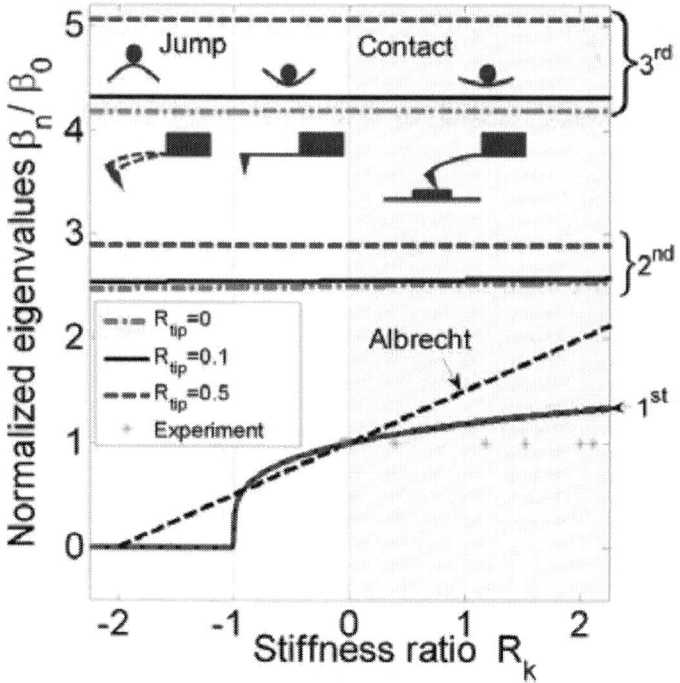

Figure 1. The first three normalized eigenvalues β_n/β_0 varying with stiffness ratio R_k and tip-mass ratio R_{tip}. $R_k > 0$ represents a contact, $-1 < R_k < 0$ represents a oscillation, and $R_k < -1$ represents a jump. Experimental data, denoted by *, are from Gotsmann et al. [11].

Giessibl applied to the atomic resolution of his studies, but the equilibrium of the quasi-static conditions was not considered [13–15]. f_0 remains constant in his papers. However, his experimental frequency shifts were still well matched with the theoretical prediction. Because the probes employed in these experiments were sufficiently stiff, $R_k \to 0^-$, so f_0 was constant. In this paper, the combination of eigen frequencies obtained from the quasi-static equilibrium and Giessibl's theory leads to the resonance frequency in vacuum:

$$f_{vac}(R_k, R_{tip}, A_0) = f_n(R_k, R_{tip})[1 + \frac{1}{\pi k_0 A_0^2} \int_{-A_0}^{A_0} k_{ts}(z-x)\sqrt{A_0^2 - x^2}\, dx]$$

(4)

If $k_{rs}(z)$ is assumed to be Lennard-Jones' formula, the integration could be solved to be a function of amplitude. Figure 2a shows the frequency shift of a tungsten tip on a KCl (100) surface in which Equation (4) is applied to compare with experimental data from Giessibl's study [16]. Model data was generated with an amplitude $A_0 = 0.15$ nm, a stiffness of $k_0 = 1800$ N/m, length 2.4 mm, width 130 µm, thickness 214 µm, and eigenfrequency set to $f_1 = 25,068$ Hz.

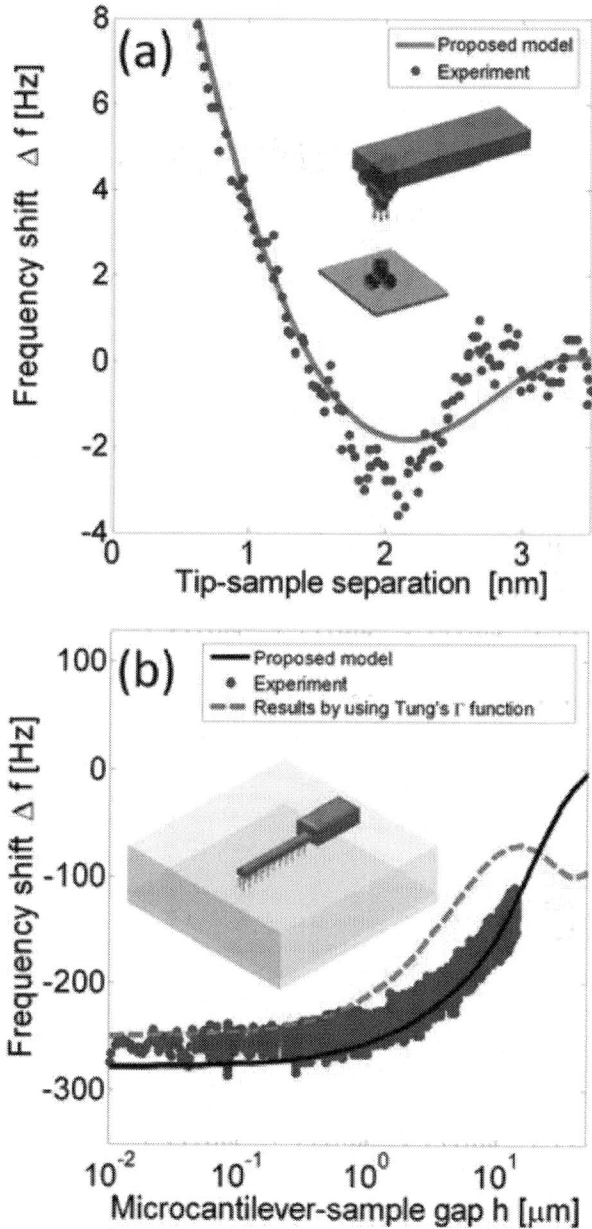

Figure 2. Frequency shift versus separation for: (**a**) a vacuum case with prediction by the proposed method (solid line) and experimental data (dots) [16], and (**b**) a liquid case with prediction by proposed method (solid line), experimental data (dots) [2], and results (dashed line) by the proposed method but with Γ_r replaced by approximation [21].

2.3. Hydrodynamic Effect

The modeled frequency shift takes into account hydrodynamic effects when the probe is immersed in liquid. Sader [9] extended the boundary integral technique of Tuck [17] and presented an explicit semi-analytical theory to solve for an oscillating cantilever immersed in a viscous fluid nearby the specimen surface. This paper extended Sader's work to FM AFM probe. It assumes that the fluid was incompressible and the oscillation amplitude was small. The fluid flow around the microcantilever is governed by the incompressible unsteady Stokes equation:

$$\rho_f \frac{\partial \mathbf{v}}{\partial t} = -\nabla P + \eta \nabla^2 \mathbf{v}$$

(5)

where v, P, and η are vector velocity, pressure, and viscosity of the fluid, respectively; and the Reynolds number of the flow is $R_e = \pi \rho_f f_h b^2 / 2\eta$ [18], and the Reynolds number of a soft AFM probe of width 30 μm oscillating in water is in range [0.1, 20]. Fluid velocities are restricted to be normal to the solid surface, whereas no-slip boundary conditions are enforced at interfaces. The hydrodynamic load per unit is obtained from pressure differences between the top and bottom of the beam, $p_h = \eta \dot{w}(x,t) \int_{-b/2}^{b/2} \Delta P(\xi) d\xi$, in which $\dot{w}(x, t)$ is the normal velocity and $\Delta P(\xi)$ is the pressure jump across the beam. Let the Fourier transform be $\widetilde{X} = \int_{-\infty}^{\infty} X e^{iwt} dt$. Taking the Fourier transform of Equation (1) yields $EI\widetilde{w}''' - m\omega^2 \widetilde{w} + i\omega c\widetilde{w} = \tilde{p}$. $\tilde{p}$ is the hydrodynamic loading component. Examining the Fourier transformed equations of Equation (5) and setting $\nabla \cdot \mathbf{v} = 0$ give the general form of the hydrodynamic loading component as $\tilde{p} = \pi^3 f_n^2 \rho_f b^2 \Gamma(Re, b, h) \tilde{w}$, where h is the gap between the probe and the surface and $\Gamma(R_e, b, h) = i \int_{-b/2}^{b/2} \Delta P(\xi) d\xi / (\pi R_e)$ is the complex hydrodynamic function [19]. Its real part Γ_r represents the effect of added mass on the surrounding fluid, and its imaginary part Γ_i represents the damping effect of the fluid. The hydrodynamic load equation is formally exact in the limit of $0 < L/b \ll 1$. Substituting $\tilde{p}$ into the Fourier-transformed motion equation leads $f_h(R_e, b, h) = f_{vac} [1 + \pi \rho_f b^2 \Gamma_r(R_e, b, h)/4m]^{-1/2}$. Note the other formula for the frequency function can refer to the study by Naik et al. [20]. Thus f_{vac} replaced by Equation (4) gives the frequency:

$$f_h(R_k, R_{np}, A_0, Re, b, h) = f_n(R_k, R_{np}) \frac{1 + \dfrac{1}{\pi k_0 A_0^2} \int_{-A_0}^{A_0} k_{ts}(z-x)\sqrt{A_0^2 - x^2}\, dx}{[1 + \dfrac{\pi \rho_f b^2}{4m} \Gamma_r(Re, b, h)]^{1/2}}$$

(6)

This is the desired equation, representing frequency as a function of the eigenfrequency and proportional to the atomic effect and inversely proportional to the hydrodynamic effect. From an engineering point of view, the interaction force curve can be assumed and the probe properties and the amplitude of the probe are given by users. This frequency function works while the hydrodynamic function Γ_r is given. Here, Γ_r has an approximation function and will be discussed later. Note that Γ_r is a function of the Reynolds number R_e, which itself is a function of f_h. Applying the numerical method allows f_h to be easily solved. As a result, Equation (6) is useful in commercial devices. This equation is different to the equation provided by Sader et al. [1]. It contains the quasi-static equilibrium condition and can predict the performance of the soft AFM probes. Furthermore, the inverse method for calculating the tip-sample forces in liquid from the experimental frequency shifts can be realized by applying Giessibl's technique [16] along with Equation (6).

2.4. Hydrodynamic Function

The details of the hydrodynamic function $\Gamma_r(R_e,b,h)$ can be obtained from the works of Green, Sader, and van Eysden [9,19] who provided closed-form analytical expressions for the entries of all submatrices except four complicated terms, (A_1,A_2,C_1,C_3). In this paper, these four terms were computed by applying numerical Gauss-Legendre quadrature. But it was a time consuming procedure. Tung et al. [21] calculated the hydrodynamic function and provided the polynomial approximation function by curve fitting in the domain R_e $\in[10^{-2},10^4]$. These fitting results were found to have errors on the order of a few percent, which might shift the frequency a few tens of hertz (the eigenfrequency is around one hundred kilohertz).

Figure 2b shows the frequency shift of a microcantilever immersed in water plotted against the gap between the probe and the surface. The dimensions of the microcantilever (Nanosensors EFM cantilever) are length 225 μm, width 23 μm, thickness 3 μm, a tip height of 13 μm, a oscillation amplitude of 1 nm, and the resonance frequency in water is 13.14 kHz. The solid line plots results obtained from Equation (6) directly by the semi-analytical approach [9]. The dashed line represents results from Equation (6) but with the hydrodynamic function derived from Tung's polynomial approximation [21]. The dots are experimental data obtained from Sader et al. [2]. The dashed line deviates when the gap is larger than 10 μm because Tung's approximation is outside the convergent region. Tung's approximation saves time and is useful in predicting the shift tendency of the resonance frequency. When the gap is much smaller than the microcantilever width, a change in the gap has less of an effect on the resonance frequency due to the hydrodynamic effect. Comparing Figure 2a with 2b, the atomic force affects the frequency shift by several hertz in a region a few nanometers in the near-field, and the hydrodynamic affects the frequency shift by hundreds of hertz in the region around a hundred micrometers. As a result, while the tip height is a few micrometers, these two effects are coupled.

2.5. Tip Heights and Frequency Shifts

Figure 3a shows frequency shift versus tip-sample separation for various tip heights, set at 0.1, 1, and 10 μm. A case of a silicon tip and a polystyrene surface is studied with probe dimensions of length 250 μm, width 35 μm, and thickness 5.7 μm.

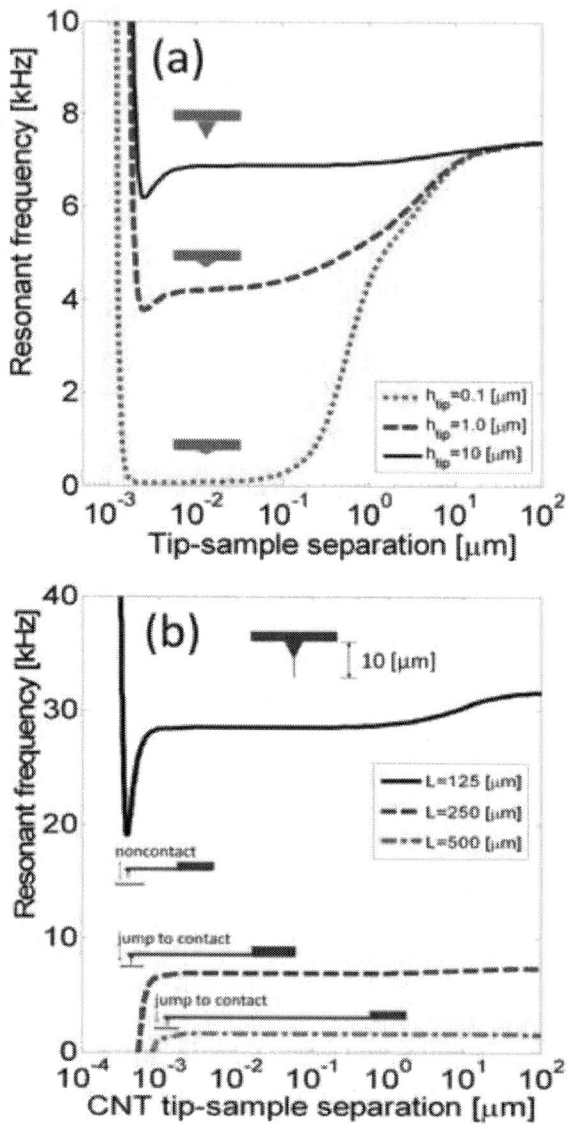

Figure 3. Frequency versus tip-sample separation for (**a**) probes with various tip heights, and (**b**) a CNT-tip with various microcantilever lengths.

The resonance frequency in water is 232.5 kHz. The Lennard-Jones interaction force is assumed to be $F(z) = A_1R/180z^8 - A_2R/6z^2$, in which R = 20 nm, A_1 and A_2 are the Hamaker constants, 0.838873×10^{-70} J m^6 and 1.15072×10^{-19} J, respectively [22]. The results indicate that the $h_{tip} = 0.1$ μm case presents the hydrodynamic affects from far field to the near field, tip-sample separation between 0.1–10 μm. Hydrodynamic effects are effectively constant when the tip-sample separation is less than 0.1 μm. Moreover, typical tips are around ten micrometers, and the frequency shifts due to the hydrodynamic force are on the order of hundreds of hertz, as shown $h_{tip} = 10$ μm. Hence, high tips feature reducing of the hydrodynamic and presenting the atomic force effects.

Recently, tips extended by a carbon nanotube (CNT) positioned on the tip have been used to measure high-aspect ratio samples. The CNT can be hundreds of nanometers to a few micrometers long, and a few hundred nanometers long is suggested to avoid buckling. In the following calculation, the total height of a CNT-tip is set to 10 μm, and the rest of the probe geometry is the same as in previous calculations. The carbon-carbon interaction potential is $F(z) = 4\varepsilon\pi n\sigma^2[-(\sigma/z)^{10}/5 + (\sigma/z)^4/2]$, where the parameters are assumed to be $\varepsilon = 4.751 \times 10^{-22}$ eV, $n = 0.114$ Å^{-3}, and $\sigma = 3.407$ Å [23]. To study the jumps, the microcantilever length was set sequentially to L = 125, 250, and 500 μm (k_0 distribution ratio is 1/8:1:8) as shown in Figure 3b. Since the CNT-tip has a rapid interaction, the stiffness ratio becomes less than −1 in the first two cases, and that represents CNT-tip jumps. The frequency shifts and critical points of the jumps can also be analytically obtained by the proposed method.

3. CONCLUSIONS

A frequency function for AFM in liquid has been introduced, with the frequency shift derived as a function of six parameters: stiffness ratio, mass ratio, oscillation amplitude, Reynolds number, probe width, and tip height. This proposed frequency function is in closed form, which is easy for engineers to implement in AFM devices. It can predict the jump to contact behavior. Accordingly, probes with stiffness ratios smaller than −1 could be applied; that is soft probes could be easily used to increase the probe sensitivity. By application of the frequency function, the probe with high tips demonstrates the effect of coupling between atomic and hydrodynamic forces, and soft probes with CNT-tips show jumping behavior in accord with the predictions.

ACKNOWLEDGMENTS

The author wishes to thank National Science Council of the Republic of China, Taiwan, for financial support under contract NSC-101-2221-E-390-003.

1. REFERENCE

2. Fukuma, T.; Jarvis, S.P. Biological applications of FM-AFM in liquid environment. In Noncontact Atomic Force Microscopy; Morita, S., Giessibl, F.J., Wiesendanger, R., Eds.; Springer-Verlag: Berlini/Heidelberg, Germany, 2009; pp. 329–345.

3. Sader, J.E.; Uchihashi, T.; Higgins, M.J.; Farrell, A.; Nakayama, Y.; Jarvis, S.P. Quantitative force measurements using frequency modulation atomic force microscopy—Theoretical foundations. Nanotechnology **2005**, 16, S94–S101.

4. Adamcik, J.; Berquand, A.; Mezzenga, R. Single-step direct measurement of amyloid fibrils stiffness by peak force quantitative nanomechanical atomic force microscopy. Appl. Phys. Lett. **2011**, 98, 193701–193703.

5. Xu, X.; Carrasco, C.; de Pablo, P.J.; Gomez-Herrero, J.; Raman, A. Unmasking imaging forces on soft biological samples in liquids when using dynamic atomic force microscopy: A case study on viral capsids. Biophys. J. **2008**, 95, 2520–2528.

6. Melcher, J.; Carrasco, C.; Xu, X.; Carrascosa, J.L.; Gomez-Herrero, J.; Jose de Pablo, P.; Raman, A. Origins of phase contrast in the atomic force microscope in liquids. PNAS **2009**, 106, 13655–13660.

7. Kodera, N.; Yamamoto, D.; Ishikawa, R.; Ando, T. Video imaging of walking myosin V by high-speed atomic force microscopy. Nature **2010**, 468, 72–76.

8. Shih, P.J. Tip-jump response of an amplitude-modulated atomic force microscope. Sensors **2012**, 12, 6666–6684. [Google Scholar]

9. Giessibl, F.J. Forces and frequency shifts in atomic-resolution dynamic-force microscopy. Phys. Rev. B **1997**, 56, 16010–16015.

10. Green, C.P.; Sader, J.E. Small amplitude oscillations of a thin beam immersed in a viscous fluid near a solid surface. Phys. Fluids **2005**, 17.

11. Clough, R.W.; Penzien, J. Dynamics of Structures, 3rd ed.; Computers & Structures, Inc: Berkeley, CA, USA, 2003; pp. 370–373.

12. Gotsmann, B.; Anczykowski, B.; Seidel, C.; Fuchs, H. Determination of tip-sample interaction forces from measured dynamic force spectroscopy curves. Appl. Surf. Sci. **1999**, 140, 314–319.

13. Albrecht, T.R.; Gütter, P.; Horne, D.; Rugar, D. Frequency modulation detection using high-Q cantilevers for enhanced force microscope sensitivity. J. Appl. Phys. **1991**, 69.

14. Giessibl, F.J. Atomic resolution of the silicon (111)-(7 × 7) surface by atomic force microscopy. Science **1995**, 267, 68–71.

15. Hembacher, S.; Giessibl, F.J.; Mannhart, J. Force microscopy with light-atom probes. Science **2004**, 305, 380–383.

16. Welker, J.; Giessibl, F.J. Revealing the angular symmetry of chemical bonds by atomic force microscopy. Science **2012**, 336, 444–449.

17. Giessibl, F.J. A direct method to calculate tip-sample forces from frequency shifts in frequency-modulation atomic force microscopy. Appl. Phys. Lett. **2001**, 78.

18. Tuck, E.O. Calculation of unsteady flows due to small motions of cylinders in a viscous fluid. J. Eng. Math. **1969**, 3, 29–44.

19. Sader, J.E. Frequency response of cantilever beams immersed in viscous fluids with applications to the atomic force microscope. J. Appl. Phys. **1998**, 84.

20. Van Eysden, C.A.; Sader, J.E. Resonant frequencies of a rectangular cantilever beam immersed in a fluid. J. Appl. Phys. **2006**, 100.

21. Naik, T.; Longmire, E.K.; Mantell, S.C. Dynamic response of a cantilever in liquid near a solid wall. Sens. Actuators A **2003**, 102, 240–254.

22. Tung, R.C.; Jana, A.; Raman, A. Hydrodynamic loading of microcantilevers oscillating near rigid walls. J. Appl. Phys. **2008**, 104.

23. Rützel, S.; Lee, S.I.; Raman, A. Nonlinear dynamics of atomic-force-microscope probes driven in Lennard-Jones potentials. Proc. R. Soc. Lond. A **2003**, 459, 1925–1948.

24. Kutana, A.; Giapis, K.P.; Chen, J.Y.; Collier, C.P. Amplitude response of single-wall carbon nanotube probes during tapping mode atomic force microscopy: Modeling and experiment. Nano Lett. **2006**, 6, 1669–1673.

CHAPTER 8

Elucidation and Identification of Double-Tip Effects in Atomic Force Microscopy Studies of Biological Structures

Yong Chen[1,2]

[1]State Key Laboratory of Food Science and Technology, Nanchang University, Nanchang, China; [2]Institute for Advanced Study, Nanchang University, Nanchang, China.

ABSTRACT

While atomic force microscopy (AFM) has been increasingly applied to life science, artifactual measurements or images can occur during nanoscale analyses of cell components and biomolecules. Tip-sample convolution effect is the most common mechanism responsible for causing artifacts. Some deconvolution-based methods or algorithms have been developed to reconstruct the specimen surface or the tip geometry. Double-tip or double-probe effect can also induce artifactual images by a different mechanism from that of convolution effect. However, an objective method for identifying the double-tip/probe-induced artifactual images is still absent. To fill this important gap, we made use of our expertise of AFM to analyze artifactual double-tip images of cell structures and biomolecules, such as linear DNA, during AFM scanning and imaging. Mathematical models were then generated to elucidate the artifactual double-tip effects and images develop during AFM imaging of cell structures and biomolecules. Based on these models, computational formulas were created to measure and identify potential double-tip AFM images. Such formulas proved to be useful for identification of double-tip images of cell structures and DNA molecules. The present studies provide a useful methodology to evaluate double-tip effects and images. Our results can serve as a foundation to design computerbased automatic detection of double-tip AFM images during nanoscale measuring and imaging of biomolecules and even non-biological materials or structures, and then personal experience is not needed any longer to evaluate artifactual images induced by the double-tip/probe effect.

Keywords: Nanobiotechnology; Atomic Force Microscopy (AFM); Double-Tip Artifact; Biological Specimens; Linear DNA

1. INTRODUCTION

Nanotechnology has been emerging as a powerful tool for the studies of life science [1]. One of the important research endeavors is to understand nanoscale structures and life events through the nano-measuring and imaging of cells [2-6] or thin sections of them [7], cellular organelles [8], proteins [9-13] , polysaccharide [14], DNA [15] and others using the instruments such as scanning tunneling microscopy (STM), near field scanning optical microscope (NSOM) [16-18] and atomic force microscope (AFM) [19-21]. It is important to note that artifactual measurements or images can occur during nanoscale analyses of cell components and biomolecules using the current-generation instruments. Artifactual or "ghost" images due to errors in STM and AFM probing have been described by some investigators since the invention of scanning probe techniques [22-25] . Such artifactual images appear to be associated with steep corrugations or sharp structural features on a sample surface [26], and with the occurrence of an "extra-tip" on the scanning tipsample interface (double-tip or multiple-tip effect and tip asymmetry).

Since the artifactual measurements or images derived from nanoscale operation can certainly jeopardize our understanding of the nano-world, the mechanisms by which such artifactual images develop have been intensively studied and well known [27]. Tip-sample geometric convolution effect is the most common mechanism responsible for causing artifacts especially when the size of an imaged object or feature is on the same scale as the radius of curvature of the AFM tip. At present, some methods or algorithms for reconstruction of specimen surfaces or tip geometries have been developed mainly by deconvolution [28,29].

The mechanism of double-tip/probe effect is different from that of convolution effect (Figure 1) [30]. Tip convolution occurs when the curvature radius of the tip of a single-tip AFM probe is on the same size scale as the imaged object of feature, and the tip images the object of feature only one time during the imaging process; for double-tip effect, however, the two tips of a double-tip probe scan the surface of an object separately, forming two images (a real image and a ghost image) of the object at two different but adjacent locations (Figure 1) [30]. Convolution effect may occur during the imaging process of each of the two tips if the size of the imaged object is on the same scale as the curvature radiuses of the two tips.

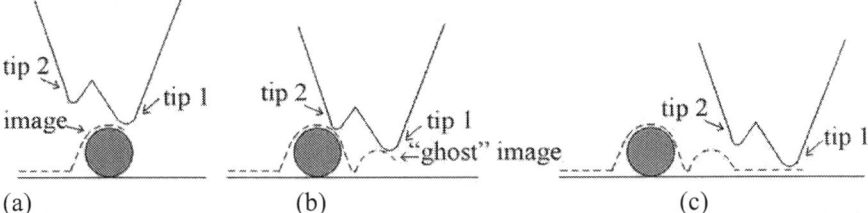

Figure 1. Schematics of the mechanism of double-tip effect. The double-tip AFM probe actually images the object twice, getting an image first at the location of the object by tip 1 (a) and then a "ghost" image nearby by tip 2 (b). During the imaging process of each tip, convolution effect may occur if the size of the object is on the same scale as the curvature radiuses of the two tips.

Due to the distinction in mechanism, the previous well-developed, deconvolution-based methods or algorithms may be unapplicable to the artifactual images induced by double-tip/probe effect. Unfortunately, up to now, there have been no objective methods to facilitate the identification or detection of potential artifacts caused by double-tip/probe effect. To fill this important gap, we made use of our AFM expertise to explore a methodology to detect potential artifactual images derived double-tip/probe effect by analyzing cell structures and DNA or proteins. Generally, it is hard to identify the artifacts caused by double-tip/probe effect especially for the linear, tangling structures or molecules. To analyze the array pattern of real and ghost images makes it possible to identify these artifacts since the distance and angle between the two tips of a double-tip AFM probe are all fixed.

2. MATERIALS AND METHODS

2.1. Human MSC Cultures

Bone marrow (BM) from human fetus thighbone was aspirated into 10 ml syringes within 4 h of harvest. The marrow samples were washed and centrifuged twice (1000 rpm for 5 min) in mesenchymal stem cell (MSC) growth medium (MSCGM; Osiris Therapeutics, USA) before suspension in fresh MSCGM. 70% Percoll (Sigma, USA) was added into the suspension, and then centrifuged at 1500 rpm for 15 min. The suspension with MSCs was added to MSCGM medium supplemented with 15% FBS. The number of mononuclear cells was determined with a hemocytometer and plated in 25-cm^2 tissue culture flasks at a density of approximately 1×10^5 BM mononuclear cells/cm^2. Cells were incubated at 37°C with 5% fully humidified CO_2. The first medium change occurred after 24 h and then every 3 - 4 days thereafter. Then they were processed by trypsinization (0.25% trypsin and 1 mM EDTA for 5 min at 37°C) and re-expanded to confluency. After a second passage, the cells were counted and assessed for viability by means of Trypan Blue dye exclusion. Before AFM imaging, MSCs were incubated on the surface of clear cover slips treated with poly-L-lysine at 37°C with 5% CO_2 at saturated humidity.

2.2. Preparation of the Cell and Biomolecule Samples for AFM

Microscope cover slips were cleaned by means of standard surfactant and rinsed with distilled water. Before AFM imaging, sterilized cover slips were placed in cell culture dishes containing freshly passaged MSC cells. The cells were cultured in conventional culture media in all cases and incubated at 37°C with 5% CO_2. After 2 days, the cover slips with MSCs were taken from the medium directly. After air-fixing rapidly by waving the slip vigorously, a 1% glutaraldehyde solution was then added for several minutes to rigidify the cells. The cells were washed with PBS and then air-dried for 10 min. The sample was then examined under the optical microscope and was imaged by AFM.

Calf thymus DNA (Sigma) was dissolved at 1 mg/ml in Milli-Q water. A 3 µl drop of the DNA solutions was deposited onto a freshly cleaved mica surface. Approximately 1 min later, the residual solution on the mica was carefully removed by a slice of filter paper, followed by air-drying. Other DNA solutions were maintained at room temperature for several days, and then sampled. At this time, most DNA molecules were hydrolyzed into spherical particles. Collagen type I (Sigma) were dissolved in acetic acid, were purposely mixed with unpurified water containing particles, were deposited onto mica surface, and then were scanned by AFM probe.

2.3. Atomic Force Microscopy

The cover slip was mounted on an AFM stage, and the integral video camera was used to locate the regions of interest. The samples were observed with an AutoProbe CP AFM (Thermomicroscopes, USA) in tapping mode. Microfabricated silicon nitrite cantilevers (Park Scientific Instruments) with sharp tips with a tip radius of curvature 10 nm and a force constant of approximately 2.8 N/m were used. Observation was carried out in air at room temperature. The scan speed of the tip was 1 Hz. The AFM images were planar leveled using the software (Thermomicroscopes Proscan Image Processing Software Version 2.1) provided with the instrument. Using the line analysis function of the software, widths and heights of the regions of interest were determined, and the height profiles of cross sections were obtained. Images of spheres and linear molecules or structures were measured for various line parameters such as horizontal and vertical lines connecting the real and ghost molecule, and the line-related angles using the line-measuring AFM software.

2.4. Statistical Analyses

Data are expressed as the mean ± SD. ANOVA and student t test analyses were undertaken to determine the extent to which Φ and x values obtained by measurements resemble those Φ and x values calculated by mathematic formulas.

3. RESULTS AND DISCUSSION

3.1. Analyses of Double-Tip Images of Cell Structures and Biomolecules

Artifacts induced by double-tip or double-probe effect have been briefly described [31-34] and the underlying mechanism has been clarified [30]. However, an effective method for objectively identifying these artifacts is needed. As an initial effort to dissect and identify double-tip effects in AFM practice, we sought to analyze and define double-tip images of cell structures and biomolecules. A mesenchymal stem cell (MSC) with fibrillar branches was scanned by AFM and assessed for double-tip images of cellular structures. First glance of the AFM topographic image of MSC on the cover slip simply revealed the occurrence of extensive and dense fibrillar structures around the cell (Figure 2(A)). Careful examination allowed the identification of double-tip images for some of fibrils (marked 1-6, Figure 2(A)), although many tiny fibrillar branches did not show ghost (false) images in both the low-magnified and high-magnified (bottom left) topographic images. The double-tip images were readily illustrated by the sequential heightprofile analyses of topographic images in six different cross sections along one of the fibrillar branches (Figure 2(A)). Evidently, the width and height of the ghost fibril images (a short arrow) were smaller than those of the real fibril images (a long arrow). Moreover, with the real fibril image getting smaller, the height of its ghost fibril image also became smaller and lower, and even disappeared (arrowhead) when the height of the real branch reached around 30 nm. The double-tip phenomenon of cell structures was more readily revealed in high-magnified topography than in the low-magnified image (Figure 2(B)). Several morphological changes that occurred as a result of double-tip effect were shown in the magnified image of the cell. First, the repetition of branches (green doted areas) was noticeable. Secondly, edges of cell were extended (pink dotted areas). Finally, large protuberances on membrane surface were doubled (blue circles).

We then sought to analyze double-tip phenomena in imaging studies of DNA molecules and single collagen fibrils. DNA specimens containing both linear single-chain DNA and hydrolyzed DNA spheres were imaged by a high-quality AFM tip. Initial AFM scanning showed a real image of linear single-chain DNA and individual spheres of hydrolyzed DNA (Figures 3(a) and (b)). However, the subsequent scanning of the upper area containing a contaminated particle (black arrow) by the same AFM tip resulted in the development of a double-tip and the occurrence of double-tip images of the both linear DNA and hydrolyzed DNA spheres (Figure 3(b)). The single-chain DNA became double-chain DNA in the double-tip image; the hydrolyzed DNA spheres were duplicated. In some cases, these false double chain-like DNA molecules were comprised of a real single-chain DNA and its paralleled artifactual single-chain DNA (ghost chain). In other cases, the double-tip effect could generate the image of a widened DNA chain (arrowheads in Figure 3(b)), which looked like a fused double-chain DNA molecule (see the mathematic models below for explanation). Interestingly, double-tip images of partially hydrolyzed linear DNA could look like unwinding double helix DNA (Figure 3(c)).

Similarly, double-tip effect on individual protein molecules was also seen. Double-tip images of collagen fibrils were identified during AFM scanning and imaging of the fibrils and the defined particles that could derive a double tip (Figure 3(d)). It should be pointed out that, without these referring particles, the double-tip effect would be difficult to recognize or distinguish. These referring particles in unpurified water were purposely mixed with collagen fibrils and were deposited on mica. The AFM-tip scanning of these particles appeared to derive a double tip and the double-tip images of real and ghost fibrils (Figure 3(d)).

Next, we did mathematical analyses of the double-tip images (Figure 4), based on which four mathematic models were developed (Figure 5). Then, mathematical measurements were performed to support the models (Figure 6). Based on the four mathematic models, finally, we created formulas (Figure 7) potentially useful for calculating and identifying a potential double-tip image.

3.2. Mathematical Analyses of Double-Tip Images of Cell Structures and Biomolecules

Mathematical measurements of double-tip images should be an important step for dissecting the development of double-tip effects during AFM scanning and imaging. We used the AFM line-measuring software to measure important angles and lines connecting ghost and real molecules in a same double-tip topography. Our studies indicates that double-tip images of both spherical and

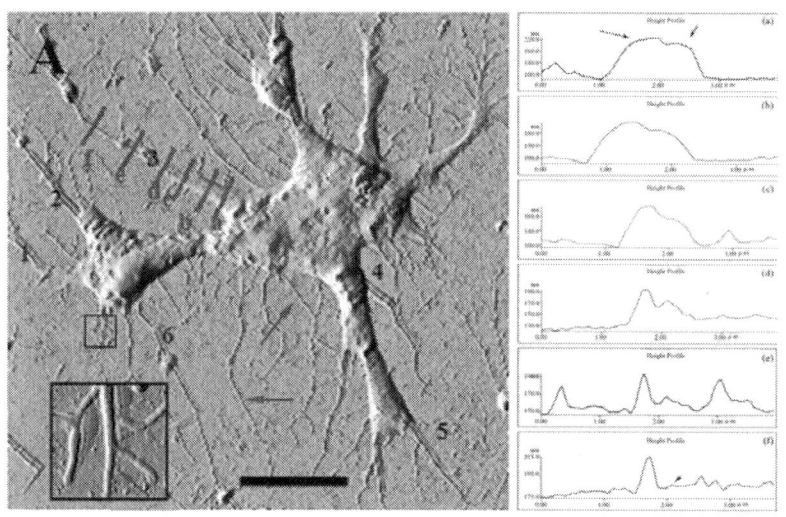

(A)

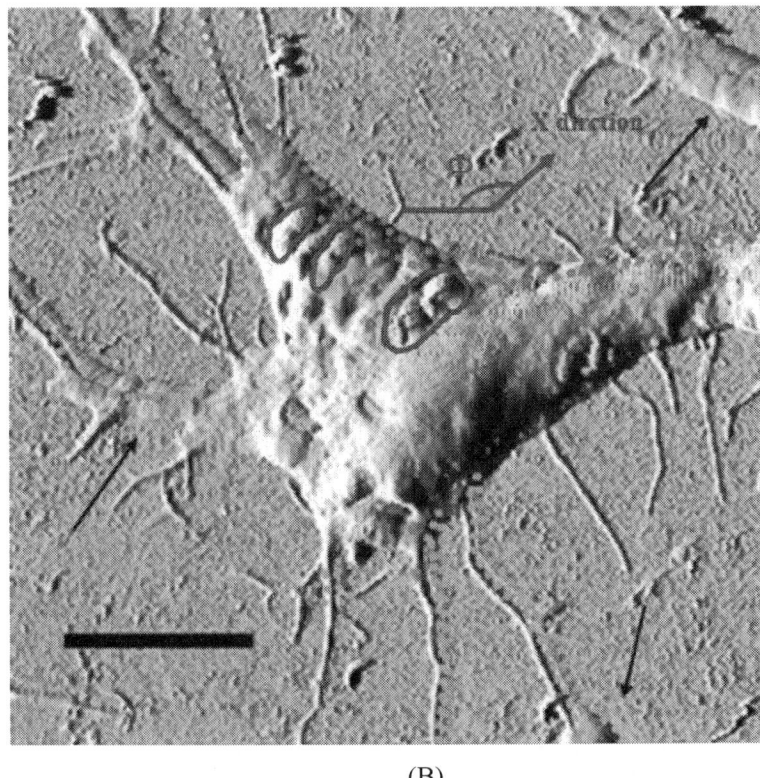

(B)

Figure 2. Analyses of double-tip images of cell structures. (A) The left panel revealed double-tip images (marked 1-6) of some fibrillar branches in a single MSC. Blue arrows indicated true images but not double-tip (ghost) images. The absence of double-tip images of a branch was also seen in the high-magnified AFM topographic image (bottom left). The right panel showed the double-tip images of a fibrillar branch by analyzing height profiles in a series of cross sections of the fibrillar branch. The cross sections a-f corresponded to the blue lines marked by letters a-f, along the branch of the cell in the left topography. The long arrow indicated the real image, and the short arrow meant the false (ghost) image induced by double-tip effect. Note that the real images were wider and higher than the ghost images. In addition, the ghost images became smaller with the branch being smaller, and no ghost was found when the height of the branch was around or less than 30 nm (the arrowhead in section F. Scan sizes of the lowand high-magnified images are 50 µm × 50 µm and 2.5 µm × 2.5 µm, respectively. Bar: 10 µm; (B) Artifactual double-tip effects on cell morphology and sizes in the higher-magnified AFM topography. Note the repeated branches and dilated edges of cell as highlighted by green and pink dashed circles. Measurement errors were easy to occur in some parts (arrows) where the ghost images overlapped with the real images. Furthermore, as shown by blue circles, some structures on cell membrane surface might also show structural ghosts that were difficult to be distinguished. Scan size: 20 µm × 20 µm. Bar: 5 µm.

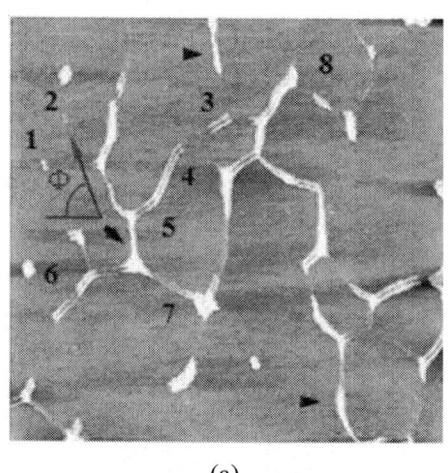

(a)

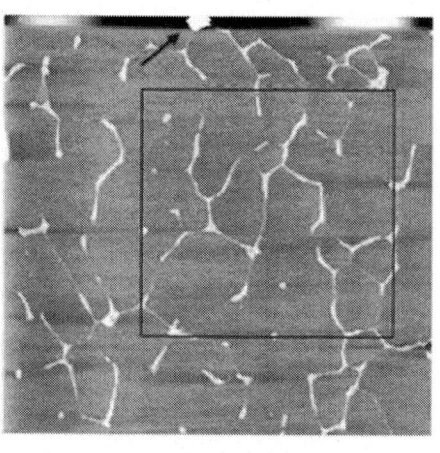

(b)

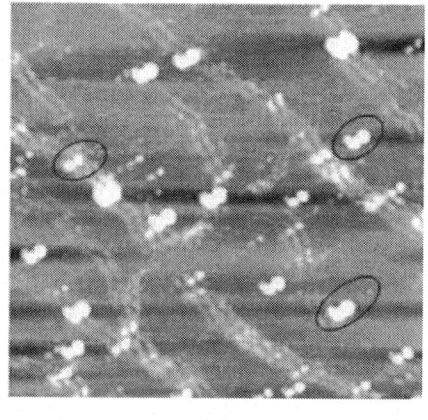

(c)

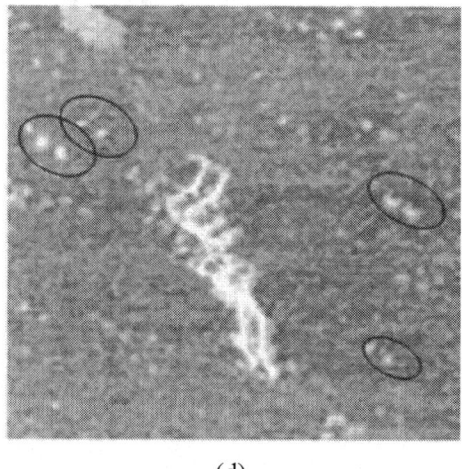

(d)

Figure 3. Analyses of double-tip images of biomolecules. (a) The control true topographic image of DNA spheres and linear DNA, which was scanned by a high-quality tip from bottom to top (the fast scan direction is X). Note that there was a particle on the top (arrow) that might induce a double-tip during the subsequent scanning. Scan size: 5 μm × 5 μm; (b) The double-tip images of DNA spheres and linear DNA in the subsequent scanning of the same square area in Figure 3(a) using the same AFM tip (subsequently contaminated by the particle mentioned in Figure 3(a)). Note that in the magnified square the single DNA spheres 1 and 2 (and others) now became double spheres in the double-tip images. Single-chain linear DNA now looked like doublechain DNA-like structures due to the double-tip effect. The single linear DNA could also exhibit a widened single-chain (arrowheads), which looked like being fused together. Scan size: 3 μm × 3 μm; (c) The double-tip images of DNA spheres (in circles) and partially-hydrolyzed linear DNA molecules. The partially-hydrolyzed DNA could wrongly be interpreted as unwinding double helix DNA. Scan size: 1 μm × 1 μm; (d) The double-tip images of known particles and bundles of collagen fibrils. The particles existed in unpurified water and were purposely mixed with collagen fibrils before AFM scanning and imaging. If one did not know the contaminated particles (in circles), the doubleimages of single collagen fibrils would be difficult to recognize (purified water does not contain the practices). Scan size: 1 μm × 1 μm.

Linear molecules share the same angle (Φ) which is formed by a horizontal line a and a line x connecting real and ghost molecules (Figure 4). For double-tip images of spherical fragments or structures, the x line is the same to the line c that vertically connects the real and ghost spheres in the same double-tip topography (Figure 4(a) and Figure 3(b)). For double-tip images of linear molecules or structures, the x appeared to be similar in values for each pair of ghost and real molecules, but different in values from the c (Figures 4(b), 4(c) and 3(b)). The Φ is related to the angle ω formed by a real linear molecule and an Φ-related line x. The measurements of these important angles and lines raises the possibility that mathematic models can be generated to dissect and identify double-tip images in AFM operation.

3.3. Mathematic Models for Dissecting Double-Tip Images in AFM Scanning of Cells or Biomolecules

Four mathematic models were developed to understand the mechanical and optical mechanisms for the encountered ghost images derived from double-tip effects: 1) Double-tip image model of spherical molecules (Figure 5(a)). When two or more spherical molecules with different sizes were scanned by an AFM double-tip, double-tip images of these spherical molecules have following common features. The ghost spheres are as big as or smaller than real molecules. Importantly, Φ, x, or c value is the same for all pairs of real and ghost spheres in the same double-tip images. That is $x1 = x2$; $c1 = c2$; $\Phi_1 = \Phi_2$; $c = x$ (Figure 4(a)). The ghost images of spheres are relatively easy to recognize; 2) Double-tip image model of curving linear molecules (Figure 5(b)). This model proposes that in the double-tip image of curving linear molecules, linear ghosts are paralleled to real linear molecules. Sometimes, such double-tip effects can make the curving linear molecules look like unwinding double helix DNA (Figures 3(c) and 5(b)). The x values are almost invariable, i.e. $x1 = x2 = x3 = x4$ (blue lines in Figure 5(b)). However, c values ($c1 - c4$) are different to each other and to x in the same double-tip image (red lines in Figures 5(b) and (d)); 3) Double-tip image model of a global cell (Figure 5(c)). The double-tip effect can enlarge the circular shape of the global cell along the x line that forms the angle Φ (Figure 5(c)). Similarly, double-tip images of the cell structures share similar x ($x1 = x2$) but different c lines ($c1 \neq c2$) (Figure 5(c)). If double-tip effect involves a global cell with branches, the ghost images of cellular branches can be explained by the model 2) as described above; 4) Double-tip image model of multiple short linear fragments or molecules (Figure 5(d)). This model helps to dissect complex double-tip images of short fragments of individual linear DNA or protein molecules in AFM scanning. Each pair of ghost and real short fragments of linear molecules has a different c but shares a same x or Φ in a given double-tip AFM topography. Importantly, two critical points, in which short fragments of linear molecules are positioned, can be defined to determine how c and x values for the images evolve in different imaging positions. When the angle ω is at $90°$, the short fragment of a real linear molecule is positioned in a critical point 1 and is perpendicular to the x that connects the ghost and real linear molecule (Figure 5(d)). In this case, the c equals to the x, and reaches its maximum value, $c = x$, $[c = x\sin(180° - \omega); c = x\sin(180° - 90°); c = x\sin90°;$ thus $c = x]$. When the ω is at $180°$ or $0°$, the short fragment of a real linear molecule is positioned to the critical point 2. Thus, the c value is 0, indicating that a ghost linear molecule overlaps with or gets close to the real linear molecule. Within the range of $0° < \omega < 90°$, an increase in the ω value is associated with an increase in c value (Figure 5(d)), which is consistent with the increases in $c1 - c4$ values for distances connecting ghost and real linear molecules (Figure 5(d)). In contrast, in the range of $90° < \omega < 180°$ an increase in ω is associated with a reduction in c values (Figure 5(d)). Such a prediction is practically true for the decreases in $c5 - c7$ values (Figure 5(d)). This can help to explain why some ghost images are close to real images and others are far from their real images in various segments of the curving linear biomolecules shown in Figures 3(b), 3(c) and 5(b). The reason why some ghost and real images of linear molecules are fused as single widen linear molecules (shown by arrowheads in Figure 3(b)) is that the real linear molecules are close to the critical point 2 ($c = 0$), with minimum distance between the ghost and real linear molecules.

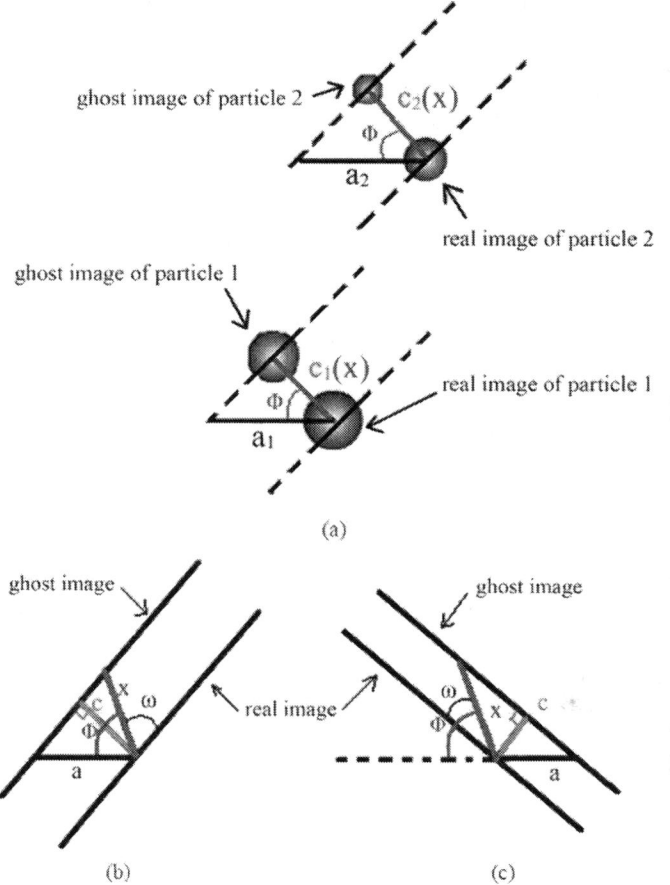

Figure 4. Mathematical analyses of double-tip images of cell structures and biomolecules. (a) Parameters for measuring double-tip images of spheres. The angle (Φ) is formed by a horizontal line a and a line x connecting real and ghost molecules. Note that the line x is the same as the line c ($x = c$) in the double-tip image of spheres. Here, $\cos\Phi_1 = c1/a1$, and $\cos\Phi_2 = c2/a2$, then $\Phi_1 = \arccos(c1/a1)$, and $\Phi_2 = \arccos(c2/a2)$. Using formula 2 in Figure 7(a) (see below), the following can be expected: $x = c/\{\cos[\phi - \arccos(c/a)]\}$, $x_1 = c1$, and $x_2 = c2$. Since c1 is the same as c2 ($c1 = c2$), it can be concluded: $x1 = x2 = c1 = c2$ and $\Phi_1 = \Phi_2$. (b) and (c) Parameters for measuring double-tip images of linear molecules in different positions or orientations. Note that the position, in which the double-tip images are located, is important for determining which group of formulas are used to calculate x and Φ (Figures 6(a) and 7).

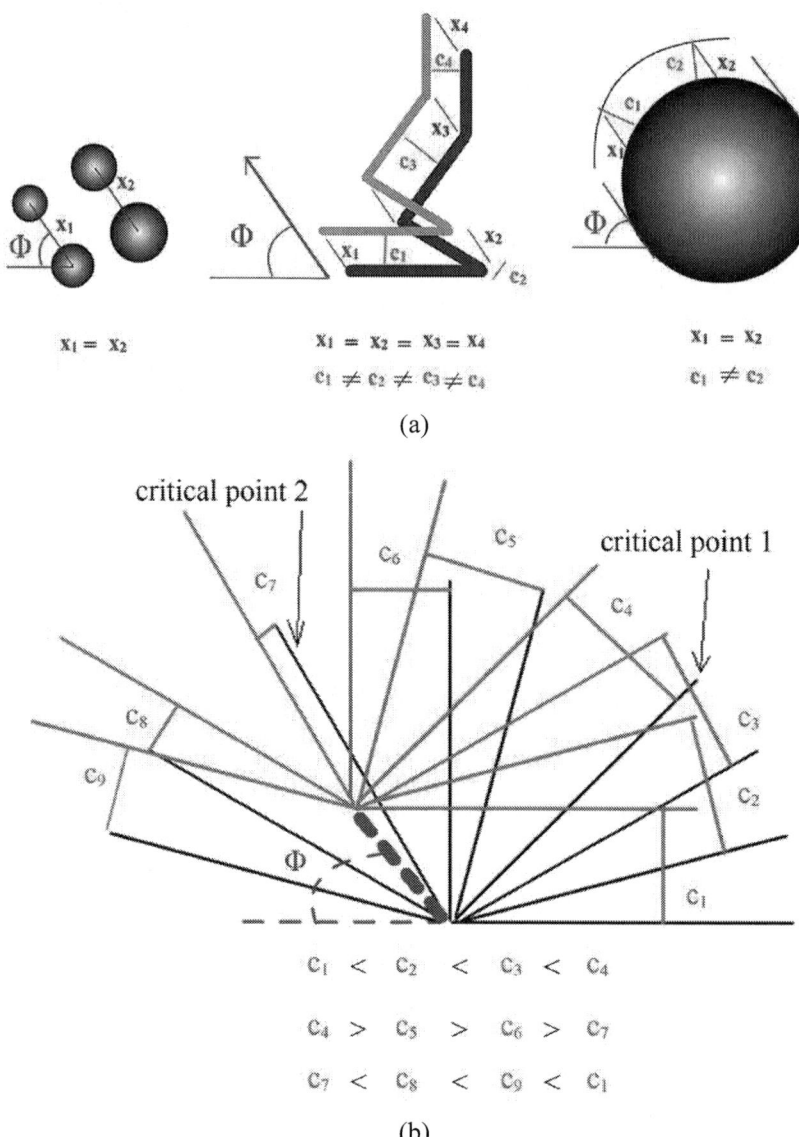

Figure 5. Four mathematic models for dissecting artifactual images caused by double-tip effect. (a) Double-tip image model of spherical molecules or structures; (b) Double-tip image model of linear curving molecules or structures. The black and gray wide lines represented the real and ghost images of the linear curving molecule, respectively; (c) Double-tip image of the spherical cell; (d) Double-tip image of multiple fragments of linear molecules. The black and pink lines meant the real and ghost images of fragments of linear molecules or cell structures, respectively. Indicated were two critical points, in which c values evolve with changes in ω (the angle formed by an x line and a real fragment).

3.4. Mathematical Measurements and Formulas for Identifying Double-Tip Images in AFM Scanning of Cells or Biomolecules

The detailed dissecting of double-tip effects in our models should make it possible to measure and identify potential double-tip images. Our mathematic models described above suggest that potential double-tip images of biomolecules share a same x but different c value. It is likely that measurements of x and c values could facilitate the recognition and identification of double-tip images of cell structures and biomolecules during AFM scanning. To prove this concept, we measured a, c, Φ and x in the defined double-tip images as shown in Figure 2. Since spherical and linear molecules or structures shared an invariant x or Φ in the same double-tip topography, we first measured the x or Φ values of ghost spheres 1 and 2 at Figure 2(d), and then made use of the Φ to measure the corresponding x, and c values for linear double-tip images 3 through 8 using the AFM linemeasuring software. The measured x values for the spherical double-tip images 1 and 2 at Figure 2(d) were very close to the c values measured by the cross-section line that vertically connects the real and ghost spheres (Figure 5(a)), confirming the scenario x = c in the double-tip image of spheres (Figure 3(a)). The measurements of linear images using the Φ derived from spherical images indicated that all the linear double-tip images 3 through 8 had very similar x values but had different c values (Table 1 and Figure 5(a)). Similarly, when cell fibril ghosts were measured for c and x values at 23 different sites of double-tip images in the Figure 2(a), we found these double-tip images had the similar x but different c values (Figure 5(b)). These results suggest that mathematic measuring of x and c values is useful for recognition and identification of double-tip images at AFM scanning and imaging.

Given that the measuring of x relies on Φ, and that it is quite challenging to precisely measure the Φ if no spheres with obvious double-tip images are present in the same topography, it would be important to develop an objective and non-bias methodology to identify doubletip images. For this purpose, we created formulas to calculate and identify a potential double-tip image based on four mathematic models as described above. We then tested these formulas for the capacity to characterize and identify double-tip images. These mathematic formulas allow one to calculate x values for individual images of spheres/particles and linear molecules or their fragments (Figure 7(a)). While c and a values can be measured using the AFM line-measuring software, a Φ value can be calculated based on the formula 1. Then, the known Φ value allows one to calculate an x value using the formula 2 (Figure 7(a)). Since calculations of Φ and x using the formulas require any two random sites of potential double-tip images, three different modifications of the equations are further developed based on the each image's position relative to the angle ω (Figures 7(b)-(d), and Figures 4(b) and (c)). Using these equations, suspected double-tip images of cell structures or molecules at multiple sites can be measured for a and c values, and calculated for Φ and x values. Double-tip images can then be determined based on the invariant x and different c values for the multiple sites in a same topography.

Table 1. Mathematic measurements indicated that doubletip images of DNA molecules had similar x values but different c values.

Parameters #DNA Fragments	a (nm)	c (nm)	Φ	x (nm)
1*			70°	43.25
2			70°	42.95
3**	70.31	41.51	70°	43.22
4	35.16	31.51	70°	43.56
5	60.59	40.50	70°	43.66
6	46.88	36.92	70°	44.84
7	70.31	30.31	70°	43.28
8	46.88	26.26	70°	44.75
Mean	55.02 ± 14.32	34.50 ± 6.09	70°	43.89 ± 0.72

#DNA fragments 1 through 8 corresponded to those DNA molecules marked 1 - 8 in **Figure 3(b)**; *Fragments 1 and 2 were hydrolyzed DNA spheres as marked in the **Figure 3(b)**. For spherical molecules, the x equals to the c (see the legends for **Figures 4(a)** and **6(a)**); **The Φ value detected in spherical DNA fragments 1 and 2 was used to measure x values for the linear DNA fragments 3 - 8, since in the same double-tip topography spherical and linear molecules had a same Φ and x.

To test the authenticity and utility of these mathematic formulas for identifying potential double-tip images of spherical and linear molecules or their fragments, we took advantage of defined double-tip images of spherical structures or particle and linear molecules in the Figure 3(b) to measure a and c, and then calculate Φ and x at random two-site combination of double-tip images of DNA spheres and linear DNA molecules (Figure 3(b)). The results indicate that the x values are almost the same, whereas c values are completely different for the any two sites of linear DNA molecules or fragments (Table 2). The Φ and x values calculated from the formulas were indeed quite similar to those measured Φ and x values shown in Table 1. The calculation of Φ and x in 6 randomly-selected sites of cell fibril ghosts also yielded similar Φ and x values as measured in Figure 6(b) (Table 3). The data showing the "same" x but different c values are consistent with the observation of the defined double-tip images of spheres and linear molecules or structures as shown in Figures 2 and 3, suggesting that our mathematic formulas can be used to identify any pairs of real/ghost spheres and linear molecules.

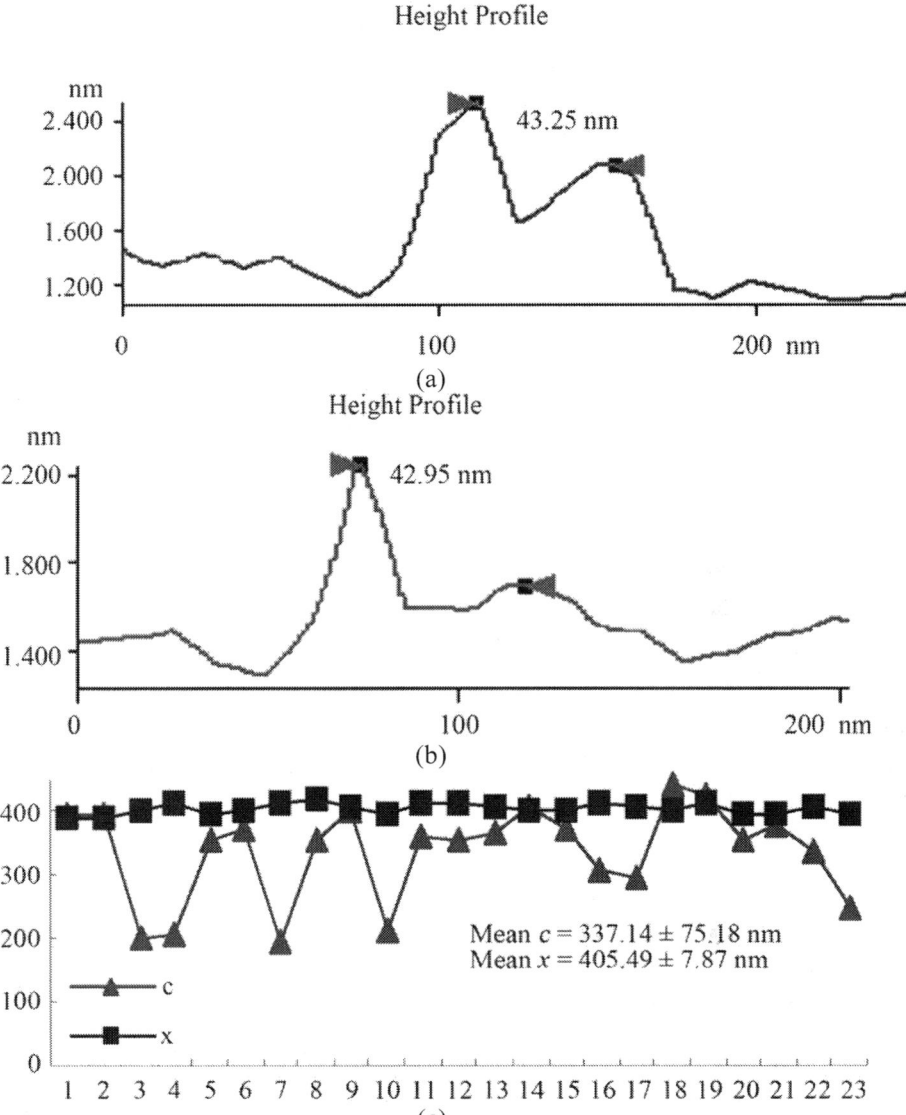

Figure 6. Mathematic measurements indicated that doubletip images of spheres and linear molecules or structures had similar x values but different c values. (a) The measured x values were similar to the c values, which were determined based on the measurements through height profile analyses of the cross section of both the real DNA sphere and ghost sphere 1 (left) and 2 (right) as marked in Figure 3(b) (see Table 1). This is consistent with what is described (x = c for double-tip spheres, Figure 4(a)); (b) Mathematic measurements of 23 sites of double-tip images of cell fibrils showed similar x values but different c values. The Φ derived from double-tip spherical structures were used to measure x and c for 23 different sites of double-tip images of cell fibrils in Figure 2(a).

$$\begin{cases} \tan \Phi = \dfrac{\pm c_1 \cdot c_2 \cdot (a_2 \pm a_1)}{a_1 \cdot c_1 \cdot \sqrt{a_2^2 - c_2^2} - a_2 \cdot c_2 \cdot \sqrt{a_1^2 - c_1^2}} \cdots\cdots (1) \\[3mm] x = c_1 \Big/ \cos\left[\Phi - \arccos\left(\pm\dfrac{c_1}{a_1}\right)\right] \cdots\cdots (2) \end{cases}$$

(a)

$$\begin{cases} \tan \Phi = \dfrac{c_1 \cdot c_2 \cdot (a_2 - a_1)}{a_1 \cdot c_1 \cdot \sqrt{a_2^2 - c_2^2} - a_2 \cdot c_2 \cdot \sqrt{a_1^2 - c_1^2}} \\[3mm] x = c_1 \Big/ \cos\left(\Phi - \arccos\dfrac{c_1}{a_1}\right) \end{cases}$$

(b)

$$\begin{cases} \tan \Phi = \dfrac{c_1 \cdot c_2 \cdot (a_1 - a_2)}{a_1 \cdot c_1 \cdot \sqrt{a_2^2 - c_2^2} - a_2 \cdot c_2 \cdot \sqrt{a_1^2 - c_1^2}} \\[3mm] x = c_1 \Big/ \cos\left[\Phi - \arccos\left(-\dfrac{c_1}{a_1}\right)\right] \end{cases}$$

(c)

$$\begin{cases} \tan \Phi = \dfrac{c_1 \cdot c_2 \cdot (a_1 + a_2)}{a_1 \cdot c_1 \cdot \sqrt{a_2^2 - c_2^2} - a_2 \cdot c_2 \cdot \sqrt{a_1^2 - c_1^2}} \\[3mm] x = c_1 \Big/ \cos\left(\Phi - \arccos\dfrac{c_1}{a_1}\right) \end{cases}$$

(d)

Figure 7. Mathematic formulas can be used to identify double-tip images of spherical and linear molecules or structures. (a) Basic formulas for calculating Φ and x values for suspected double-tip images; (b) Modified formula 1. This equation group was used to calculate Φ and x values when the random two sites of images for calculation were both orientated in the positions ($0° < \omega < 90°$), as shown at Figure 4(b); (c) Modified formula 2. This equation group was used to calculate Φ and x values when the random two sites of images for calculation were both orientated in the positions ($90° < \omega < 180°$) as shown at Figure 4(c); (d) Modified formula 3. This equation group was used to calculate Φ and x values when one of the random two sites of images for calculation was orientated in the positions ($0° < \omega < 90°$) as shown at Figure 4(b) and the other was in the positions ($90° < \omega < 180°$) as shown in Figure 4(c). The ω is an angle formed by the x line and the imaging line of a real linear molecule or fragment (Figure 4).

Thus, the direct measurement and mathematic formulation provide a useful methodology to evaluate double-tip images. Our results can serve as a foundation to design computer-based automatic detection of double-tip/probe AFM images during nanoscale measuring and imaging of cell structures and biomolecules. However, for practical purposes of this methodology, it is necessary to recruit some specialists in other fields like computer science to figure out a practically implementable form for AFM instruments.

4. ACKNOWLEDGMENTS

This work was supported by the National Natural Science Foundation of China (No. 30900340), the Open Project Program of State Key Laboratory of Food Science and Technology, Nanchang University (No. SKLF-KF-2010011), the Natural Science Foundation of Jiangxi Province (No. 2010GZN0138), the Scientific Research Foundation for Returned Overseas Chinese Scholar of State Education Ministry, and the Scientific Research Fund of Jiangxi Provincial Education Depart

Table 2. Mathematic formulas (Figure 7) made it possible to calculate Φ and x values at random two sites of imaged molecules and to identify double-tip images of linear molecules as defined in Figure 3(b) (Mean $\Phi = 70.17 \pm 2.23$; Mean x = 43.18 ± 1.16).

Φ \ x^*	3^*	4	5	6	7	8
3		43.08	42.27	42.98	43.24	43.50
4	69.33		43.74	43.43	43.42	44.05
5	64.69	70.26		43.94	43.55	43.91
6	68.83	69.83	70.88		43.43	43.90
7	70.05	69.81	69.64	69.81		39.32
8	71.20	70.66	70.79	70.80	75.97	

*The individual sites 3 through 8 of double-tip images corresponded to those linear DNA molecules marked in **Figure 3(b)**. The sites 1 and 2 were double-tip images of spherical DNA fragments and, in this case, the c equals to the x (see **Figure 4(a)**, **Figure 7(a)** and **Table 1**); *Formulas in Figure 7 (b) were used to calculate x and Φ for any two random sites of imaged fragments 3 - 6 as marked in **Figure 3(b)**. The formulas in **Figure 7(c)** were used to calculate x and Φ for the two random sites of fragments 7 and 8. The formulas in **Figure 7(d)** were used to calculate x and Φ for one of fragments 3 - 6 and one of the fragments 7 - 8 (see **Figures 7(a)-(d)** for details); Statistical analyses indicated that Φ and x values measured at **Table 1** significantly resemble those Φ and x values calculated in **Table 2** by mathematic formulas.

Table 3. Mathematic formulas (Figure 7) made it possible to calculate Φ and x values at random two sites of imaged cell structures and to identify double-tip images of cell branches as defined in Figure 2(a) (Mean Φ = 124.68 ± 0.15; Mean x = 404.18 ± 12.93).

Φ \ x	1*	2	3	4	5	6
1		386.81	411.67	389.35	392.14	396.14
2	124.49		392.85	412.73	410.35	404.77
3	124.78	124.81		402.46	422.97	421.34
4	124.55	124.57	125.08		415.16	418.31
5	124.60	124.58	124.70	124.58		385.65
6	124.65	124.63	124.71	124.61	124.80	

*The formulas in **Figure 7(c)** were chosen to calculate Φ and x for any two random sites of six double-tip images of cell fibrils in **Figure 2(a)**; Statistical analyses indicated that Φ and x values measured at **Figure 6(b)** significantly resemble those Φ and x values calculated in **Table 3** by mathematic formulas.

REFERENCES

1. R. F. Service, "Nanotechnology: Biology Offers Nanotechs a Helping Hand," Science, Vol. 298, No. 5602, 2002, pp. 2322-2323.
2. S. W. Schneider, K. C. Sritharan, J. P. Geibel, et al., "Surface Dynamics in Living Acinar Cells Imaged by Atomic Force Microscopy: Identification of Plasma Membrane Structures Involved in Exocytosis," Proceedings of the National Academy of Sciences of USA, Vol. 94, No. 1, 1997, pp. 316-321.
3. D. Fotiadis, Y. Liang, S. Filipek, et al., "Atomic-Force Microscopy: Rhodopsin Dimers in Native Disc Membranes," Nature, Vol. 421, No. 6919, 2003, pp. 127-128.
4. J. L. Alonso and W. H. Goldmann, "Feeling the Forces: Atomic Force Microscopy in Cell Biology," Life Sciences, Vol. 72, No. 23, 2003, pp. 2553-2560.
5. J. A. Dvorak, "The Application of Atomic Force Microscopy to the Study of Living Vertebrate Cells in Culture," Methods, Vol. 29, No. 1, 2003, pp. 86-96.

6. Y. Chen, J. Cai, "Membrane Deformation of Unfixed Erythrocytes in Air with Time Lapse Investigated by Tapping Mode Atomic Force Microscopy," Micron, Vol. 37, No. 4, 2006, pp. 339-346.

7. Y. Chen, J. Cai, C. Wang, et al., "Atomic Force Microscopy Imaging and 3-D Reconstructions of Serial Thin Sections of a Single Cell and Its Interior Structures," Ultramicroscopy, Vol. 103, No. 3, 2005, pp. 173-182. L. F. Jimenez-Garcia and R. Fragoso-Soriano, "Atomic Force Microscopy of the Cell Nucleus," Journal of Structural Biology, Vol. 129, No. 2-3, 2000, pp. 218-222.

8. J. G. Forbes and G. H. Lorimer, "Structural Biology: Unraveling a Membrane Protein," Science, Vol. 288, No. 5463, 2000, pp. 63-64. G. H. Thomas, "New Routes to Membrane Protein Structures. Practical Course: Current Methods in Membrane Protein Research," EMBO Report, Vol. 2, No. 3, 2001, pp. 187-191.

9. T. Osada, A. Itoh and A. Ikai, "Mapping of the ReceptorAssociated Protein (RAP) Binding Proteins on Living Fibroblast Cells Using an Atomic Force Microscope," Ultramicroscopy, Vol. 97, No. 1-4, 2003, pp. 353-357.

10. Y. Yang, H. Wang and D. A. Erie, "Quantitative Characterization of Biomolecular Assemblies and Interactions Using Atomic Force Microscopy," Methods, Vol. 29, No. 2, 2003, pp. 175-187.

11. Y. Chen, J. Cai, Q. Xu and Z. W. Chen, "Atomic Force Bio-Analytics of Polymerization and Aggregation of Phycoerythrin-Conjugated Immunoglobulin G Molecules," Molecular Immunology, Vol. 41, No. 12, 2004, pp. 1247- 1252.

12. P. E. Marszalek, H. Li and J. M. Fernandez, "Fingerprinting Polysaccharides with Single-Molecule Atomic Force Microscopy," Nature Biotechnology, Vol. 19, No. 3, 2001, pp. 258-262.

13. J. Tamayo and M. Miles, "Scanning Probe Microscopy for Chromosomal Research," Archives of Histology and Cytology, Vol. 65, No. 5, 2002, pp. 369-376.

14. Y. Chen, L. Shao, Z. Ali, J. Cai and Z. W. Chen, "NSOM/ QD-Based Nanoscale Immunofluorescence Imaging of Antigen-Specific T-Cell Receptor Responses during an in Vivo Clonal Vg2Vd2 T-Cell Expansion," Blood, Vol. 111, No. 8, 2008, pp. 4220-4232.

15. Y. Chen, J. Qin and Z. W. Chen, "Fluorescence-Topographic NSOM Directly Visualizes Peak-Valley Polarities of GM1/GM3 Rafts in Cell Membrane Fluctuations," The Journal of Lipid Research, Vol. 49, No. 10, 2008, pp. 2268-2275.

16. Y. Chen, J. Qin, J. Cai and Z. W. Chen, "Cold Induces Microand Nano-Scale Reorganization of Lipid Raft Markers at Mounds of Cell-Membrane Fluctuations," PLoS One, Vol. 4, No. 4, 2009, p. e5386.

17. S. Bunk, "Better Microscopes Will Be Instrumental in Nanotechnology Development," Nature, Vol. 410, No. 6824, 2001, pp. 127-129.

18. K. Keren, M. Krueger, R. Gilad, et al., "Sequence-Specific Molecular Lithography on Single DNA Molecules," Science, Vol. 297, No. 5578, 2002, pp. 72-75.

19. J. K. Horber and M. J. Miles, "Scanning Probe Evolution in Biology," Science, Vol. 302, No. 5647, 2003, pp. 1002- 1005.

20. K. L. Westra, A. W. Mitchell and D. J. Thomson, "Tip Artifacts in Atomic-Force Microscope Imaging of ThinFilm Surfaces," Journal of Applied Physics, Vol. 74, No. 5, 1993, pp. 3608-3610.

21. D. Nyyssonen, L. Landstein and E. Coombs, "2-Dimensional Atomic Force Microprobe Trench Metrology System," Journal of Vacuum Science & Technology B, Vol. 9, No. 6, 1991, pp. 3612-3616.

22. D. J. Keller and C. C. Chou, "Imaging Steep, High Structures by Scanning Force Microscopy with Electron-Beam Deposited Tips," Surface Science, Vol. 358, No. 1-3, 1992, pp. 333-339.

23. P. Grutter, W. Zimmermannedling and D. Brodbeck, "Tip Artifacts of Microfabricated Force Sensors for Atomic Force Microscopy," Applied Physics Letters, Vol. 60, No. 22, 1992, pp. 2741-2743.

24. S. N. Magonov, A. Y. Gorenberg and H. J. Cantow, "Atomic Force Microscopy on Polymers and Polymer RelatedCompounds," Polymer Bulletin, Vol. 28, No. 5, 1992, pp. 577-584.

25. U. D. Schwarz, H. Haefke, P. Reimann and H. J. Guntherodt, "Tip Artifacts in Scanning Force Microscopy," Journal of Microscopy, Vol. 173, No. 3, 1994, pp. 183- 197.

26. J. S. Villarrubia, "Morphological Estimation of Tip Geometry for Scanned Probe Microscopy," Surface Science, Vol. 321, No. 3, 1994, pp. 287-300.

27. J. S. Villarrubia, "Algorithms for Scanned Probe Microscope Image Simulation, Surface Reconstruction, and Tip Estimation," Journal of Research of the National Institute of Standards and Technology, Vol. 102, No. 4, 1997, pp. 425-454.

28. Y. Chen, J. Cai, M. Liu, et al., "Research on DoubleProbe, Doubleand Triple-Tip Effects during Atomic Force Microscopy Scanning," Scanning, Vol. 26, No. 4, 2004, pp. 155-161.

29. N. C. Santos, E. Ter-Ovanesyan, J. A. Zasadzinski and M. A. Castanho, "Reconstitution of Phospholipid Bilayer by an Atomic Force Microscope Tip," Biophysical Journal, Vol. 75, No. 4, 1998, pp. 2119-2120.

30. N. H. Thomson, B. L. Smith, N. Almqvist, et al., "Oriented, Active Escherichia coli RNA Polymerase: An Atomic Force Microscope Study," Biophysical Journal, Vol. 76, No. 2, 1999, pp. 1024-1033.

31. F. J. Giessibl, S. Hembacher, H. Bielefeldt and J. Mannhart, "Subatomic Features on the Silicon (111)-(7 × 7) Surface Observed by Atomic Force Microscopy," Science, Vol. 289, No. 5478, 2000, pp. 422-426.

32. J. Jass, T. Tjarnhage and G. Puu, "From Liposomes to Supported, Planar Bilayer Structures on Hydrophilic and Hydrophobic Surfaces: An Atomic Force Microscopy Study," Biophysical Journal, Vol. 79, No. 6, 2000, pp. 3153- 3163.

CHAPTER 9

Ultrafast Electron Microscopy for Chemistry, Biology and Material Science

Sergei A. Aseyev[1], Peter M. Weber[2], Anatoli A. Ischenko[3] *

[1]Institute of Spectroscopy of the Russian Academy of Sciences, Troitsk, Russia; [2]Department of Chemistry, Brown University, Providence, USA; [3]Moscow M. V. Lomonosov State University of Fine Chemical Technologies, Moscow, Russia.

ABSTRACT

For the past thirty years, intense efforts have been made to record atomic scale movies that reveal the movement of atoms in molecules, the fast dynamical processes in biological tissues and cells, and the changes in the structure of a solid confined to nano-scale volumes. A combination of sub-nanometer spatial resolution with picosecond or even femtosecond temporal resolution is required for such atomic movies. Additional important information can be obtained when the energy of the electron beam transmitted through the sample is measured. The four dimensional (4D) spatially and temporally resolved ultrafast electron microscopy method is made possible by the extremely high detection efficiency that is reached in 4D electron microscopy. Using ultra-short electron bunches for the visualization of biological tissue can also improve the spatial resolution compared to conventional electron microscopes, thereby enabling the study of complex biological samples of relevance to the life sciences. Of particular interest to a broad audience is the possibility to create a video, and in the future, a real atomic movie, using 4D electron tomography.

Keywords: Ultrafast Electron Microscopy; Dynamic Processes; Structural Dynamics; Atomic Movie; Femtosecond Temporal Resolution; Atomic Spatial Resolution; Electron Tomography; Spectral-Spatial-Temporal Resolution

1. INTRODUCTION

Since the 1980s, intense efforts have been made to record an atomic movie showing the movements of atoms within molecules, the fast dynamical processes in biological tissues and cells, and the changes in the structure of a solid in a nano-scale volume [1-3]. Such an atomic movie requires a combination of sub-nanometer spatial resolution with picosecond or even femtosecond temporal resolution.

The aim of any microscopy is to study the structure, composition, and a variety of physical and chemical characteristics of samples (usually solids) on a very small length scale, with linear dimensions of irregularities that are on the order of micrometers or nanometers. While in most microscopies, photon beams (in optical microscopy) or corpusclar beams (for example in electron microscopy) are used as the probes, tunneling microscopies use a very sharp tip that is scanned across surfaces. Electron microscopy and optical microscopy use light and a focused electron beam for imaging, respectively. Beyond those, other microscopic methods may use ions, protons, positrons, or neutrons, or acoustic or microwave radiation in addition to other, less common, methods [3-6]. In each of them, the specificity of the interaction of a beam of the particles (or photons) and the molecules or atoms of a sample yield unique and rather useful information about the structure, composition and microscopic inhomogeneities of the sample, and the nature of their intermolecular interactions [3,4,7]. For example, the slow neutrons in neutron microscopy and spectroscopy barely interact with the electrons, but interact strongly with the nuclei of the atoms.

There are two main approaches to achieve imaging in classical microscopy: 1) In transmission mode, a large area of the sample is illuminated by a beam of light or particles. Specifically designed lenses project the transmitted beam onto a screen or detector to form an image of the structure of the sample. 2) In reflection mode, the beam reflected from the sample, or secondary particles or photons generated by the incident beam in the sample, form an image of the sample surface or a thin region underneath the surface [8].

Because of absorption of the optical radiation by matter, transmission optical microscopy is limited to thin samples, typically less than 1 mm thick. In electron microscopy, the sample has to be much thinner, because the electrons interact with matter more strongly than light. In transmission electron microscopy it is therefore possible to get images only of very thin slices, with film thicknesses of much less than 1 micrometer. Such samples have to be prepared using special ultramicrotomes that can produce film thicknesses down to a few tens of nanometers.

There are two different ways to obtain an image [9,10]: 1) "all at once", when the sample is irradiated entirely with light or electrons and the transmitted or the reflected beam is detected at once using, for example, a film, photographic plate, photosensitive recorder or digital camera; 2) in "scanning microscopy" mode, where the sample is scanned by a beam focused to a spot of rather small diameter, and the resulting image is assembled point by point.

2. TRANSMISSION AND SCANNING ELECTRON MICROSCOPY

In conventional scanning microscopy, images are formed by a beam of charged particles or photons that is focused to a very small diameter. This probe is moved across the surface of an object, and signal of any physical nature, but in particular reflected particles or photons, is detected and the image obtained "point by point". The method is quite analogous to the generation of an image using a scanning electron beam in a traditional cathode ray tube (CRT), or in modulating the brightness of each luminous point in a LCD monitor [11]. Scanning microscopy is a rather convenient tool to study the morphology and topography of various geometric objects, to investigate elemental compositions, to measure the electric and magnetic fields in micro-volumes, (the so-called "micro-fields", as introduced by Prof. GV Spivak), or to test electrical parameters of different semiconductor crystals (e.g., lifetime and diffusion length of charge carriers).

In scanning electron microscopy, the scan of the electron beam should be synchronous with a sweep of the probe on a sample. Like in a TV image, the raster image consists of individual small dots or pixels, the brightness and make-up of which, formed after signal amplification, define the image. Thus, the elements of the image appear one after the other, as if we move our fingers to spell a complicated book of life, but not all at once, as in the usual optical or transmission electron microscopes. This "pointwise" map of the object greatly facilitates the processing and interpretation of images [12].

In a transmission electron microscope, the electrons illuminate and pass through a very thin sample. An image of the sample is formed by means of electrostatic or magnetic lenses on the fluorescent screen or equivalent position sensitive detector [13]. In 1931, the first such device was demonstrated by M. Knoll and E. Ruska, who imaged a wire mesh using an incident electron beam and axially symmetric magnetic electron lenses with a narrow annular gap and a sharp maximum of the magnetic field on the axis. The first commercial transmission electron microscope was developed and released by "Siemens" in 1939 under the supervision of E. Ruska, who received the Nobel Prize for Physics in 1986 for his invention of the electron microscope.

The transmission electron microscope (TEM) is the now the most common tool, designed to study very thin samples with thicknesses of about 10 - 100 nm at electron accelerating voltages of up to 200 kV [14]. Its scheme is similar to the optical system of an optical light microscope in that it creates an image of the sample based on the transmitted radiation beam. Figure 1 compares the main units of the transmission electron microscope to those of the optical microscope and the scanning electron microscope.

The transmission electron microscope consists of an electron gun and electromagnetic lenses. Figure 1 shows the path of the rays in a microscope with three levels of zoom and one condenser lens, which is used to pre-focus of the electron beam. The two common modes of its operation are the imaging mode and the recording of micro-diffraction patterns. In imaging mode, the plane of a sample conjugates to the screen, where it is assumed that the sample is very thin compared to the focal length of the imaging objective lens. In the diffraction mode, the object of observation conjugates to the back focal plane of the objective lens, where the distribution of the amplitudes of the electron beam, diffracted off the sample, corresponds to the Fraunhofer's diffraction of the electrons. The resulting picture is observed on a remote screen. It is possible to determine the composition of the sample by analyzing the diffraction pattern [16].

In the phrase "micro-diffraction", the word "micro" implies that the diffraction signal is observed only from a rather small area of the sample. The area is chosen either by a special selector diaphragm, installed in the plane of the first intermediate image (see Figure 1), or by irradiation of the desired part of the sample with a sharplyfocused electron beam with a diameter of about 1 micrometer or less.

In imaging mode, the so-called aperture diaphragm is installed in the focal plane of the object lens in order to limit the aperture of the beam, i.e. the opening angle of the cone of rays emerging from the sample. This helps to decrease the spherical aberration, a basic imaging error associated with the projection of a point in the sample to a point of the image.

Modern microscopes are usually equipped with a twolens condenser, which allows for a rather high electron flux to a small area of the object with characteristic size of 1 - 5 microns. This eliminates the growth of the films due to hydrocarbon polymerization of oil vapors as a result of electron bombardment of the unobserved sample areas in microscope where the column is pumped by diffusion pumps.

According to statistical data, tens of thousands of transmission microscopes with accelerating voltages of 80 - 100 kV have been produced. Hundreds of devices with an accelerating voltage of between 200 kV and 500 kV were created for the study of thicker objects, with a thickness of up to 10 microns. And there are only a few dozen microscopes intended for the voltage range from 1 MV to 3.5 MV.

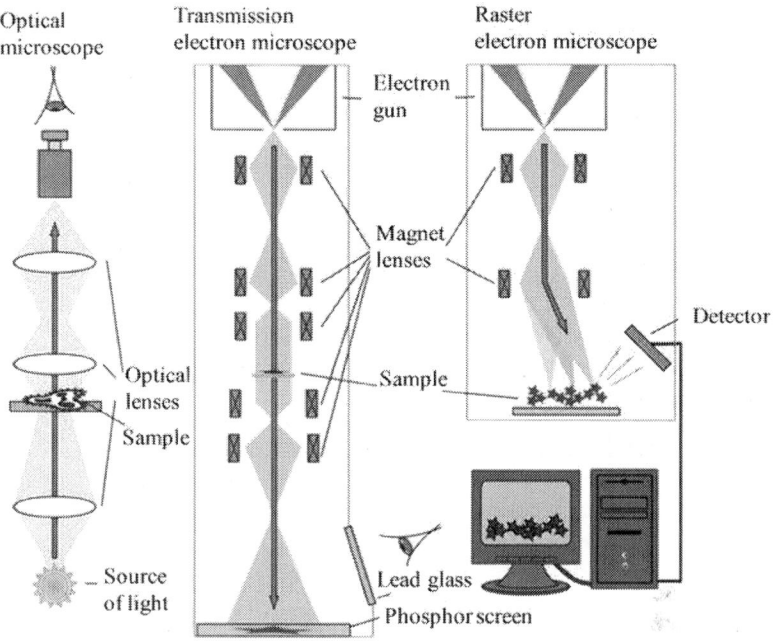

Figure1. Schematic of the optical, transmission and scanning electron microscopes [15].

For a typical transmission electron microscope with an accelerating voltage of 100 kV, the de Broglie wavelength of the electron is 3.7 pm. This is much smaller than the diameter of a hydrogen atom, which equals approximately 100 pm. This would seem to make it easy to observe any atomic particle, as long as the resolution of the electron microscope is on the order of the electron wavelength as is the case in optical microscopes. But in fact, it is extremely difficult to observe individual atoms because of the poor quality of the electron lenses compared to the light-optical lenses: at the same angular size of the light and electron beams, the electron lenses create almost of two orders of magnitude more aberrations and image distortion [13].

At a kinetic energy of 100 eV, the de Broglie wavelength of the electron is equal to 0.1 nm, i.e. on the order of atomic dimensions. Electrons with such energies are strongly diffracted by atomic lattices. This property is used in low-energy electron diffraction to investigate the structure and composition of surfaces and near-surface layers, because the penetration depth of the electrons is very small.

3. STROBOSCOPIC ELECTRON MICROSCOPY

Electron microscopy takes advantage of two of the three main properties of the electron: the small de-Broglie wavelength of the electron, which enables very high spatial resolution; and the electric charge, which allows to control its movement by electric and magnetic fields. About 60 years ago, the third property of the electron, its small mass (i.e., low inertia) began to be used in socalled stroboscopic electron microscopy for the study of periodic processes [17]. The essence of the method is quite simple: a sample with periodically time-varying characteristics, with a fixed rate of change, should be irradiated by short bunches of electrons with the same frequency and fixed phase. The synchronicity freezes the image, which corresponds to the state of the sample at the time of arrival of the electron pulses.

In the early 60's, in the Soviet Union as a result of the work of Prof. G.V. Spivak, and in other countries, stroboscopic techniques were developed for all major types of electron microscopes: translucent, raster, emission and mirror types [18,19]. These developments allowed the exploration of processes in rapidly varying thin films (transmission electron microscopy) and in the surface layers of solids (raster, emission and mirror microscope): re-polarization of various ferroelectrics; distinctions of the domain structure during re-magnetization of thin magnetic films; heterogeneity of alternating fields of magnetic heads; and local defects of p-n junctions during their fast switching between the locked and open states [20]. Additionally, from a theoretical point of view, it is possible to study the microstructure of solution-gels in periodic reactions such as the Belousov-Zhabotinsky reaction using electrons with low kinetic energies, less than 100 eV, and to investigate the periodic sorptiondesorption processes of catalysis in a similar fashion [20].

The initial work on stroboscopic microscopy was performed with microand nanosecond time resolution. Later, several laboratories started to investigate periodic processes with picosecond temporal resolution, usually using stroboscopic scanning electron microscopy. The main difficulties arose from the gating, the interruption, or the modulation of the electron beam. Generally, one uses one of two methods: first, it is possible to change of the intensity of the beam by locking or unlocking it with appropriate voltage pulses applied across the cathode or across the modulator in the electron gun of the microscope; or secondly, one can deflect a continuous electron beam by an electric or magnetic field in the vicinity of a skipping aperture of small diameter. The blanked, locked or unlocked electron pulses should be properly frequency and phase synchronized with the signals on the sample in order to avoid a loss of resolution.

The larger the duty cycle of the blanking pulses, i.e. the higher the ratio of the repetition period to the duration of the front of the electron pulse, the better the temporal resolution of the stroboscopic microscope, but the lower the brightness of the image. Consequently, each application requires a careful assessment of the best trade-offs. Moreover, a very short front of the pulse causes the resolution of the microscope to deteriorate because of non-ideal edges of the pulse, the appearance of chromatic aberration in transmission electron microscopy and the manifestation of the Bursch effect, which describes the transition of the longitudinal component of the electron velocity to a transverse component due to Coulomb repulsion.

4. DYNAMIC TRANSMISSION ELECTRON MICROSCOPY

Transmission electron microscopy, with its rather wide range of its tools, has long been a powerful technique in many fields of science and technology. While allowing to reach a spatial resolution on the sub-nanometer scale, conventional transmission electron microscopy does not permit an ultrashort temporal resolution, i.e. a time resolution in the picosecond or femtosecond regime. In contrast, optical microscopy using fluorescent probes, such as green fluorescent proteins, made it possible to visualize the processes occurring in vitro [2]. However, despite the high temporal resolution of optical methods, their spatial resolution is usually limited by the wavelength to the range of 200 - 800 nm. A review article [21] and a recently published monograph [2] describe the development of the methods of 4D, dynamic transmission electron microscopy (DTEM) and ultrafast electron microscopy (UEM), respectively.

The images and diffraction patterns in the work [22] were obtained at an accelerating voltage of 120 keV for non-organic materials (single crystals of gold, amorphous carbon and polycrystalline aluminum) and for biological cells of the intestine of rats [23]. In those studies, the gated beam contains on the average only one electron per pulse, with the total dose being limited to a few electrons per $Å^2$. Note, that these electron pulses were completely controlled in space and in time. Conceptually, this works is based on the methods of time resolved electron diffraction (TRED) or ultrafast electron diffraction (UED) [1,24-26,78,112], ultrafast electron crystallography (UEC) [23] and the other DTEM studies [21], but with one important difference-namely, the implementation of synchronized pulses of single electrons to form an image. The sketch of this 4D DTEM is shown in Figure 2.

This single-electron approach differs quite remarkably from the one used in the work [27], which used single pulses with -10^8 electrons and pulse durations of -20 ns.

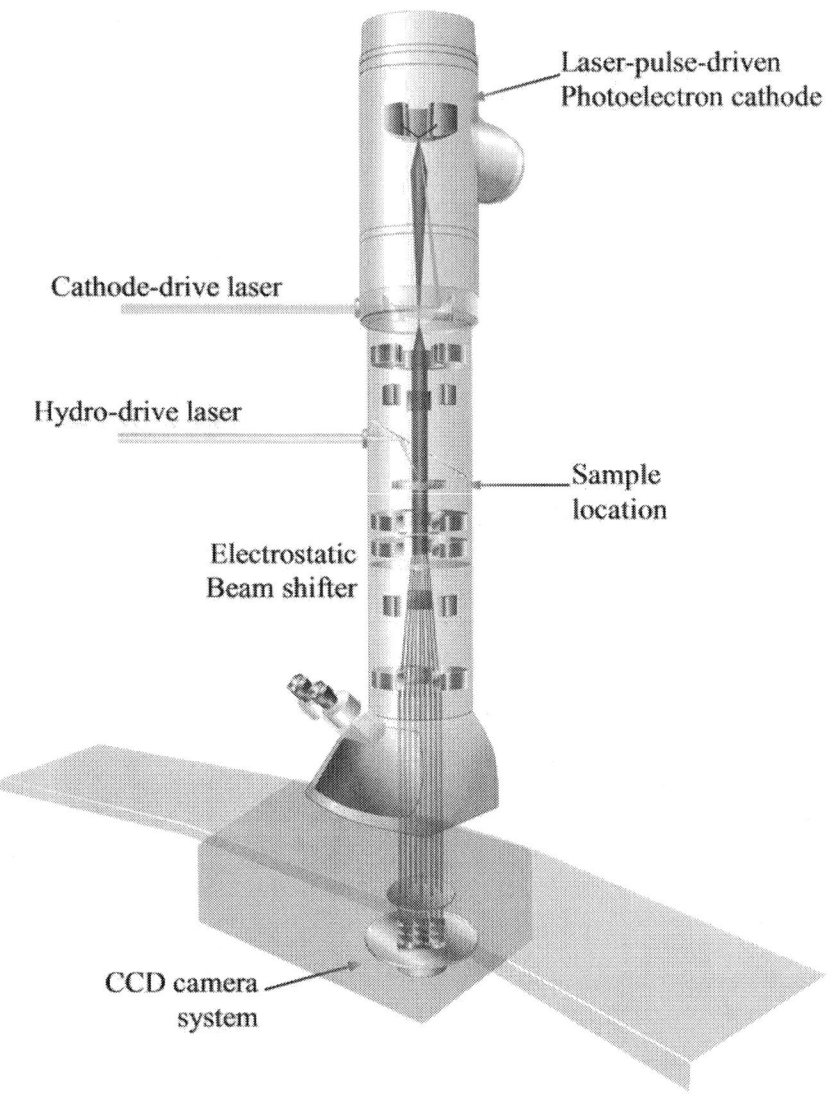

Figure 2. The schematic of a dynamic transmission electron microscope (DTEM) [21].

Those pulses, used to study laser-induced melting of different metals, contain so many electrons that high temporal resolution is not possible. Moreover, as noted in [28], the uncertainty in the spatial resolution in these experiments due to statistical interferences was of the order of micrometers, because the temporal window for imaging was on a nanosecond scale. It was found, that the use of amplified femtosecond light pulses, with much higher intensity of about 10^{12} W/cm^2, led, at first, to the formation of an electron beam that was insensitive to the focusing system of the microscope, and, secondly, to a significant lengthening of the pulse, which was due to electron-electron Coulomb repulsion.

The single-electron per pulse approach of 4D UEM [22,29] was used to obtain images from polycrystalline and single crystals [2]. To verify the authenticity of the experiment, some images were collected when the femtosecond pump laser pulses were blocked. In this case, the pattern was not observed, confirming that the time resolved signal generated in the electron microscope was indeed induced optically, and that the contribution of thermal electrons that are continuously generated by the cathode can be neglected [29]. It is useful to remember that electron microscopes can operate in UEM and DTEM mode. To reach atomic scale spatial resolution, the UEM diffraction regime should be used by introducing an intermediate lens to select the back focal plane of the lens as the object.

5. CONTROLLING THE MOTIONS OF FREE ELECTRONS BY FEMTOSECOND LIGHT PULSES

Femtosecond lasers play a key role in 4D electron microscopy. Ultrashort pulses are necessary for the optical excitation of ultrafast processes in the sample and to form the ultrashort probe electron pulses. Additionally, pulsed laser radiation makes it possible to measure the temporal characteristics of the photoelectron bunches, to implement additional compression of electron pulses, and to generate electron bunches with durations that may potentially reach into the attosecond regime.

The last statements need to be amplified on, because they imply that is possible to control free electrons with a laser beam. At first glance, this statement conflicts with the well-known fact that free electrons cannot absorb the electromagnetic (EM) radiation because the laws of conservation of energy and momentum cannot simultaneously be satisfied. Nevertheless, free electrons can scatter EM radiation. From a quantum mechanical point of view, this is the result of stimulated Compton scattering in a strong laser field.

The possibility to control the translational degrees of freedom of charged particles, in particular of free electrons, by a spatially inhomogeneous electromagnetic field was demonstrated theoretically by Gaponov and Miller in the mid 1950's [30]. This mechanism leads to the ejection of charged particles from the strong field. In the case of high-power laser radiation, the GaponovMiller, or ponderomotive (gradient) force (PF) [31], which is defined as the spatial gradient of the so-called ponderomotive potential U_p, can reach sufficiently high values to form a basis for the effective control of electron pulses in vacuum.

For non-relativistic electrons, the concept of the ponderomotive potential in a spatially inhomogeneous electromagnetic field $E = E_o(r)\sin(\omega t)$ can be introduced by averaging the Hamiltonian H over the fast oscillations with frequencies w and 2w:

$$H = \left\langle \left(p + eE_o(r)\sin(\omega t)/\omega \right)^2 \right\rangle / 2m_e$$
$$= p^2/2m_e + \left[eE_o(r) \right]^2 / \left(4m_e\omega^2 \right)$$
$$= p^2/2m_e + U_p$$

Here, m_e and e are the mass and the charge of the electron and $\langle \cdots \rangle$ denotes an averaging over optical cycles. As a result, the expression for the PF can be written as the following expression:

$$F^{(\text{pond})}(r,t) = -\left[e^2\lambda^2/64\pi^2 m_e\varepsilon_0 c^3 \right]\nabla I(r,t)$$

Here e_0 is the dielectric constant, c the speed of light and λ the wavelength of the laser radiation. It follows that for a tightly focused laser pulse with an intensity of $10^{15} \, W/cm^2$ in the center of a focal spot of 2 μm diameter (at the level of $1/e$) and λ=800nm, the force, $F^{(\text{pond})} \approx 10^{-11} N$, is approximately equal to the strength of the interaction between two electrons separated by a distance of 5 nm. In this example, the ponderomotive potential is $U_p \approx 60eV$.

It is evident that ultrashort laser radiation allows to control free electrons in vacuum on a femtosecond scale. This makes possible both the creation of ultrashort photoelectron bunches and the measurement of their duration [32,33]. The determination of the temporal characteristics of a pulsed electron beam is based on scanning the time delay between the laser radiation that forms the photoelectric bunch and the tightly focused laser beam that changes the velocity distribution of the photoelectrons as a result of the PF. Such a method compares favorably with a standard streak camera in its final characteristics and can be used for femtosecond electron beams [33].

In a streak camera, the electrons are deflected by a high-speed and high-voltage electric field with an amplitude of about 2 - 5 kV before reaching a positionsensitive detector. The rate of change of the high electric field determines the temporal resolution of the device. The currently best time resolution is in the subpicosecond range and reaches a value of −300 fs. In addition to this, it should be noted that the femtosecond temporal synchronization of the high-voltage electrical pulse in the streak camera and the femtosecond laser pulse that forms the ultrashort photoelectron bunch poses serious technical challenges.

The study of the behavior of the free electrons in spatially inhomogeneous electromagnetic fields began shortly after the appearance of the theory of Gaponov and Miller and the first experiments were devoted mainly to the possibility of creating traps using microwave technology [34]. Here, special attention was focused on the observation of the passage of the electrons with a certain kinetic energy through the ponderomotive potential, which allowed, for example, to determine the value of U_p [34,35]. The first demonstration of scattering of the photoelectrons by the ponderomotive potential, created by an intense sub-nanosecond laser pulse, was done in the work [36]. Here the multi-photon ionization of Xe, especially bleeding in the vacuum system, was used to prepare the pulsed photoelectron beam with kinetic energies less than 5 eV. This experiment demonstrated the control of low-energy photoelectrons by an optical ponderomotive potential with the height of about 10 eV.

Special attention should be given to the measurement of the value of the PF with which tightly-focused femtosecond laser radiation acts on the electrons propagating in vacuum. This may be useful for in situ spacetime diagnostics of the laser fields of high intensity, and can be used to validate the measurement of ultrashort electron pulse durations based on the irradiation of the electron bunches with the laser pulses of high intensity. Here it should be noted that the use of laser radiation with intensities of more than $10^{14} \ \text{W/cm}^2$ usually is accompanied by the photoionization of the residual gas in the vacuum system and thus by a possible deformation of the electron trajectories that pass through a cloud of charged particles.

As was noted above, electrons are ejected from their field-free paths by a strong electromagnetic field. This is illustrated in Figure 3, which shows two counterpropagating tightly-focused laser beams forming a standing wave from which the electron pulse is deflected. Such a scheme has a dual purpose. First, it allows to determine the duration of the initial electron bunch by scanning the timing of the laser pulses that create the standing wave, and using a position-sensitive scheme to detect the rejected, but in general, diffused, electron beam components. Secondly, the deflected electron pulses potentially have a shorter time duration compared to the incident electron pulses.

Two important remarks should be made here. The scheme in Figure 3 requires the spatial-temporal matching of two femtosecond laser pulses from two directions and a (sub) picosecond electron bunch from another di- rection. The use of a single laser beam for the electron scattering would, of course, be simpler. However, in the standing wave the spatial inhomogeneity of the electromagnetic field is about $\lambda/2$, a rather small value that is technically difficult to obtain with a single amplified femtosecond laser beam. For example, a parabolic mirror can usually focus 0.8-micron amplified femtosecond laser radiation to a spot diameter of $d_{1/e} \sim 6$ microns. Therefore, for a given laser pulse energy, laser wavelength and focusing conditions, the scattering of electrons in a pulsed standing wave will be more pronounced.

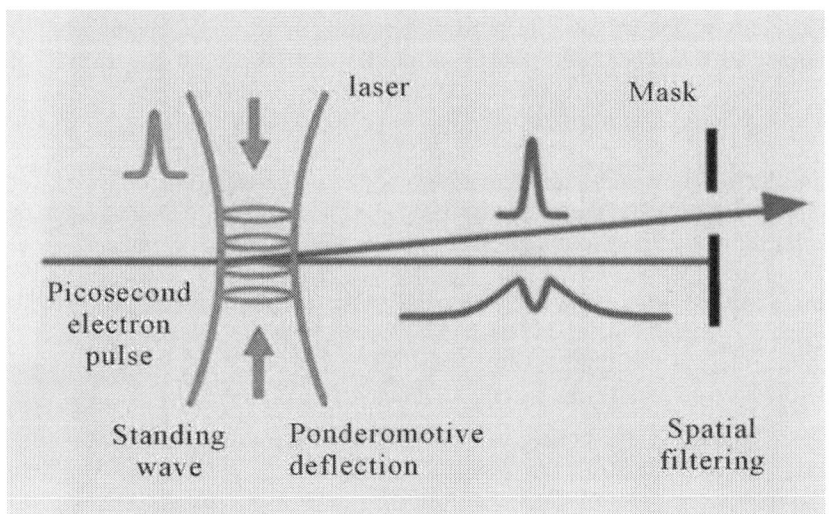

Figure 3. Deflection of an electron beam as a result of the optical ponderomotive force generated by an ultrashort laser pulse [2].

Another important consideration relates to the ultimate possibilities of this approach. This is important because it is rather interesting to know the accuracy of the electron pulse duration measurement as well as the duration of the deflected electron bunch. These characteristics are directly related to the transit time of the fast electrons through the laser focal spot, $\tau_{fin} \approx \tau_{tr} = d_{1/e}/v_e$ where v_e is the speed of the electron beam. Using $d_{1/e} = 6$ μm and $v_e \approx 10^8$ m/s for electrons with kinetic energy of 30 keV, we find $\tau_{tr} \approx 60$ fs. Note that the use of a standing wave formed by two laser beams (Figure 3) will not lead to a decrease in τ_{fin}, and hence the accuracy of measurement for this example will remain of the order of 100 fs. Therefore, to obtain a femtosecond electron bunch with $\tau \sim 10\text{-}50$ fs and its full diagnostics it is important to ensure the tight focusing of intense laser radiation, and/or to use electrons with higher kinetic energy. Of course, in order to deflect these ultrafast electrons it is necessary to implement intense laser fields.

An alternative way of creating ultrashort electron bunches also employs a standing light wave, but in a collinear geometry [2]. This approach is shown schematically in Figure 4.

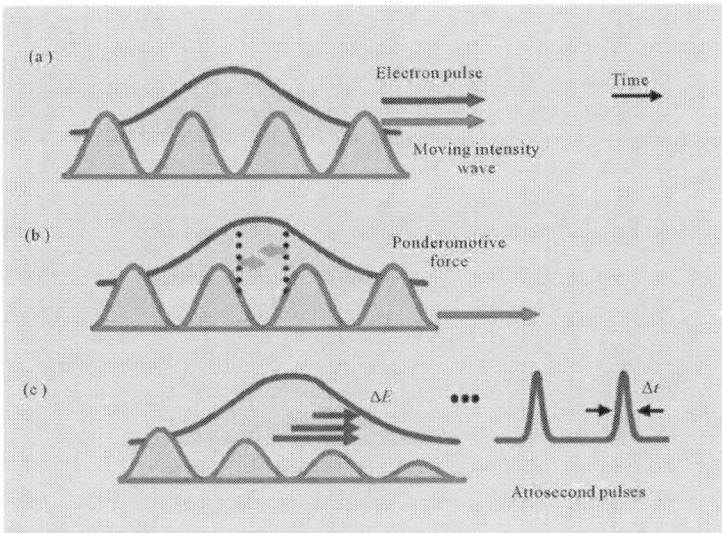

Figure 4. Schematic diagram of the creation of attosecond electron pulses. (a) An electron bunch with sub-picosecond duration is irradiated by two counter-propagating laser beams (forward and after the bunch) with different wavelengths. In the reference frame of the electrons, the wavelengths are the same as a result of the Doppler effect, and a standing light wave is formed; (b) Ponderomotive forces push electrons away from the strong laser field; (c) As a result, the compression leads to the formation of electron bunches with attosecond duration. Strictly speaking, the envelope duration of the initial electron bunch remains constant, but importantly, the attosecond spikes appear inside the envelope [2].

In this context it should be recalled that on the basis of the uncertainty principle, $\Delta E \Delta t \geq \hbar$, the existence of ultrashort, in this case attosecond, bursts needs a wide spectrum of energy DE$_e$. For example, a bunch with 100 as (10^{-16} s) duration has at least $\Delta E_e \approx 6eV$. The initial picosecond photoelectron pulse does not exhibit such spectral properties. As shown in Figure 4, the required broadening is due to the work of the ponderomotive forces [2]. Let us illustrate this with a simple estimate, by writing down the expression for the work in the traditional sense, where the energy gain of the electrons is equal to $\Delta E_e = F^{(pond)} \tau_p v_e$ (τ_p-the laser pulse duration), and the force is approximately defined as $F^{(pond)} \approx 2U_p / \lambda^*$ (λ^*-wavelength of the laser radiation in the reference frame of the fast electron). In the conditions of the numerical experiment [2] $v_e \approx 10^8$ m/s $\lambda^* = 370$nm, $U_p \approx 0.2eV$, and $\tau_p = 300$fs, one finds $\Delta E_e \approx 32$ eV. The calculated value DE$_e$ exceeds by 2 orders of magnitudes the spectral width of "normal" sub-picosecond electron pulse, emitted from the solid photocathode under the action of femtosecond laser radiation.

As considered in the monograph [2], this scheme is rather elegant. Even so, it is important to emphasize that the measurement of the duration of the formed attosecond electron bursts is a quite a challenging experimental problem. It is likely that it belongs to the class of experiments in which it is easier to create an ultrashort bunch than to measure it.

6. ULTRAFAST ELECTRON MICROSCOPY

An early antecedent of modern methods to investigate fast processes is the English photographer Eadweard Muybridge, who used multiple cameras to temporally resolve fast motions. For example, in 1871 Muybridge captured the moment when a galloping horse separated all four hooves from the surface of the earth (Figure 5).

Of interest to the scientific world today would be movies showing the movement of atoms in molecules, the fast dynamical processes in biological systems, or the changes in the geometrical structures of nanoscale devices. To estimate the magnitudes of time scales involved, we can extrapolate from the observation of macroscopic objects by using as the maximum velocity the speed of sound. Thus, a time resolution of 1 µs suffices to freeze in time motions of objects with spatial dimensions of 1 mm. In a system with a characteristic length scales of nanometers, a picosecond time resolution is appropriate. And, for molecular systems, femtosecond time resolution is necessary to capture the motions of individual atoms and functional groups.

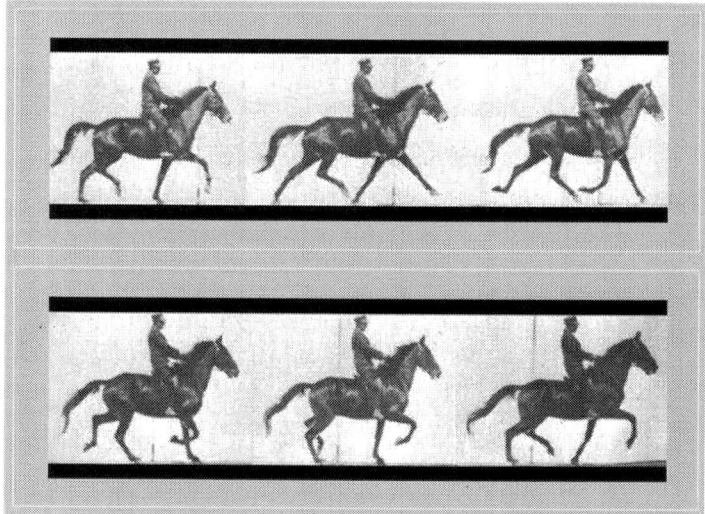

Figure 5. One of the first movies in history. These images were obtained by E. Muybridge in 1871. The time delay between the exposures was about 50 ms [37].

The ultrafast electron microscope built at the California Institute of Technology [2] is designed to explore a class of problems that occur on picosecond or subpicosecond time scales. A schematic illustration of this instrument is shown in Figure 6.

It is convenient to conceptualize the setup as consisting of three main sections: 1) the formation and acceleration of the sub-picosecond pulsed electron beam; 2) the interaction unit, in which the sample is excited by femtosecond laser pulses and its spatial-temporal structure probed by ultrashort electron bunches; and 3) the position-sensitive detection of the electrons after their interaction with the sample. The use of electron beams of subpicosecond duration imposes strict requirements on the detector, because the probe pulse cannot contain a large number of electrons. As it will be illustrated below, the main cause of the electron pulse temporal spreading is the space-charge interaction between the electrons within each pulse on account of their Coulomb repulsion. Therefore, the detection efficiency for electrons in this microscope should be as high as possible. Much preparatory work was done to achieve this [2]. As a result, the detection efficiency for the time-resolved instrument is almost an order of magnitude higher than that of a standard electron microscope. This also provides an interesting opportunity to significantly reduce the exposure (the total number of electrons passing through the sample to form an image on the detector), which is particularly important for imaging biological samples that are very sensitive to radiation induced damage.

The ultrashort electron bunch is the essence of 4D electron microscopy. Therefore, the question of its duration should be given special attention. The duration of the photoemission of electrons from the surface of a solid cathode and therefore the initial duration of the photoelectron bunch are determined by the duration of the laser pulse. But during the propagation from the photocathode to the target, the electron beam is stretched in time. Here we consider three major factors that lead to a smearing of the electron bunch.

First, the acceleration of photoelectrons in a static electric field near the cathode causes so-called time-offlight chromatic aberration for the electrons (TFCA). The temporal resolution limit is given by:

$\Delta\tau_{EF} = \sqrt{2m_e \Delta E_e}\,/eF$ where DE$_e$ is the initial spread of the kinetic energy of the electrons as they emerge from the cathode and F is the electric field in the acceleration gap.

The expression for TFCA can be obtained by solving a quadratic equation that describes the uniformly accelerated motion of the electrons in the acceleration gap of the length of l:

$$v_0 t + \left(eF/m_e\right)t^2/2 = l$$

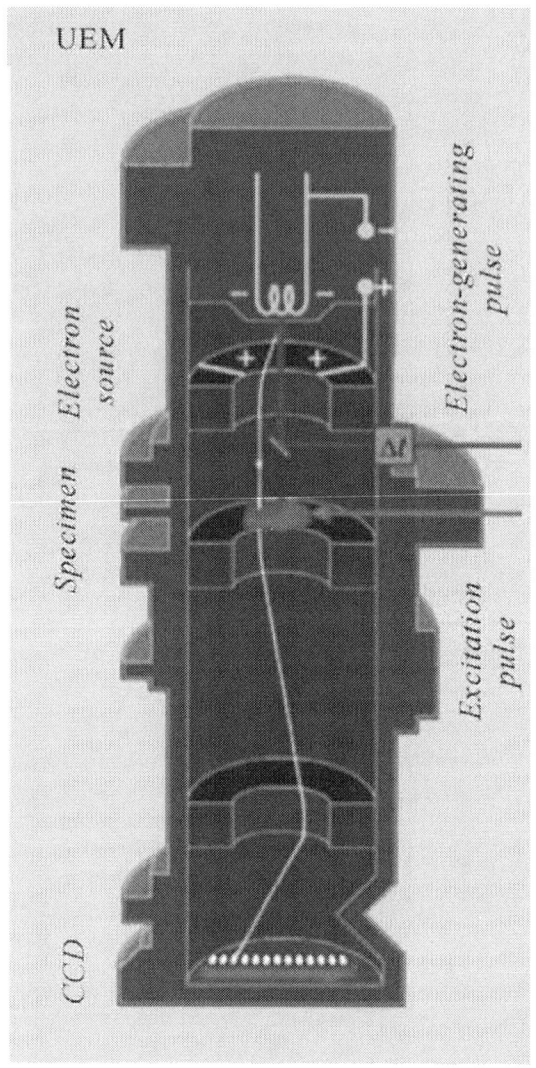

Figure 6. Schematic illustration of an ultrafast electron microscope. A commercial electron microscope, operating in a continuous mode, can be taken as a basis in order to create such a device. To adapt it to 4D microscopy requires a femtosecond laser, which provides a pulsed excitation of the sample (corresponding to a laser beam shown in red), and allowing for the preparation of the photoelectron probe pulse of ultrashort duration (ray marked in blue) [2].

Differentiating the both sides of the equation with respect to time t, we get:

$$\Delta v_0 + \left(eF/m_e \right) \Delta t \approx 0$$

from which immediately follows the expression for TFCA. Let us consider a numerical example. At DE$_e$ = 0.3 eV and $F \approx 10^8$ V/m, which approximately corresponds to the maximum value of the field (higher values lead to vacuum breakdown even with special electrodes that underwent thorough mechanical, electronic and ionic treatment), this contribution is $\Delta t_{EF} \approx 60$ fs.

Secondly, the movement of electrons through the field free flight tube with slightly different kinetic energies is accompanied by an elongation of the time of flight, which is expressed as

$$\Delta \tau_{TOF} = t_{TOF} \left(\Delta E_e / 2K \right)$$

Here, t$_{TOF}$ is the time of flight of the electrons with kinetic energy K from the photocathode to the sample. At $K = 30$ keV, $\Delta E_e = 0.3$ eV and $t_{TOF} \approx 3$ ns, corresponding to the length of 30 cm between the cathode and the target, we get $\Delta t_{TOF} \approx 15$ fs.

Thirdly, the Coulomb repulsion between the electrons within a pulse causes a swelling of the photoelectron bunch. To assess the effect of this mechanism on the temporal properties of the beam, we consider a simple model, in which the electron bunch is travelling through a field-free region. (Here the field-free region is taken for the simplicity.) The total energy of the bunch, consisting of N electrons, is conserved:

$$V + W = 1/2 \left(\sum e^2 / 4\pi\varepsilon_0 \left| r_i - r_j \right| \right) + 1/2 \sum m_e v_i^2$$

Therefore, the characteristic value of the distribution of the velocities, dv, resulting from the electrons pushing apart each other can be qualitatively estimated by using the expression

$$m_e \delta v^2 / 2 \approx e^2 N / \left(16\pi\varepsilon_0 \delta r \right)$$

where dr is the initial size of the photoelectron cloud. In this model, the velocity spread of the electrons causes the elongation of the order of $\Delta \tau_C \approx t_{TOF} \, \delta v / v_e$, which can be rewritten as:

$$\Delta \tau_C = \left(t_{TOF} / v_e \right) \sqrt{e^2 N / \left(8\pi\varepsilon_0 \delta r m_e \right)}$$

At $v_e \approx 10^8$ m/s, $N \approx 10^4$, $t_{TOF} \approx 3$ ns and $\delta r \approx 10$ mm, corresponding to the size of the focal spot of the tightly focused laser beam, this contribution in the framework of the simple model is equal to $\Delta \tau_C \approx 5$ ps. This shows that the Coulomb repulsion of the electrons in the electron bunch is the main factor limiting the ability of ultrafast electron microscopy. Naturally, the use of ultra-high accelerating voltages, such as 500 kV instead of 30 kV, can reduce the swelling of the electron bunch due to Coulomb repulsion, but can also lead to destruction of the sample. Use of very fast electrons, in the MeV range, can further reduce the space-charge broadening because of the onset of relativistic effects [110,111]. It should also be noted that a system based on a TOF electrostatic mirror, a so-called reflectron, can be used to re-compress the electron bunches after their original spreading [26,38]. At the present time, a radio frequency compression scheme is already in use for this purpose.

A serious challenges in the field of 4D electron microscopy remains the task of observing an image or a diffraction pattern produced by a single, ultrashort electron pulse. The solution to this problem is rather important, because it would enable the study of ultrafast irreversible processes. An early breadkthrough has been achieved by a collaboration of scientists at Brown University and at SLAC, who used ultrashort electron pulses with 5.4 MeV of energy, generated using the Gun Test Facility at SLAC, to record single-shot diffraction patterns of 160 nm thick aluminum foils [110,111]. Their work showed that MeV electrons should make it possible to achieve sub-100 fs time resolution. Subsequently, physicists in Japan were able to collect diffraction patterns of thin gold films with single subpicosecond electron bunches [39]. They applied a magnetic sector in order to compress the electron beam in the experiment [39].

Non-Thermal vs Thermal Melting of a Solid Irradiated by a Femtosecond Laser

The study of ultrafast phase transitions in condensed matter is one of the most interesting fields of 4D electron microscopy, viz. ultrafast electron crystallography. To induce an ultrafast phase transition such as the melting of a solid, it is necessary to insert energy into the lattice within a small period of time and to reach the critical temperature as quickly as possible. The most suitable tool to accomplish this is a femtosecond laser pulse. In metals and semiconductors, the photons excite the electrons (and the holes) to high-lying electronic states, resulting in a change of the charge density. Initially, the absorbed energy is stored in electronic degrees of freedom. With the decrease of electron density between the atoms, the bonds in the system become weaker, and at some degree of excitation, the lattice is no longer fixed. Such processes may lead to non-thermal melting of a solid, provided that they are fast compared to the relaxation of the electron energy into the lattice vibrations (phonons). Preliminary studies of aluminum by a femtosecond laser pump-probe have shown that in this material, the transition from a solid to a liquid phase occurs within 500 fs, i.e. faster than the energy transfer from the excited electrons to the phonons [40].

The direct observation of melting of Al with subpicosecond temporal resolution has been achieved by ultrafast electron diffraction [41]. A 20-nm thick film of polycrystalline aluminum was used as the sample in this experiment. The results are shown in Figure 7.

Figure 7 clearly demonstrates, that the lattice is heated and remains in the solid phase with a distinct diffraction pattern on a time scale of up to 1.5 - 2.5 ps after the excitation by laser pulse, which is consistent with the thermal melting. Current research continues in this direction.

7. PHASE TRANSITIONS IN NANOPARTICLES

Vanadium dioxide, VO_2, undergoes a phase transition of the first type from a low-temperature monoclinic phase (M) into a high-temperature tetragonal phase of rutile (R) at $-67°C$. This phase transition has been the subject of intensive research since its discovery almost half a century ago. The work [29] describes the first study of this material using 4D UEM. Because single electron probe pulses were used, Coulomb (space-charge) repulsion was absent. The experiment demonstrated the possibility of obtaining a sequence of images (i.e. a movie) with atomic-scale spatial resolution combined with an ultrashort temporal resolution. In particular, it was shown that it was possible to investigate the ultrafast metalinsulator phase transitions in vanadium dioxide. A structural phase transition in VO_2 nanoparticles is exhibited in the diffraction patterns (at atomic scale) and UEM images (at nanometer scale) with a characteristic hysteresis with a temporal resolution on the order of 100 fs (Figure 8).

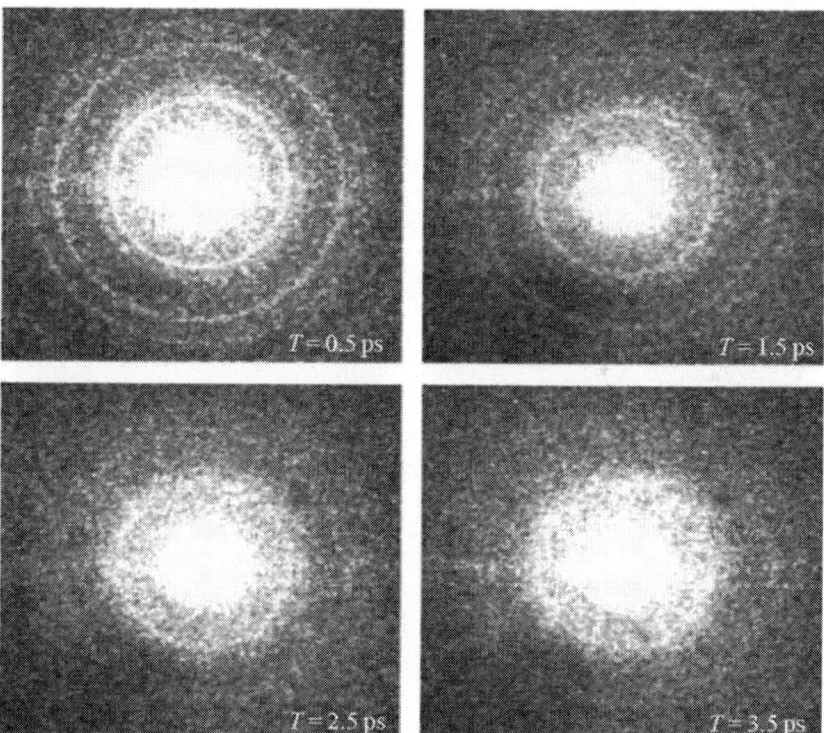

Figure 7. Electron diffraction patterns from a film of polycrystalline Al, recorded at different times after the excitation by a femtosecond laser pulse as labelled. The energy density of the laser pulse is about 70 mJ/cm^2 [41].

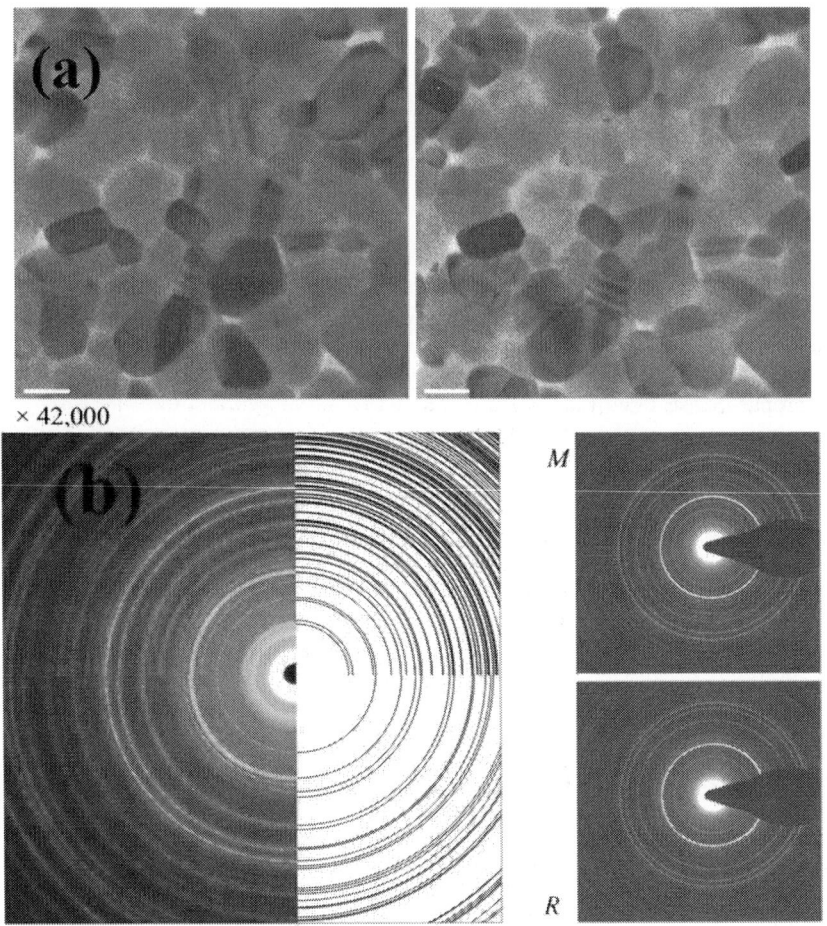

Figure 8. (a) Images corresponding to a phase transition in the films of VO_2 (left) and after the phase transition (right), were collected in UEM with the magnification of 42,000 (the bar corresponds to 100 nm). It should be noted that these images could not be observed when the generation of ultrashort photoelectrons was locked. (b) The diffraction patterns, corresponding to the phase transition in the films of VO_2 (right) and after the phase transition (left), were obtained by UEM. The diffraction patterns of two phases (the monoclinic phase M and the high-temperature phase of tetragonal rutile, R), the experimentally observed (left side of (b)) and constructed by calculations (right side of (b)). The analysis was described in the paper [29].

4D UEM with femtosecond laser irradiation of the sample in the near IR allowed to record the movement of atoms during the phase transition in all three dimensions and in time. This study showed that at first, the vanadium atoms were separated from each other for a short time period, and then began to move towards their final positions.

8. VISUALIZATION OF MECHANICAL DISPLACEMENTS OF A CANTILEVER

The development of ultrafast electron microscopy made possible the direct observation of vibrations of nanostructures in the ultrasound range, covering the range from kHz to GHz [42]. In this connection it should be recalled that the methods of optical interferometry cannot ensure complete three-dimensional spatial resolution, while the possibilities of standard electron microscopy are usually limited to a temporal resolution of about 30 ms.

The work [42] used a pseudo-one-dimensional molecular material, crystal Cu (TCNQ), which exhibits highly anisotropic electrical and optical properties, to prepare a nanoand microcantilever with a thickness of 300 nm and 2 mm, respectively. Opto-mechanical motions can be initiated in such a crystal by a charge transfer from the radical $(TCNQ)^-$ to copper Cu^+.

In the experiment, the charge transfer occurred as a result of irradiation of the crystal by nanosecond laser pulses with a wavelength of 671 nm, a pulse intensity of about $160 \, mJ/cm^2$ and a pulse repetition rate of 100 Hz. This exposure induced resonant cantilever vibrations in the megahertz frequency range. The tiny displacements of the tip were visualized using the 120 keV photoelectrons (Figure 9).

The results so obtained allowed to determine the Young's modulus, measure the resonance frequency of the sample, and to calculate the energy stored due to the opto-mechanical expansion of the crystal Cu (TCNQ).

9. 4D MICROSCOPY OF FATTY ACIDS— TOWARD THE STUDY OF CELL TISSUE MATERIALS

The study of organic and biological structures is an extremely important field of electron microscopy. However, the investigation of such objects entails special demands on the microscope. This is primarily due to the increased sensitivity of the bio-structures to possible irreversible deformation and damage that can be caused by a beam of fast electrons. Therefore, for successful experiments in this field it is necessary to ensure an extremely high detection efficiency of the electrons passing through the sample. Note that the efficiency of the detector is directly related to the time of exposure. In this connection it is helpful that the high detection efficiency in the 4D microscopy allows working with a smaller number of electrons in the bunch.

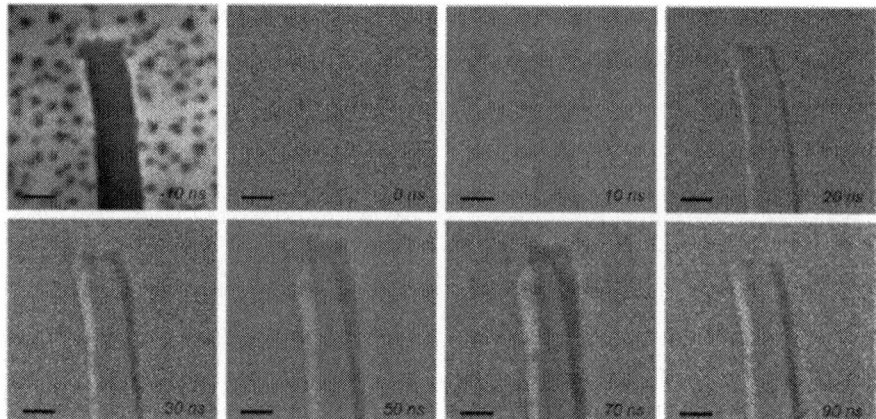

Figure 9. 4D electron micrographs of a nanoscale cantilever. The frames show difference images, i.e. time dependent images where a reference frame is subtracted out. The image recorded at negative delay (−10 ns) relative to the 671 nm pump laser pulse served as a reference frame. A scale bar in the lower left corner corresponds to 200 nm [42].

Amongst the most common biological structures are membrane-type layers of fatty acids. In the laboratory, two-dimensional crystals can be created from long fatty acid hydrocarbon chains. It is possible to use the Langmuir-Blodgett technique to prepare the monomolecular layers of organic material on a solid substrate. It is rather interesting to observe the transition from a two-dimensional to a three-dimensional crystal of the long molecular chains and to visualize the self-assembly processes.

$C_{20}H_{42}$ chains with a length of 95Å on a silicon substrate were considered in computer simulations [2]. In the model, an ultrafast pulsed heating of the substrate leads to further changes in the inter-atomic distances of $C_{20}H_{42}$, reflecting the coherent motion of the atoms in the molecular chain. Each step of the calculation corresponds to a time interval of 0.5 fs, while the total number of steps reaches 200,000. Model calculations have shown an increase in the length of the fragment -CH_2-CH_2-CH_2- that is located in the immediate vicinity of the silicon surface, by about −0.08 Å during 5 ps.

The 4D UEM method should allow us to study more complex biological samples that are directly related to the life sciences. It should be recalled that the use of ultra-short electron bunches for the visualization of biological tissues can improve the spatial resolution of the device compared to conventional electron microscopy. In some ways the situation is similar to making a snapshot by a photographic camera with a long exposure time: the resulting image becomes quite fuzzy (Figure 10), but admittedly, can lead to a new artistic perception of reality.

The 4D electron microscopy succeeded in imaging stained cellular material in rat tissue (Figure 11), and of crystalline proteins, or cells in the vitreous bodies.

10. MUSICAL NANOSCALE INSTRUMENTS: FROM THE DRUM TO THE HARP AND THE PIANO

Ultrathin plates of graphite, gold, or other material, subjected to pulsed laser heating, can experience an ultrafast deformation (bending) as a result of anisotropic expansion of the atomic structure. As it turned out, the vibrations that accompany such deformations may be surprisingly interesting.

Electron bunches at 200 keV were used to visualize mechanical vibrations of graphite nano-sheets (the membranes) with thicknesses of 75 nm. Vibrations with submillisecond temporal intervals resulted from exposure to pulsed laser radiation with a wavelength of 532 nm and a repetition rate of 5 kHz [2]. The energy density of the focused laser pulses reached a value of about 7 mJ/cm^2.

Figure 10. Photograph of a city at night, made with a long exposure time. (The figure is carried over from [43]).

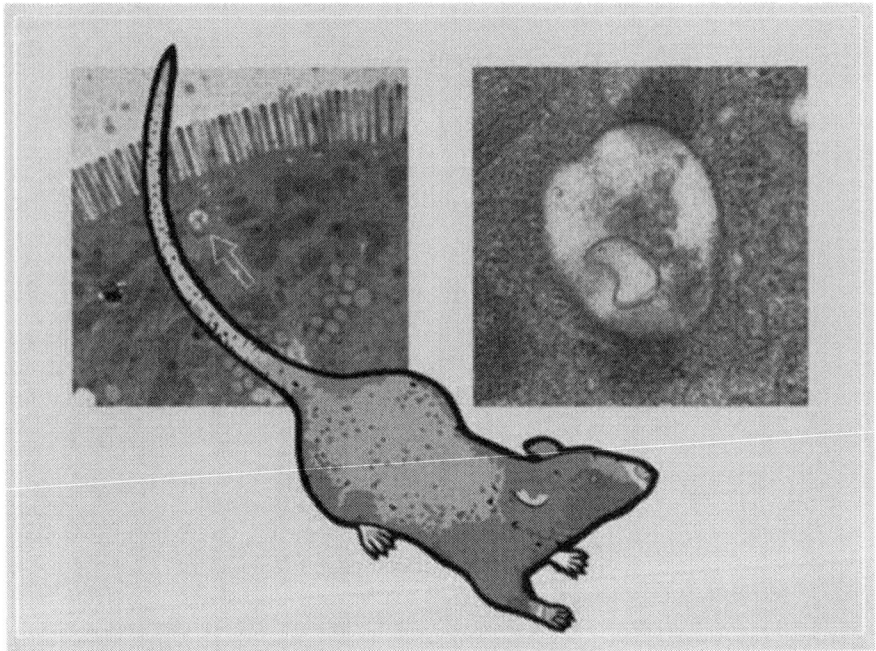

Figure 11. Images of stained cells of rat-related tract, obtained using ultrafast electron microscopy. The micrograph shows the micro-villas in the intestinal epithelium of newborn rats and numerous small vesicles (blisters) in the cytoplasm. An enlarged image of the vesicles, pointed by the arrow on the left frame, is given on the right [2].

Immediately following the pulsed laser heating of the membrane, individual carbon atoms were found to vibrate in a random order. But after a few tens of microseconds, the corresponding modes of the membrane began to be synchronized (Figure 12), resembling a "drumbeat" with a frequency of around 1 MHz.

The processing of the images (Figure 12) used a cross-correlation function defined as

$$\gamma(t';t)$$

$$=\left[\Sigma_{x,y}C_{x,y}(t)C_{x,y}(t')\right]\Big/\sqrt{\left[\Sigma_{x,y}C_{x,y}(t)^2\,S_{x,y}C_{x,y}(t')^2\right]}$$

where the contrast is

$$C_{x,y}(t)=\left[N_{x,y}(t)-\langle N(t)\rangle\right]\Big/\langle N(t)\rangle$$

Here, $N_{x,y}(t)$ is the number of counts corresponding to the picture element (pixel) with the coordinates of (x, y), and $\langle N(t) \rangle$ is the average number of the counts. 2000 images, collected with an interval of 50 ns, were used in Figure 12. The micrograph of the graphite membrane is represented in the lower left corner.

The success with the graphite nano-sheets inspired the search for other musical nano-instruments. A harp and a piano, constructed at the nanoscale, were made using arrays of cantilevers (Figure 13). They were prepared by micro-structuring of multilayer workpieces of $Ni/Ti/Si_3N_4$ using sharply-focused ion beams [44]. The workpieces consisted of 30 nm layers of nickel and titanium, consistently applied to 15 nm film of Si_3N_4. For the piano, the cantilevers had almost the same lengths, approximately 4.6 μm, but their width varied from about 400 nm to 2.3 μm. In the harp, the lengths and widths of individual elements varied in the range of 1.2 - 9.1 μm and −300 - 600 nm, respectively.

In the experiment, a sample (the harp or the piano) was heated by nanosecond 519 nm (or 532 nm) pulses from a Nd:YAG laser with a pulse repetition rate of 1 kHz. The energy density per pulse reached 2 mJ/cm². As a result of the heating of the sample, consisting of the layers with different coefficients of thermal expansion, the mechanical deformation of the cantilever triggered its vibrations. The mechanical movements were observed using the photoelectrons that were obtained by illuminating the photocathode with the fourth harmonic of a Nd:YAG laser. The results for individual cantilevers p1, p5, h1 and h5 (for convenience they are shown in Figure 13) are presented in Figure 14. The combined results of the observations are given in Figure 15.

In the next series of the experiments, the femtosecond laser with MHz pulse repetition rate was used "to play an instrument." Here a fiber laser generator with pulse repetition frequency of about 25 MHz was used. The energy density per pulse was in the range of 20 - 30 mJ/cm².

Beyond their academic interest, the musical nanosystems can have very specific applications. For example, the authors [44] note that such layered nanostructures can be used to precisely measure the temperature of a device with a spatial resolution of 10 μm in combination with a microsecond temporal resolution.

11. 4D ELECTRON TOMOGRAPHY

Electron tomography began to develop as early as the 1960-ies [45]. Progress in this field was mainly limited by the development of computers, because it is important to process a large number of 2-dimensional electron diffraction patterns and combine them in a single movie. A humorous illustration to these words is given in two images of (Figure 16), taken from the book [45].

The use of modern laser technology has allowed to add to electron tomography the fourth dimension, time, and thus to create a completely unique device (Figure 17).

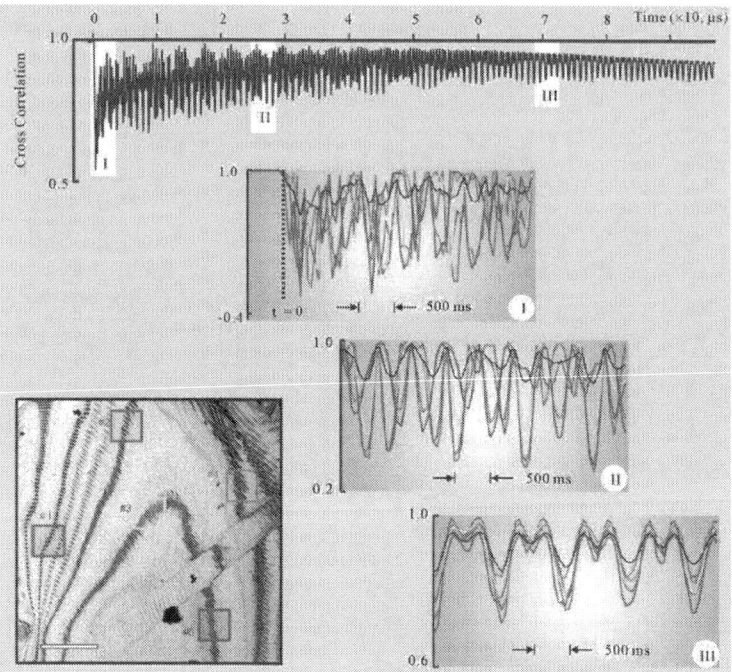

Figure 12. Time-dependence of the cross-correlation function. The scale in the lower left corner corresponds to 5 microns. Five areas with different initial dynamics are marked in red, yellow, green, blue and purple. The transition from chaotic behavior to coherent oscillations is shown in the panels I-II-III. They correspond to different temporal intervals as indicated in the figure [44].

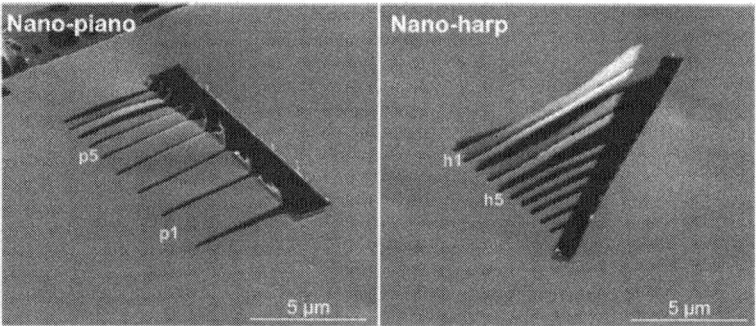

Figure 13. Images of musical nano-instruments. These frames were obtained with a scanning electron microscope. Note that two of the thinnest cantilevers, p7 and p8 on the left, and two pairs of cantilevers, h1/h2 and h3/h4 on the right, are partially melted in the central part as a result of micro-structuring by the focused ion beam [44].

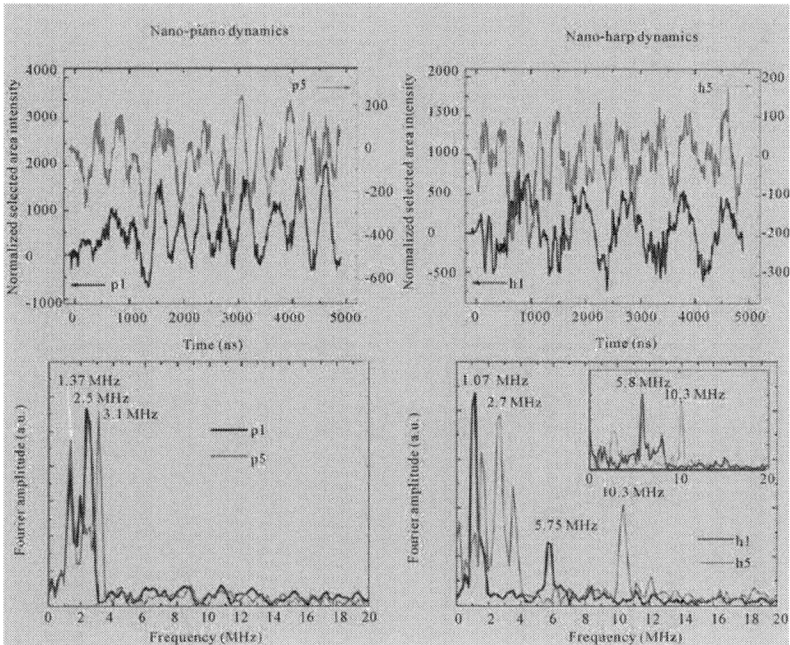

Figure 14. The oscillations of the cantilevers were caused by the pulsed heating of the sample (top). The results of the fast Fourier transform are presented below. The experimental data obtained in the tilted position of the sample with the angle of 30° are given below (insert in the right). The tilted position made it possible to record the movements of the cantilevers, performed outside of the plane of the nano-harp [44].

The temporal behavior of carbon nanotubes was studied by 4D tomography in the work [46]. The sample was twisted in the form of a bracelet. The measured length of the nanotube was L ≈ 4.4 microns. The images in Figures 18 and 19 were obtained using the 200 keV electron pulses. Femtosecond laser excitation led to the heating of the nanotube and caused the subsequent structural changes that began on a picosecond temporal scale.

Of particular interest to a broad audience is that it is possible to create a video, and in the future, a real atomic movie, using the results of 4D electron tomography. It should be emphasized that the pioneering experiment [46] did not induce irreversible changes in the structure of the nanotube, because the total dose received by the carbon sample during the experiment was about 2 orders of magnitude less than the value at which the irreversible deformation occurs. This is a tribute to the extremely high detection efficiency reached in 4D electron microscopy.

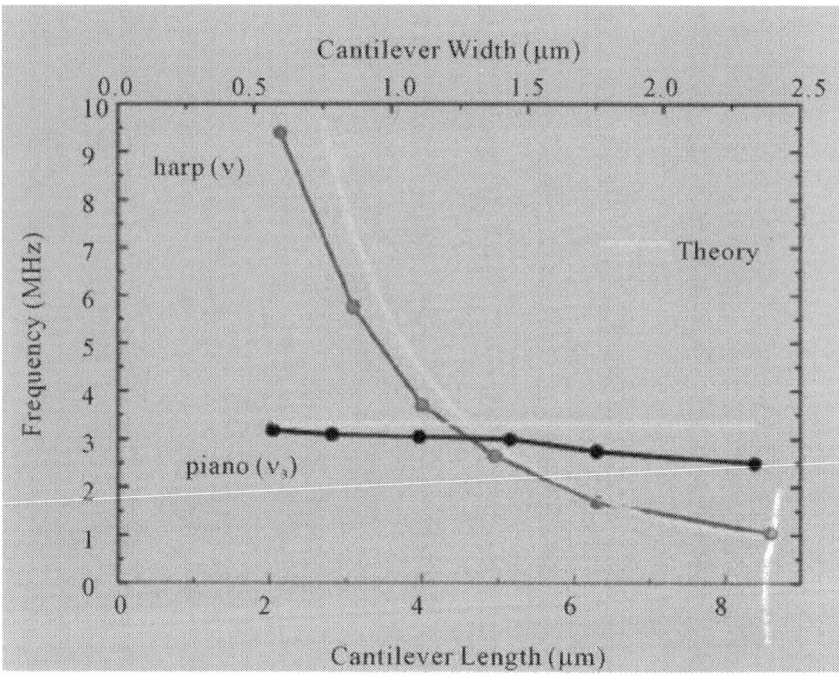

Figure 15. The dependence of the oscillation frequency on the size of the respective cantilever. Only the vibrations that occur out of the plane of musical nano-instruments are considered in this case [44].

Figure 16. The humorous evolution of 3D tomography [45].

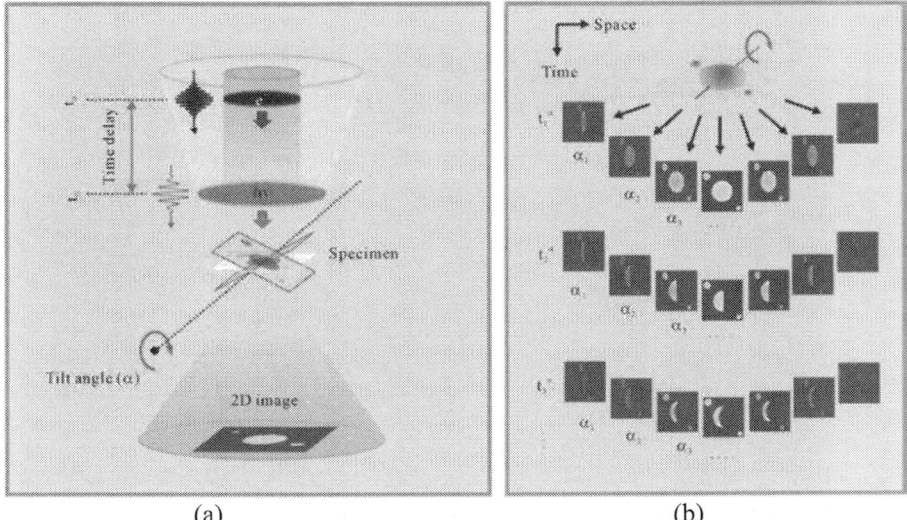

(a)   (b)

Figure 17. (a) Schematic diagram of 4D electron tomography. (b) A series of two-dimensional electronic images taken at different angles a and time delays. In [46] the angle range from −58° to +58° with increments of 1°. The total number of projections reaches 4000 [46].

12. ELECTRON MICROSCOPY WITH HIGH SPECTRAL, SPATIAL AND TEMPORAL RESOLUTION

The combination of spatial nano-resolution and subpicosecond, eventually femtosecond temporal resolution, forms the basis of the atomic movie. Naturally, when the energy of the electron beam, transmitted through the sample, is measured in a similar geometry, additional and very important information is obtained. Note that for most modern research laboratories, such a study would pose quite a serious experimental challenge. Therefore, the precedent in this field plays an important role.

This idea was realized in the work [47] using a single metal nanostructure. In this experiment, the silver particle of triangular shape with a characteristic length of 130 nm and thickness of 20 nm, placed on a substrate of graphene (Figure 20), was irradiated by femtosecond laser pulses with a photon energy of 2.4 eV. The plasma oscillations in silver, excited by the optical radiation, were probed by an ultrashort electron beam with a diameter of 10 nm. The electron beam could be moved over the surface of the sample. Here the energy gain of the electrons, passed through the sample, was measured in addition to the spatial and temporal characteristics.

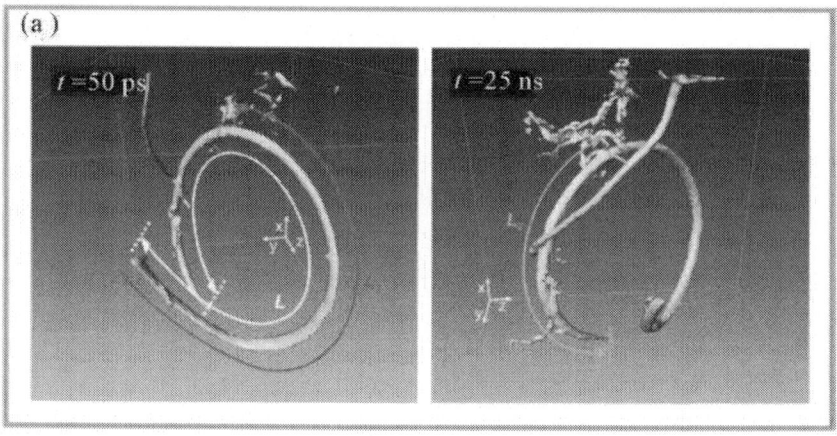

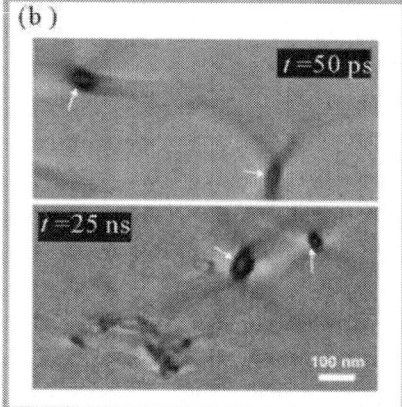

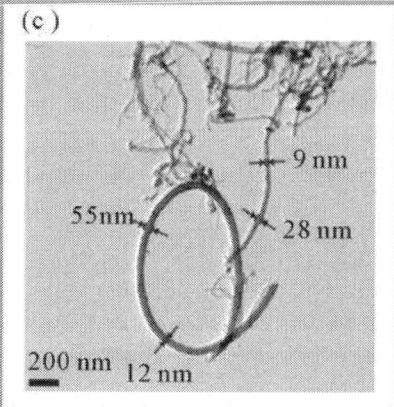

Figure 18. (a) 3D images of multi-walled carbon nanotubes at different times. The lengths of the respective segments around a fixed area are denoted as L1 and L2; (b) Cross section of the 3D images. Two-dimensional slices in the plane xy are shown. Slice thickness is 4.6 nm. Dark areas of the arrows point to the nanometer-thick walls. According to these micrographs, the spatial resolution of the method is sufficient for the correct imaging of channels with a diameter of −10 nm; (c) An image, taken with a transmission electron microscope [46].

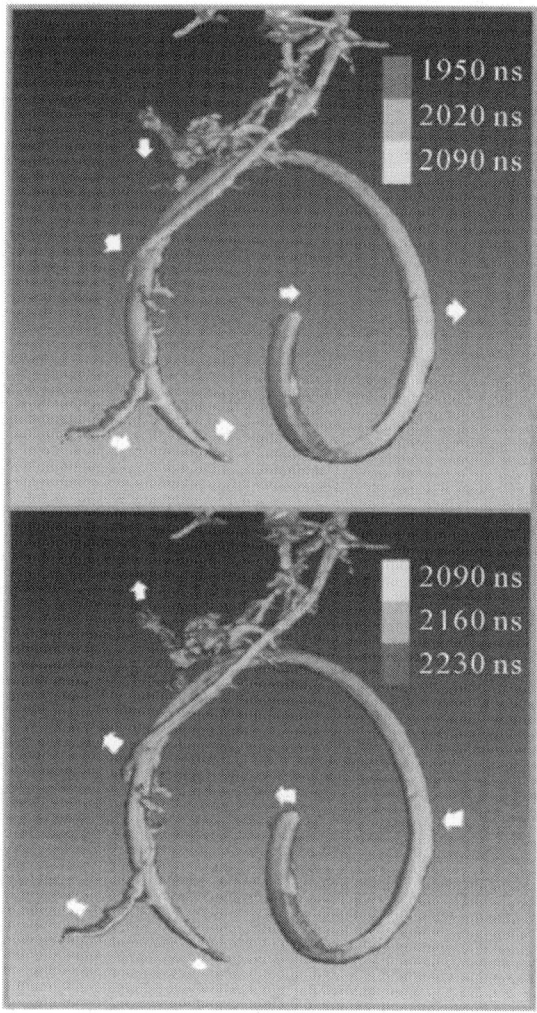

Figure 19. Mechanical vibrations of a carbon bracelet. The arrows indicate the direction of the individual sections of a carbon nanotube [46].

It is important to emphasize that due to the field localized on the surface of a metal particle, the electrons can not only lose kinetic energy, as would be the case with a standard transmission electron microscope, but also to acquire energy (Figure 21). In principle, this process can be controlled by changing the wavelength of the laser radiation. The accuracy of this procedure is determined by the spectral width of the laser pulse and could reach values of about 1 meV.

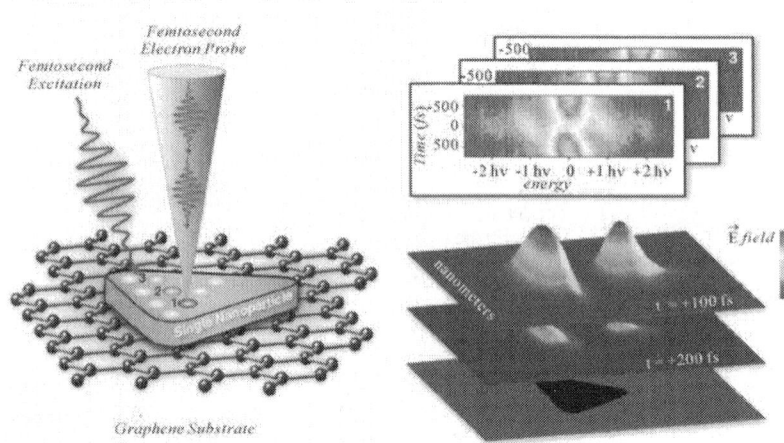

Figure 20. The process of ultrafast spectral imaging. A 10-nm electron beam was scanned over the silver particle that was previously excited by a femtosecond laser pulse. For each position of the probe, the change of the kinetic energy of the electrons (viz. electron spectrum) was measured as a function of the delay between the exciting, optical, and probing, electronic, pulses. The increase in the energy of the electrons is measured in units of the photon energy, being equal to hn = 2.4 eV [47].

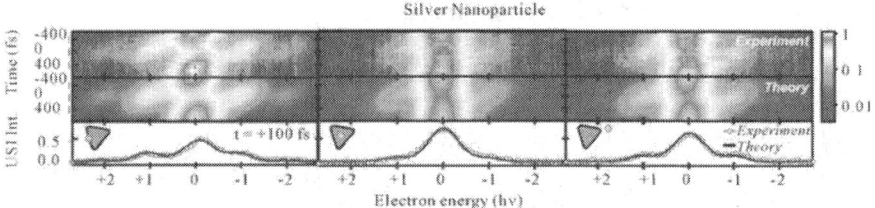

Figure 21. Spectral and temporal characteristics of silver nanoparticles. The electron spectrum, obtained at a delay of 100 fs after the femtosecond optical pulse, is shown below. It is schematically indicated by the position of the electron nanobeam with respect to the triangular silver particle, which is depicted in gray [47].

The results of the measurement of the plasmic texture, generated on the surface of the silver particle, is shown in Figure 22.

It is clearly seen in Figure 22 that the field of plasma oscillations, excited by the laser radiation, is concentrated near the vertices (the cusps) of the triangular particle. There is a fairly good agreement with theory. Qualitatively similar results were obtained near the sharp edge of a copper surface irradiated by the laser [47].

13. CONCLUSIONS

Electron microscopy and diffraction with a high temporal resolution have opened the possibility of directly observing processes that occur in non-steady states of the studied substances. A temporal resolution on the order of 100 fs corresponds to the transition of a quantum system through an energy barrier of a potential surface, describing a chemical reaction in the processes of breaking and forming new bonds of the interacting agents. The advances thus opened the possibility of investigating the coherent nuclear dynamics of molecular systems and condensed matter [24,25,48-51].

In the past two decades it has been possible to observe the motions of nuclei in time intervals corresponding to the period of the oscillation of the nuclei. The observed coherent changes in the nuclear system at such time steps determine the fundamental transition from the standard kinetics to the dynamics of the phase trajectory of a single molecule, the molecular quantum state tomography [50- 53].

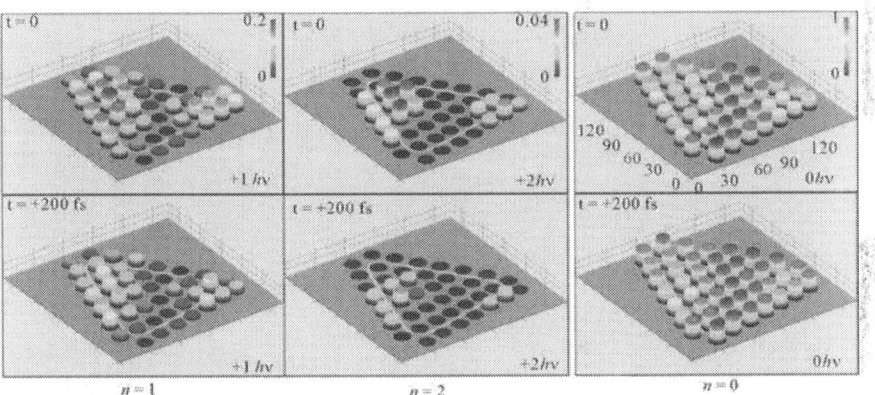

Figure 22. Plasma oscillations, optically excited in the silver nanostructure, were visualized by a 10 - nm electron beam. Here, the electron signal is proportional to the height of the corresponding cylinder. For each panel, the energy, acquired as a result of the interaction of the electrons with nanolocalized fields, is shown in the lower right corner [47].

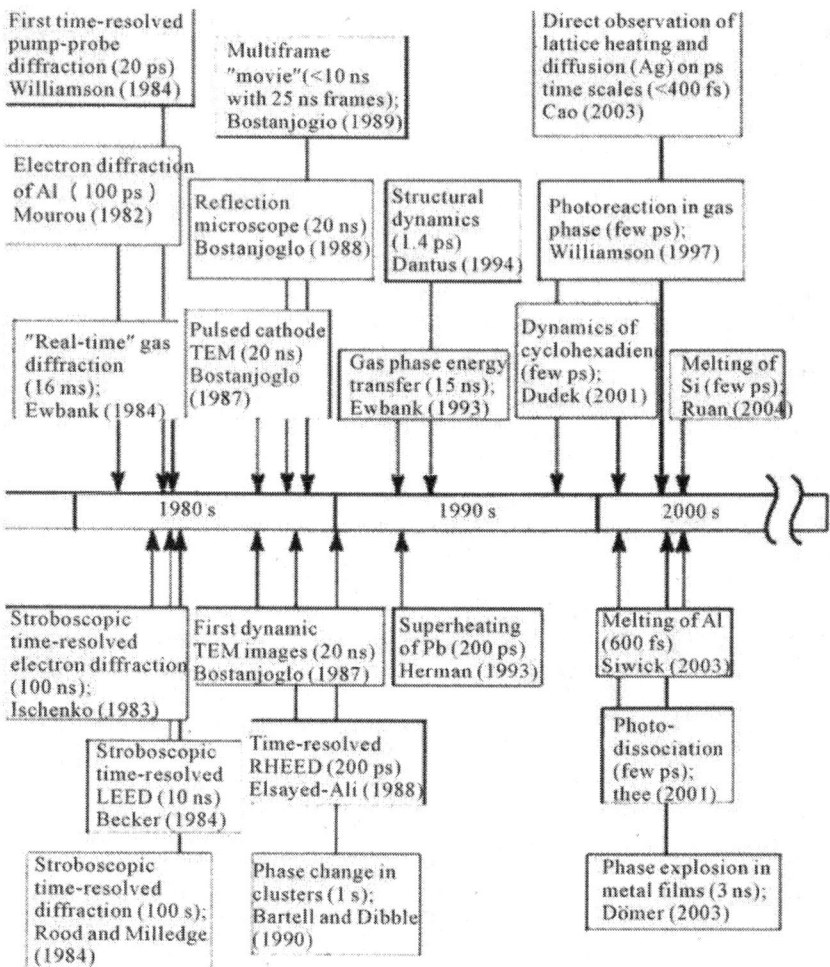

Figure 23. Chronology of the basic works (the first steps) in Dynamic Electron Microscopy, done in the first two decades (1982-2004) [21]. The parallel development of UED, UEM, UEC and DTEM until 2010 is described in [2]. The references [58-109] demonstrate the worldwide experimental activity in this field.

At the present time, the method of ultrafast diffraction continues its rapid development. The methods of ultrafast electron crystallography and electron microscopy with temporal resolution from microto subpicoseconds provide great opportunities to the study of the 4D structural dynamics (see the review articles [21,54] and the monograph [2]). Very recent advances in the formation of ultrashort electron pulses allow us to reach an attosecond temporal resolution and observe the coherent dynamics of the electrons in molecules [55-57].

The development of dynamic electron microscopy and the first steps devoted to the progress in the study of structural dynamics of ultrafast processes are illustrated in Figure 23. It is all but certain that exciting developments in this field will continue into the future.

14. ACKNOWLEDGEMENTS

Academician of the RAS, Prof. A. L. Buchachenko, Prof. E. A. Ryabov, Prof. V. N. Bagratashvili, Dr. A. E. Luk'yanov, A. A. Timofeev, Prof. V. R. Flid, and Dr. James Beckman (Therapon, Inc.) are acknowledged for the hepful discussions. This work was supported by grants: No. 2012-1.1-12-000-1009-055 (from the Ministry of Education and Science of the Russian Federation) and No. 10-02-92000-HHC_a, 11-02-00868-a, 12-02-00840-a (from the Russian Fund for Basic Research). PMW acknowledges support by the Division of Chemical Sciences, Geosciences, and Biosciences, the Office of Basic Energy Sciences, the U.S. Department of Energy, Grant No. DE-FG02-03ER15452.

REFERENCES

1. A. Ischenko, V. V. Golubkov, V. P. Spiridonov, A. V. Zgurskii, A. S. Akhmanov, M. G. Vabishevich and V. N. Bagratashvili, "A Stroboscopical Gas-Electron Diffraction Method for the Investigation of Short-Lived Molecular Species," Applied Physics B, Vol. 32, No. 3, 1983, pp. 161-163.

2. H. Zewail and J. M. Thomas, "4D Electron Microscopy. Imaging in Space and Time," Imperial College Press, London, 2010.

3. K. Weinstein, "Atomic Resolution Electron Microscopy," Soviet Physics Uspekhi, Vol. 152, No. 3, 1987, pp. 75-122.

4. J. Brandon and W. Kaplan, "Microstructure of Materials. Methods of Research and Monitoring," M. Technosphere, Bergenfield, 2004.

5. Vlasov, K. A. Yelsukov and V. Panfilov, "Microscopy Techniques," Publishing House of Moscow State Technical University named after Bauman, Moscow, 2011.

6. Vlasov, K. A. Yelsukov and I. A. Kosolapov, "Electron Microscopy," M. Bauman Bauman, Moscow, 2011.

7. Shindo and T. Oikawa, "Analytical Transmission Electron Microscopy for Materials Science," M.: Mir, Moscow, 2006.

8. Y. S. Umansky, J. N. Skakov, A. N. Ivanov and L. N. Rastorguev, "Crystallography, X-Ray and Electron Microscopy," Metallurgy, Moscow, 1982.

9. R. Kheidenraikh, "Fundamentals of Transmission Electron Microscopy," Academic Press, Waltham, 1966.

10. P. Hirsch, A. Howie, R. Nicholson, D. Pashley and M. Whelan, "Electron Microscopy of Thin Crystals," World, Moscow, 1968.

11. M. V. Locquim and M. Langeron, "Handbook of Microscopy," Butterworths, London, 1983.
12. J. Spence, "Experimental Electron Microscopy Razresheniya," M.: Science, Poole, 1986.
13. Watt, "The Principles and Practice of Electron Microscopy," Cambridge University Press, Cambridge, 1985.
14. G. Thomas and M. J. Gorindzh, "Transmission Electron Microscopy of Materials," Nauka, Moscow, 1983.
15. E.A. Gudilin, "Electron Microscopy," 2008. http://www.nanometer.ru/2008/12/21/electron_microscopy_55002.html
16. Amelinks, "Methods of Direct Observation of Dislocations," Wiley, Hoboken, 1968.
17. M. Motosuke and S. Tetsuya, "Stroboscopic Scanning Electron Microscope," US Patent Number: 4538065, 1985.
18. G. V. Spivak, V. V. Shakmanov, V. I. Petrov, A. E. Lukyanov and S. Yakunin, "On the Application of the Gates with the Deflection Plates in a Stroboscopic Electron Microscopy," Mathematical USSR Physics Series, Vol. 32, No. 7, 1968, pp. 1111-1114.
19. E. Lukyanov, V. Galstyan and G. V. Spivak, "About Stroboscopic Scanning Electron Microscopy Four-Semiconductor Structures," Technology and Electronics, No. 11, 1970, pp. 2424-2427.
20. M. L. Taheri, N. D. Browning and J. Lewellen, "Symposium on Ultrafast Electron Microscopy and Ultrafast Science, Special Section: Ultrafast Electron Microscopy," Microscopy and Microanalysis, Vol. 15, No. 4, 2009, pp. 271-271.
21. W. E. King, G. H. Campbell, A. Frank, B. Reed, J. F. Schmerge, B. J. Siwick, B. C. Stuart and P. M. Weber, "Ultrafast Electron Microscopy in Materials Science, Biology, and Chemistry," Journal of Applied Physics, Vol. 97, No. 111101, 2005, pp. 1-27.
22. V. A. Lobastov, R. Srinivasan and A. H. Zewail, "FourDimensional Ultrafast Electron Microscopy," Proceedings of the National Academy of Sciences of the United States of America, Vol. 102, No. 20, 2005, pp. 7069-7073.
23. H. Zewail, "4D Ultrafast Electron Diffraction, Crystallography, and Microscopy," Annual Review of Physical Chemistry, Vol. 57, 2006, pp. 65-103.
24. S. Williamson, G. Mourou and L. C. M. Li, "Time-Resolved Laser-Induced Phase Transformation in Aluminum," Physical Review Letters, Vol. 52, No. 26, 1984, pp. 2364-2367.
25. S. A. Akhmanov, V. N. Bagratashvili, V. V. Golubkov, A. V. Zgurskaya, A. A. Ischenko, S. A. Krikunov, V. P. Spiridonov and A. G. Tunkin,

"Generation in the EMR- 100 Electron Diffraction Apparatus the Picosecond Pulses of Fast Electrons by Photoemission in the Laser Field," Soviet Technical Physics Letters, Vol. 11, No. 3, 1985, pp. 157-161.

26. P. M. Weber, S. D. Carpenter and T. Lucza, "Reflectron Design for Femtosecond Electron Guns," SPIE Proceedings, Vol. 2521, No. 1, 1995, pp. 23-30.

27. H. Domer and O. Bostanjoglo, "High-Speed Transmission Electron Microscope," Review of Scientific Instruments, Vol. 74, No. 10, 2003, pp. 4369-4372.

28. O. Bostanjoglo, "High-Speed Electron Microscopy," Advances in Imaging and Electron Physics, Vol. 121, 2002, pp. 1-51.

29. M. S. Grinolds, V. A. Lobastov, J. Weissenrieder and A. H. Zewail, "Four-Dimensional Ultrafast Electron Microscopy of Phase Transitions," Proceedings of the National Academy of Sciences of the United States of America, Vol. 103, No. 49, 2006, pp. 18427-18431.

30. V. Gaponov and M. A. Miller, "Potential Wells for Charged Particles in a High-Frequency Electromagnetic Field," Soviet Physics—JETP, Vol. 7, No. 1, 1958, pp. 168-169.

31. T. W. B. Kibble, "Refraction of Electron Beams by Intense Electromagnetic Waves," Physical Review Letters, Vol. 16, No. 23, 1966, pp. 1054-1056.

32. T. Hebeisen, R. Ernstorfer, M. Harb, T. Dartigalongue, R. E. Jordan and R. J. D. Miller, "Femtosecond Electron Pulse Characterization Using Laser Ponderomotive Scattering," Optics Letters, Vol. 31, No. 23, 2006, pp. 3517-3520.

33. T. Hebeisen, G. Sciaini, M. Harb, R. Ernstorfer, T. Dartigalongue, S. G. Kruglik and R. J. D. Miller, "Grating Enhanced Ponderomotive Scattering for Visualization and Full Characterization of Femtosecond Electron Pulses," Optics Express, Vol. 16, No. 5, 2008, pp. 3334-3341.

34. R. Gekker, "Interaction of Strong Electromagnetic Fields with Plasmas," Clarendon Press, Oxford, 1982.

35. M. V. Fedorov, "Atomic and Free Electrons in a Strong Light Field," World Scientific, London, 1997.

36. P. H. Bucksbaum, M. Bashkansky and T. J. McIlrath, "Scattering of Electrons by Intense Coherent Light," Physical Review Letters, Vol. 58, No. 4, 1987, pp. 349- 352.

37. J. Muybridge, "Animal Locomotion, an Electrophotographic Investigation of Consecutive Phases of Animal Movement," J. B. Lippincott and Co., Philadelphia, 1887.

38. G. H. Kassier, K. Haupt, N. Erasmus, E. G. Rohwer and H. Schwoerer, "Achromatic Reflectron Compressor Design for Bright Pulses in

Femtosecond Electron Diffraction," Journal of Applied Physics, Vol. 105, No. 1, 2009, pp. 113111-1-113111-10.

39. S. Tokita, M. Hashita, S. Inoue, T. Nishoji, K. Otani and S. Skabe, "Single-Shot Femtosecond Electron Diffraction with Laser-Accelerated Electrons: Experimental Demonstration of Electron Pulse Compression," Physical Review Letters, Vol. 105, No. 21, 2010, pp. 215004-215007.

40. Guo, G. Rodriguez, A. Lobad and A. J. Taylor, "Structural Phase Transition of Aluminum Induced by Electronic Excitation," Physical Review Letters, Vol. 84, No. 19, 2000, pp. 4493-4496.

41. J. Siwick, J. R. Dwyer, R. E. Jordan and R. J. D. Miller, "An Atomic-Level View of Melting Using Femtosecond Electron Diffraction," Science, Vol. 302, No. 5649, 2003, pp. 1382-1385.

42. J. Flannigan, P. C. Samartzis, A. Yurtsever and A. H. Zewail, "Nanomechanical Motions of Cantilevers: Direct Imaging in Real Space and Time with 4D Electron Microscopy," Nano Letters, Vol. 9, No. 2, 2009, pp. 875-881. http://hqwall.net

43. S. Baskin, H. S. Parkand A. H. Zewail, "Nanomusical Systems Visualized and Controlled in 4D Electron Microscopy," Nano Letters, Vol. 11, No. 5, 2011, pp. 2183- 2191.

44. Frank, "Electron Tomography: Methods for ThreeDimensional Visualization of Structures in the Cell," Springer, New York, 2010.

45. O.-H. Kwon and A. H. Zewail, "4D Electron Tomography," Science, Vol. 328, No. 5986, 2010, pp. 1668-1673.

46. Yurtsever, R. M. van der Veen and A. H. Zewail, "Subparticle Ultrafast Spectrum Imaging in 4D Electron Microscopy," Science, Vol. 335, No. 6064, 2012, pp. 59- 64.

47. A. Ischenko, V. P. Spiridonov, L. Schäfer and J. D. Ewbank, "The Stroboscopic Gas Electron Diffraction Method for Investigation of Time-Resolved Structural Kinetics in Photoexcitation Processes," Journal of Molecular Structure, Vol. 300, 1993, pp. 115-140.

48. V. A. Lobastov, J. D. Ewbank, L. Schafer and A. A. Ischenko, "Instrumentation for Time-Resolved Electron Diffraction Spanning the Time Domain from Microseconds to Picoseconds," Review of Scientific Instruments, Vol. 69, No. 7, 1998, pp. 2633-2643.

49. D. Ewbank, L. Schafer and A. A. Ischenko, "Structural and Vibrational Kinetics of Photoexcitation Processes Using Time-Resolved Electron Diffraction," Journal of Molecular Structure, Vol. 534, No. 1-3, 2000, pp. 1-49.

50. A. Ischenko, V. N. Bagratashvili and A. S. Avilov, "Methods of Studying Coherent 4D Structural Dynamics of Free Molecules and Condensed Matter," Crystallography, Vol. 56, No. 5, 2011, pp. 805-828.

51. A. Ischenko, "The Study of Coherent Dynamics of the Nuclei by Electron Diffraction with a Time Resolution. II. Electron Scattering Coherently Excited Molecules," Proceedings of the Universities: Chemistry and Chemical Technology, Vol. 52, No. 5, 2009, pp. 62-67.

52. A. Ischenko, "The Study of Coherent Dynamics of the Nuclei by Electron Diffraction with a Time Resolution. III. Molecular Quantum State Tomography," Proceedings of the Universities: Chemistry and Chemical Technology, Vol. 52, No. 8, 2009, pp. 58-63.

53. R. Dwyer, C. T. Hebeisen, R. Ernstorfer, M. Harb, V. B. Deyirmenjian, R. E. Jordan and R. J. D. Miller, "Femtosecond Electron Diffraction: 'Making the Molecular Movie'," Philosophical Transactions of the Royal Society A, Vol. 364, No. 1840, 2006, pp. 741-778.

54. Ben-Nun, J. Cao and K. Wilson, "Ultrafast X-Ray and Electron Diffraction: Theoretical Considerations," Journal of Physical Chemistry A, Vol. 101, No. 47, 1997, pp. 8743-8761.

55. H.-C. Shao and A.F. Starace, "Detecting Electron Motion in Atoms and Molecules," Physical Review Letters, Vol. 105, No. 26, 2010, pp. 263201-263204.

56. H.-C. Shao and A. F. Starace, "Ultrafast Electron Pulse (e,2e) Processes," Bulletin of the American Physical Society 43rd Annual Meeting of the APS Division of Atomic, Molecular and Optical Physics, Vol. 57, No. 5, 2012, Article ID: N3.00008.

57. S. Williamson and G. Mourou, "Picosecond Electron Diffraction," Applied Physics Letters, Vol. 41, No. 1, 1982, pp. 44-45.

58. V. V. Golubkov, A. V. Zgurskii, A. A. Ischenko, V. I. Petrov and V. P. Spiridonov, "Pulse-Resonance Method for Detecting a Signal in a Stroboscopic Electron Microscopy," Abstracts XII All-Union Conference on Electron Microscopy, Moscow, Science, 1982, p. 62.

59. V. V. Golubkov, A. V. Zgurskii, A. A. Ischenko and V. I. Petrov, "New Methods of Signal Detection in Gas Electron Diffraction," Proceedings of the USSR Academy of Sciences, Vol. 47, No. 6, 1983, pp. 1115-1121.

60. R. S. Becker, G. S. Higashi and J. A. Golovchenko, "LowEnergy Electron Diffraction during Pulsed Laser Annealing: A Time-Resolved Surface Structural Study," Physical Review Letters, Vol. 52, No. 4, 1984, pp. 307-310. doi:10.1103/PhysRevLett.52.307

61. A. P. Rood and J. Milledge, "Combined Flash-Photolysis and Gas-Phase Electron Diffraction Studies of Small Molecules," Journal of the Chemical Society, Faraday Transactions, Vol. 2, No. 80, 1984, pp. 1145-1151.

62. J. D. Ewbank, L. Schafer, D. W. Paul, O. J. Benston and J. C. Lennox, "Real-Time Data Acquisition for Gas Electron Diffraction," Review of Scientific Instruments, Vol. 55, No. 10, 1984, pp. 1598-1603.

63. A. A. Ischenko, V. N. Bagratashvili, V. V. Golubkov, V. P. Spiridonov, A. V. Zgurskii and A. S. Akhmanov, "Observation of Electron Diffraction from Free Radicals—Products IR Multiphoton Dissociation of CF_3I Molecules by Stroboscopic Gas Electron Diffraction," Moscow University Chemistry Bulletin, Vol. 26, No. 2, 1985, pp.140-143.

64. O. Bostanjoglo, E. Endruschat, F. Heinricht, R. P. Tornow and W. Tornow, "Short-Interval Electron Microscopy and Pulsed Lasers," European Journal of Cell Biology, Vol. 44, No. 1, 1987, pp. 10-15.

65. O. Bostanjoglo, R. P. Tornow and W. Tornow, "Nanosecond-Exposure Electron Microscopy of Laser-Induced Phase Transformations," Ultramicroscopy, Vol. 21, No. 4, 1987, pp. 367-372.

66. O. Bostanjoglo and F. Heinricht, "A Laser Pulsed High Emission Thermal Electron-Gun," Journal of Physics: Conference Series, Vol. 93, No. 1, 1988, pp. 105-106.

67. H. E. Elsayed-Ali and G. A. Mourou, "Picosecond Reflection High-Energy Electron Diffraction," Applied Physics Letters, Vol. 52, No. 2, 1988, pp. 103-104.

68. O. Bostanjoglo and P. Thomsen-Schmidt, "Laser-Induced Multiple Phase Transitions in Ge and Te Films Traced by Time-Resolved TEM," Applied Surface Science, Vol. 43, No. 1-4, 1989, pp. 136-141.

69. H. E. Elsayed-Ali and J. W. Herman, "Picosecond TimeResolved Surface-Lattice Temperature Probe," Applied Physics Letters, Vol. 57, No. 15, 1990, pp. 1508-1510.

70. L. S. Bartell and T. S. Dibble, "Observation of the Time Evolution of Phase Changes in Clusters," Journal of American Chemical Society, Vol. 112, No. 2, 1990, pp. 890- 891.

71. J. C. Williamson, M. Dantus, S. B. Kim and A. H. Zewail, "Ultrafast Diffraction and Molecular Structure," Chemical Physics Letters, Vol. 196, No. 6, 1992, pp. 529-534.

72. J. D. Ewbank, J. Y. Luo, J. T. English, R. F. Liu, W. L. Faust and L. Schafer, "Time-Resolved Gas Electron Diffraction Study of the 193-nm Photolysis of 1,2-Dichloroethenes," Journal of Physical Chemistry, Vol. 97, No. 34, 1993, pp. 8745-8751.

73. J. W. Herman, H. E. Elsayed-Ali and E. A. Murphy, "Time-Resolved Structural Study of Pb(100)," Physical Review Letters, Vol. 71, No. 3, 1993, pp. 400-403.

74. M. Dantus, S. B. Kim, J. C. Williamson and A. H. Zewail, "Ultrafast Electron Diffraction. 5. Experimental Time Resolution and Applications," Journal of Physical Chemistry, Vol. 98, No. 11, 1994, pp. 2782-2796.

75. J. C. Williamson, J. Cao, H. Ihee, H. Frey and A. H. Zewail, "Clocking Transient Chemical Changes by Ultrafast Electron Diffraction," Nature, Vol. 6, No. 386, 1997, pp. 159- 162.

76. A. A. Ischenko, L. Schafer and J. D. Ewbank, "Tomography of the Molecular Quantum State by Time-Resolved Electron Diffraction," SPIE Proceedings, Vol. 3516, 1998, pp. 580- 595.

77. R. C. Dudek and P. M. Weber, "Ultrafast Diffraction Imaging of the Electrocyclic Ringopening Reaction of 1,3-Cyclohexadiene," Journal of Physical Chemistry A, Vol. 105, No. 17, 2001, pp. 4167-4171.

78. H. Ihee, V. A. Lobastov, U. M. Gomez, B. M. Goodson, R. Srinivasan and A. H. Zewail, "Direct Imaging of Transient Molecular Structures with Ultrafast Diffraction," Science, Vol. 291, No. 5503, 2001, pp. 458-462.

79. B. J. Siwick, J. R. Dwyer, R. F. Jordan and R. J. D. Miller, "Ultrafast Electron Optics: Propagation Dynamics of Femtosecond Electron Packets," Journal of Applied Physics, Vol. 92, No. 3, 2002, pp. 1643-1648.

80. [82] H. Domer and O. Bostonjoglo, "High-Speed Transmission Electron Microscope," Review of Scientific Instruments, Vol. 74, No. 10, 2003, pp. 4369-4372.

81. C.-Y. Ruan, F. Vigliotti, V. A. Lobastov, S. Chen and A. H. Zewail, "Ultrafast Electron Crystallography: Transient Structures of Molecules, Surfaces, and Phase Transitions," Proceedings of the National Academy of Sciences of the United States of America, Vol. 101, No. 5, 2004, pp. 1123-1128.

82. C.-J. Ruan, D.-S.Yang and A. H. Zewail, "Structures and Dynamics of Self-Assembled Surface Monolayers Observed by Ultrafast Electron Crystallography," Journal of the American Chemical Society, Vol. 126, No. 40, 2004, pp. 12797-12799.

83. W. E. King, G. H. Campbell, A. Frank, B. Reed, J. F. Schmerge, B. J. Siwick, B. C. Stuart and P. M. Weber, "Ultrafast Electron Microscopy in Materials Science, Biology, and Chemistry," Journal of Applied Physics, Vol. 97, No. 11, 2005, pp. 111101-111127.

84. W. E. King, M. R. Armstrong and O. Bostonjoglo, "HighSpeed Electron Microscopy," Science of Microscopy, Vol. 1, 2007, pp. 404-444.

85. O.-H. Kwon, B. Barwick, H. S. Park, J. S. Baskin and A. H. Zewail, "Nanomechanical Motions of Cantilevers: Direct Imaging in Real Space and Time with 4D Electron Microscopy," Nano Letters, Vol. 8, No. 11, 2008, pp. 3557-3562.

86. D. J. Flannigan, B. Barwick and A. H. Zewail, "Biological Imaging with 4D Ultrafast Electron Microscopy," Proceedings of the National Academy of Sciences of the United States of America, Vol. 107, No. 22, 2010, pp. 9933-9937.

87. O.-H. Kwon, H. S. Park, J. P. Baskin and A. H. Zewail, "Nonchaotic Nonlinear Motion Visualized in Complex Nanostructures by Stereographic 4D Electron Microscopy," Nano Letters, Vol. 10, No. 8, 2010, pp. 3190-3198.

88. V. Ortalan and A. H. Zewail, "4D Scanning Transmission Ultrafast Electron Microscopy: Single-Particle Imaging and Spectroscopy," Journal of the American Chemical Society, Vol. 133, No. 28, 2011, pp. 10732-10735.

89. G. Sciaini and R. J. D. Miller, "Femtosecond Electron Diffraction: Heralding the Era of Atomically Resolved Dynamics," Reports on Progress in Physics, Vol. 74, No. 9, 2011, pp. 096101-096136.

90. S. T. Park, D. J. Flannigan and A. H. Zewail, "Irreversible Chemical Reactions Visualized in Space and Time with 4D Electron Microscopy," Journal of the American Chemical Society, Vol. 133, No. 6, 2011, pp. 1730-1733.

91. O.-H. Kwon, V. Ortalan and A. H. Zewail," Macromolecular Structural Dynamics Visualized by Pulsed Dose Control in 4D Electron Microscopy," Proceedings of the National Academy of Sciences of the United States of America, Vol. 108, No. 15, 2011, pp. 6026-6031.

92. O. F. Mohammed, D.-S. Yang, S. K. Pal and A. H. Zewail," 4D Scanning Ultrafast Electron Microscopy: Visualization of Materials Surface Dynamics," Journal of the American Chemical Society, Vol. 133, No. 20, 2011, pp. 7708-7711.

93. V. Ortalan and A. H. Zewail, "4D Scanning Transmission Ultrafast Electron Microscopy: Single-Particle Imaging and Spectroscopy," Journal of the American Chemical Society, Vol. 133, No. 28, 2011, pp. 10732-10735.

94. M. M. Lin, D. Shorokhov and A. H. Zewail, "Structural Dynamics of Free Proteins in Diffraction," Journal of the American Chemical Society, Vol. 133, No. 42, 2011, pp. 17072-17086.

95. S. Schäfer, W. Liang and A. H. Zewail, "Structural Dynamics of Nanoscale Gold by Ultrafast Electron Crystallography," Chemical Physics Letters, Vol. 515, No. 4-6, 2011, pp. 278-282.

96. S. Schäfer, W. Liang and A. H. Zewail, "Structural Dynamics of Surfaces by Ultrafast Electron Crystallography: Experimantal and Multiple Scattering Theory," Journal of Chemical Physics, Vol. 135, No. 21, 2011, pp. 214201- 214215.

97. I.-R. Lee, A. Gahlmann and A. H. Zewail, "Structural Dynamics of Free Amino Acids in Diffraction," Angewandte Chemie International Edition, Vol. 51, No. 1, 2012, pp. 99-102. doi:10.1002/anie.201105803

98. A. Yurtsever, J. S. Baskin and A. H. Zewail, "Entangled Nanoparticles: Discovery by Visualization in 4D Electron Microscopy," Nano Letters, Vol. 12, No. 9, 2012, pp. 5027-5032.

99. S. T. Park and A. H. Zewail, "Relativistic Effects in Photon-Induced near-Field Electron Microscopy," Journal of Physical Chemistry A, Vol. 116, No. 46, 2012, pp. 11128- 11133.

100. A. Yurtsever, S. Schäfer and A. H. Zewail, "Ultrafast Kikuchi Diffraction: Nanoscale Stress-Atrain Dynamics of Wave-Guiding Structures," Nano Letters, Vol. 12, No. 7, 2012, pp. 3772-3777.

101. A. Yurtsever and A. H. Zewail, "Direct Visualization of Near-Fields in Nanoplasmonics and Nanophotonics," Nano Letters, Vol. 12, No. 6, 2012, pp. 3334-3338.

102. S. T. Park, D. J. Flannigan and A. H. Zewail, "4D Electron Microscopy Visualization of Anisotropic Atomic Motions in Carbon Nanotubes," Journal of the American Chemical Society, Vol. 134, No. 22, 2012, pp. 9146-9149.

103. S. T. Park, O.-H. Kwon and A. H. Zewail, "Chirped Imaging Pulses in Four-Dimensional Electron Microscopy: Femtosecond Pulsed Hole Burning," New Journal of Physics, Vol. 14, 2012, Article ID: 053046.

104. S. T. Park and A. H. Zewail, "Enhancing Image Contrast and Slicing Electron Pulses in 4D Near-Field Electron Microscopy," Chemical Physics Letters, Vol. 521, 2012, pp. 1-6.

105. D. J. Flannigan and A. H. Zewail, "4D Electron Microscopy: Principles and Applications," Accounts of Chemical Research, Vol. 45, No. 10, 2012, pp. 1828-1839.

106. D.-S. Yang, O. F. Mohammed and A. H. Zewail, "Environmental Scanning Ultrafast Electron Microscopy: Structural Dynamics of Solvation at Interfaces," Angewandte Chemie International Edition, Vol. 52, No. 10, 2013, pp. 2897-2901.

107. W. Liang, S. Schäfer and A. H. Zewail, "Ultrafast Electron Crystallography of Monolayer Adsorbates on Clean Surfaces: Structural Dynamics," Chemical Physics Letters, Vol. 542, 2012, pp. 1-7.

108. W. Liang, S. Schäfer and A. H. Zewail, "Ultrafast Electron Crystallography of Heterogeneous Structures: GoldGraphene Bilayer and Ligand-Encapsulated Nanogold on Graphene," Chemical Physics Letters, Vol. 542, 2012, pp. 8-12.

109. J. B. Hastings, F. M. Rudakov, D. H. Dowell, J. F. Schmerge, J. Cardoza, J. M. Castro, S. M. Gierman, H. Loos and P. M. Weber, "Ultrafast Time-Resolved Electron Diffraction with Megavolt Electron Beams," Applied Physics Letters, Vol. 89, No. 18, 2006, pp. 184109- 184111.

110. F. M. Rudakov, J. B. Hastings, D. H. Dowell, J. F. Schmerge and P. M. Weber, "Megavolt Electron Beams for Ultrafast Time-Resolved Electron Diffraction," In: M. D. Furnish, M. Elert, T. P. Russel and C. T. White, Eds.

Shock Compression of Condensed Matter—2005, American Institute of Physics, New York, 2006, pp. 1287-1292.

111. J. D. Geiser and P. M. Weber, "High Repetition Rate Time-Resolved Gas Phase Electron Diffraction," Proceedings of SPIE Conference on Time Resolved Electron and X-Ray Diffraction, Vol. 2521, San Diego, July 1995, pp. 136-144.

CHAPTER 10

Influence of a Thiolate Chemical Layer on GaAs (100) Biofunctionalization: An Original Approach Coupling Atomic Force Microscopy and Mass Spectrometry Methods

Alex Bienaime [1], Therese Leblois [1], Nicolas Gremaud [1], Maxime-Jean Chaudon [1], Marven El Osta [2], Delphine Pecqueur [2], Patrick Ducoroy [2] and Celine Elie-Caille [1,]*

[1] MicroNanoSciences and Systems Department, Franche-Comté Electronique, Mécanique, Thermique et Optique—Sciences et Technologies (FEMTO-ST) Institute, 32 avenue de l'Observatoire, 25044 Besançon Cedex, France

[2] Clinical and Innovation Proteomic Platform, University of Burgundy, CHU, 21000 Dijon, France

ABSTRACT

Widely used in microelectronics and optoelectronics; Gallium Arsenide (GaAs) is a III-V crystal with several interesting properties for microsystem and biosensor applications. Among these; its piezoelectric properties and the ability to directly biofunctionalize the bare surface, offer an opportunity to combine a highly sensitive transducer with a specific bio-interface; which are the two essential parts of a biosensor. To optimize the biorecognition part; it is necessary to control protein coverage and the binding affinity of the protein layer on the GaAs surface. In this paper; we investigate the potential of a specific chemical interface composed of thiolate molecules with different chain lengths; possessing hydroxyl (MUDO; for 11-mercapto-1-undecanol ($HS(CH_2)_{11}OH$)) or carboxyl (MHDA; for mercaptohexadecanoic acid ($HS(CH_2)_{15}CO_2H$)) end groups; to reconstitute a dense and homogeneous albumin (Rat Serum Albumin; RSA) protein layer on the GaAs (100) surface. The protein monolayer formation and the covalent binding existing between RSA proteins and carboxyl end groups were characterized by atomic force microscopy (AFM) analysis. Characterization in terms of topography; protein layer thickness and stability lead us to propose the 10% MHDA/MUDO interface as the optimal chemical layer to efficiently graft proteins. This analysis was coupled with in situ MALDI-TOF mass spectrometry measurements; which proved the presence of a dense and uniform grafted protein layer on the 10% MHDA/MUDO interface. We show in this study that a critical number of carboxylic docking sites (10%) is required to obtain homogeneous and dense protein coverage on GaAs. Such a protein bio-interface is of fundamental importance to ensure a highly specific and sensitive biosensor.

Keywords: GaAs; self-assembled thiolate monolayers; proteins grafting; AFM; MALDI-TOF MS

1. INTRODUCTION

In the field of biosensors, both the transducer and the bio-specific interface are considered as the cornerstones of each device. The performances of the sensors are conditioned to control the building of the interface, which requires physical processes, chemical and biochemical functionalization steps and biorecognition events in complex biological samples. The substrate plays a key role in this regard. It combines the possibility to graft biological elements onto a surface and to detect on it specific events thanks to the transducer properties. Some typical examples are gold for surface plasmon resonance (SPR) [1] and quartz for surface acoustic wave devices [2].

Among active materials, Gallium arsenide (GaAs) is of particular interest for many reasons. Firstly, GaAs is a crystal that presents interesting physical properties, especially piezoelectric and piezoresistive effects, which are used for sensor transducing. Secondly, the microfabrication processes are well known, making it possible to miniaturize components and to develop a specific interface. In this way, we designed an original piezoelectric resonant biosensor based on lateral field excitation that generates bulk acoustic waves. This original design, presenting the electronic part separated from the biological interaction reactor, makes it possible to detect a very low mass variation of captured peptides on the surface [3]. This promising sensor structure is obtained using wet etching, which is a low-cost and reproducible process that gives a specific behavior to the surface [4] thanks to adapted microstructuration. Moreover, without the requirement of an added layer [5,6], direct protein grafting is possible through a specific interaction between crystalline facets and specific amino-acid sequences [7,8,9,10,11,12], or genetically engineered proteins [13]. Another method consists of using an intermediate chemical self-assembled monolayer [14,15] (SAM), which offers the opportunity to control the orientation of a large choice of native proteins. Moreover, this "self-assembled-based process" has some advantages, namely it is easy, fast and inexpensive.

For SAM chemistry, the possibility to use either the bare substrate or the native oxidized layer offers a large choice of chemical interfaces for biofunctionalization. In particular, alkanethiol-based SAMs on III-V semiconductors are extensively used. The precursors in this field are probably Lunt et al. [16] and Nakagawa et al. [17], but their works were performed with a view to conferring specific functionalities on the surface through the modification of chemical and electrical properties of the material [18,19,20]. The most widely studied self-assembled monolayer is the octadecanethiolates monolayer [18], but various molecules could also be adsorbed onto the bare substrate thanks to their sulfhydryl groups. Among them, different types of variable-length chains (alkane, PEG, biphenyl etc.) have been used and terminal tail groups adapted ($-CH_3$, $-OH$, $-COOH$, $-NH$, $-SH$ etc.) according to the application being envisaged [21]. For biosensor applications, different strategies could be adapted [15] as seen in Dubowski et al. [22] and Adlkofer et al. [23]. These authors give us some examples of functionalization of specific chemical interfaces adapted for biological applications.

In this paper, we developed a specific chemistry composed of mixed thiolate molecules with hydroxyl and carboxyl tail groups [24] on GaAs crystals, developed originally on gold substrates for SPR measurements. This mixed layer is well known to make it possible, on the one hand, to covalently bind proteins to the material surface thanks to the activation of the carboxylate groups with amine-reactive esters [25]; and on the other hand, to reduce non-specific adsorption, thanks to the presence of inert hydroxyl tail groups. Furthermore, different chain lengths between these two molecules are used to render the –COOH tail group more easily accessible for protein grafting [20,26,27]. It was observed by several authors that a 100% monolayer of MHDA thiolates gave bad results in terms of protein graftings, especially due to heterogeneity and imperfections in the layer: repulsions between COOH groups creating sort of "holes" in the layer, interactions between COOH groups and certain substrates. Mixing the MHDA thiolates with shorter thiolates (like MUDO here) revealed to be conducive to a better MHDA orientation, repartition and efficiency on a surface. The possibility to use these molecules to functionalize the GaAs surface has previously been shown [28,29,30]. We optimized and analyzed the covalent grafting of proteins on this chemical interface thanks to an original characterization. The methods we used were atomic force microscopy (AFM) and matrix-assisted laser desorption/ionization time-of-flight mass spectrometry (MALDI-TOF MS). We wanted to engage this characterization study on several different thiolate layers reconstituted on the GaAs surface. This means that we had to use relatively high quantities of proteins. Albumin is one of the less expensive proteins, that is the reason why we chose it. Our results give comfort to what we could expect with an antibody monolayer, that would certainly give less non-specific interactions with the surface and then even more pronounced specific results. In this work, different ratios of hydroxyl and carboxyl end groups were tested to highlight their influence on protein grafting.

2. RESULTS AND DISCUSSION

2.1. Characterization of the Presence of a Mixed Thiol Layer on GaAs Surface

Figure 1 presents the fluorescence intensity obtained on the GaAs surfaces, functionalized by 10% MHDA/90% MUDO or 0% MHDA/100% MUDO.

Through this characterization, we managed to obtain a Cy5-NHS fluorescence response on functionalized GaAs samples. The fluorescence on 10% MHDA is not uniform along the whole line (0–0.7 cm), simply because the GaAs sample was not entirely covered by the NHS-fluorescent probe during the graftings. Only one part of the GaAs functionalized surface (roughly 1 cm²) was covered by the NHS fluorescence probe. This allows us to have a control area on the same sample just next to the area covered by the probe. Thus, the fluorescence proves 1) the presence of COOH groups on the mixed thiolate layer, and 2) the binding specificity of the NHS-molecule (here cy5-NHS, and then EDC-NHS used for protein grafting in the following step) to the MHDA/MUDO functionalized GaAs surface.

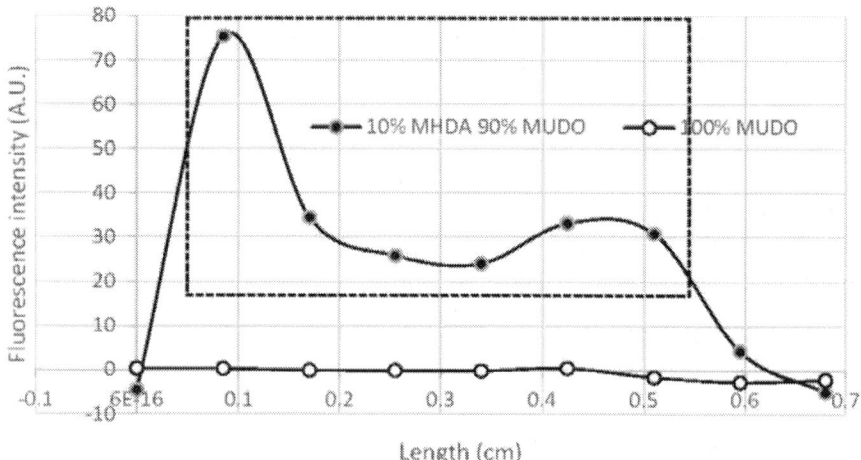

Figure 1. GaAs functionalized surfaces after incubation with Cy5-NHS fluorophore. GaAs samples functionalized by 10% mercaptohexadecanoic acid (MHDA)/90% 11-mercapto-1-undecanol (MUDO) (black dots) or 0% MHDA/100% MUDO (transparent dots). Fluorescence intensity emission at 680 nm is given in arbitrary units, while exactly the same conditions of excitation and filters were used in both cases of functionalization. The fluorescence was registered on the GaAs sample along a straight line, crossing the area functionalized by cy5-NHS (the cy5-NHS drop covers roughly 1 cm² of the GaAs sample—the fluorescence measurements gave the results in the window that is delimited by the dashed line on the graph).

2.2. Chemisorption of Mixed Thiol Layer on GaAs Surface

Before protein grafting, we checked the organization of a 10% MHDA layer on the GaAs (100) surface. Figure 2 shows the surface topography of this layer.

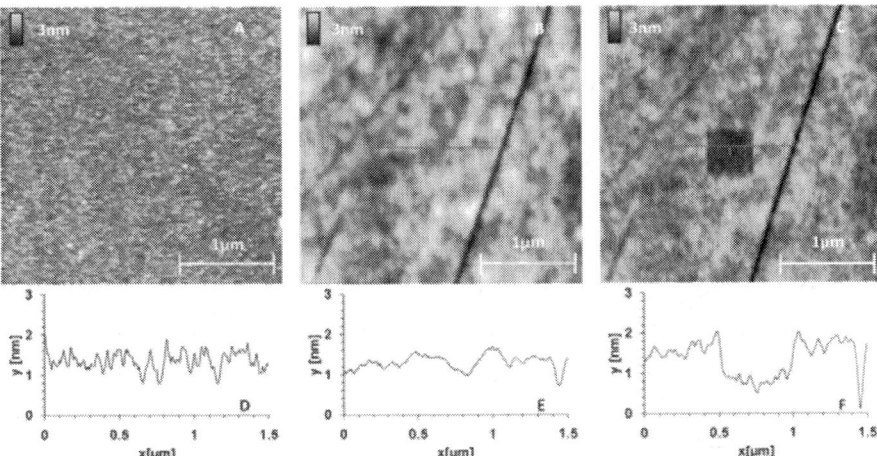

Figure 2. GaAs surface before (**A**) and after (**B,C**) thiolate functionalization (10% MHDA) and their corresponding sections (respectively **D,E,F**).

Figure 2A shows the initial surface state of a bare GaAs wafer. Figure 2B,C correspond to the same functionalized GaAs surface, without and with a 500 nm scraped zone respectively. Images of cleaned and functionalized surfaces show a very low roughness (z range 3 nm) corresponding to a highly smooth surface similar to those presented in other publications [8,23,31,32,33]. Since the GaAs sample is polished on one side (see experimental section), topography images can reveal the presence of stripes on the substrate. These stripes give the black lines on the AFM images in Figure 2B,C).

The functionalized surface is even smoother (Figure 2E), which is confirmed by the decrease in arithmetic average and root mean square roughness values (Ra and RMS respectively), from Ra = 0.23 nm/RMS = 0.32 nm for a naked surface to Ra = 0.18 nm/RMS = 0.23 nm for a chemically functionalized surface. To measure the thickness of this layer, we made a hole by scraping a square of 500 nm width with a stiff cantilever (k = 0.58 N/m) (Figure 2C). This scraping makes it possible to locally remove the layer of thiols as seen by a well-defined dark square. The section profile (Figure 2F) highlights the material removal at this place and we observed a layer thickness of approximately 1 nm. Although this method precludes precise estimation of the thickness of the mixed thiol layer and any conclusion regarding its organization or composition, the AFM results tend to demonstrate that a homogeneous mixed thiolate layer has been established. The obtained thickness suggests that this layer is organized as a monolayer because this value is in the same range as those found by Zhou and Walker [34] with single wavelength ellipsometry (SWE) measurements: 1.7 nm for MHDA and 1.4 nm for MUA (for 11-mercaptoundecanoic acid, a thiolate composed of 11 carbons, terminating with a carboxyl tail group). Then, we used complementary characterization methods like contact angle, ellipsometry and X-ray photoelectron spectroscopy (XPS) to validate the good organization and composition of this chemical layer. These results have previously been published [28].

2.3. Protein Grafting on the Functionalized GaAs Surface

The protein was covalently grafted to MHDA / MUDO mixed SAM through its amine groups using the EDC/NHS activation of COOH docking sites presented by MHDA molecules. After the incubation of this surface with proteins, followed by a washing step with a 40 mM Octyl Glucoside solution, carboxyl group deactivation was finally performed with ethanolamine. The protein covered surfaces (Figure 3D–F) and some references surfaces (Figure 3A–C) were characterized.

The images (Figure 3) give results on different chemically modified GaAs surfaces in the same experimental imaging conditions, as regards AFM tips and scan settings. Repeatability and reliability of results were controlled: three samples for each condition were taken, and measurements were performed in different places on each sample. Compared to the reference sample (Figure 3A), we clearly observed a modification of the GaAs surface after the protein grafting protocol on the different chemically modified substrates. The initial very smooth surface (z range of 5 nm) was modified and the roughness strongly increased after this step (z range of 20 nm for the other samples). Figure 4 gives RMS values as a function of the MHDA percentage. The unfunctionalized (Figure 3B) and 0% MHDA (Figure 3C) surfaces behaved similarly, presenting fine and small grains. These motifs are probably proteins or aggregates of proteins. In spite of the detergent washing, some proteins or aggregates remained absorbed on the surface. Nevertheless, the substitution of some MUDO by MHDA molecules induced a change on the surface topography and bigger motifs were observed in these cases. With the mechanical filtering induced by the tip, the observed motifs had a size between 5 and 15 nm, which is consistent with 3–8 nm RSA protein dimensions [35]. The 10% MHDA sample shows a highly rough surface compared to the others (RMS = 2.63 nm) and grafted proteins form a dense layer on the GaAs surface. The 3% MHDA covered surface presents an intermediate value of RMS. The 100% MHDA surface shows the least roughness. In this last case, it may be that the presence of supernumerary docking sites (100% MHDA) induced RSA grafting by several sites. These multiple protein-surface bonds could tend to flatten the proteins on the surface, thus smoothing out the surface (RMS = 0.98 nm).

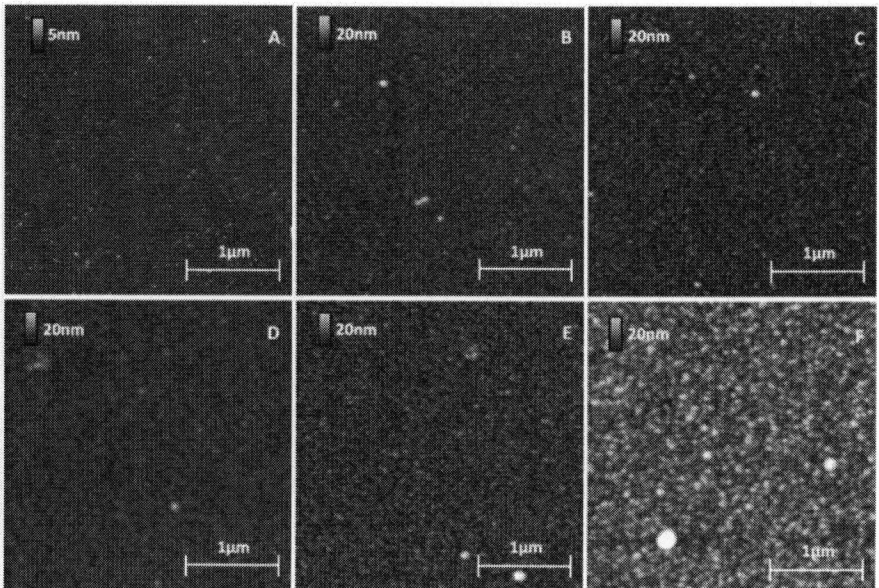

Figure 3. GaAs surface images before (**A**) and after protein grafting for several chemical interfaces: unfunctionalized (**B**); 0% MHDA (**C**); 100% MHDA (**D**); 3% MHDA (**E**); and 10% MHDA (**F**).

The corresponding interpretation of these roughness values may be due to the number of docking sites offered by the different SAM layers. Indeed, if we assume the protein diameter to be 5 nm, a complete and saturated protein monolayer would give coverage of around 50,000 proteins per μm^2. The density of the thiolate layer is approximately 5×10^6 molecules per μm^2, because the surface occupied by a thiol on the GaAs (100) surface is roughly 20 Å^2 [18,36]. The surface ratio of thiolates to protein would be 100 thiolates for one protein. Theoretically, with this complementary information, self-assembled monolayers composed of 3%, 10% or 100% MHDA should offer successively 3, 10 and 100 docking sites for one protein. At this stage, we have to keep in mind that the ratio of thiolates in the solution is certainly not respected on the surface, due to several parameters like the thiolate functional groups and their chain length. Moreover, if we consider that MHDA molecules are non-uniformly distributed and that only a part of the carboxyl groups reacts (only 10% of docking sites activated for Ding et al. [37]), we can assume that self-assembled monolayers are composed of an insufficient number of docking sites at 3% MHDA, an excess at 100% MHDA and a good compromise at 10% MHDA. Based on the AFM images, the 10% MHDA layer seems to be an ideal ratio to immobilize enough proteins to create a dense protein layer. Similar to our AFM images of RSA grafted on 10% MHDA surface, Duplan et al. [32] showed AFM images of immobilized neutravidin (molecules close to RSA in size, with a molecular weight of 60,000 Da, versus 64,500 Da for RSA) on GaAs modified by polyethylene glycol mixed thiol, and confirmed the ability to form an organized, densely packed protein layer on the GaAs surface. The authors used a ratio of 1/15 (6.66%) biotinylated docking sites distributed in a hydroxyl terminated PEG layer.

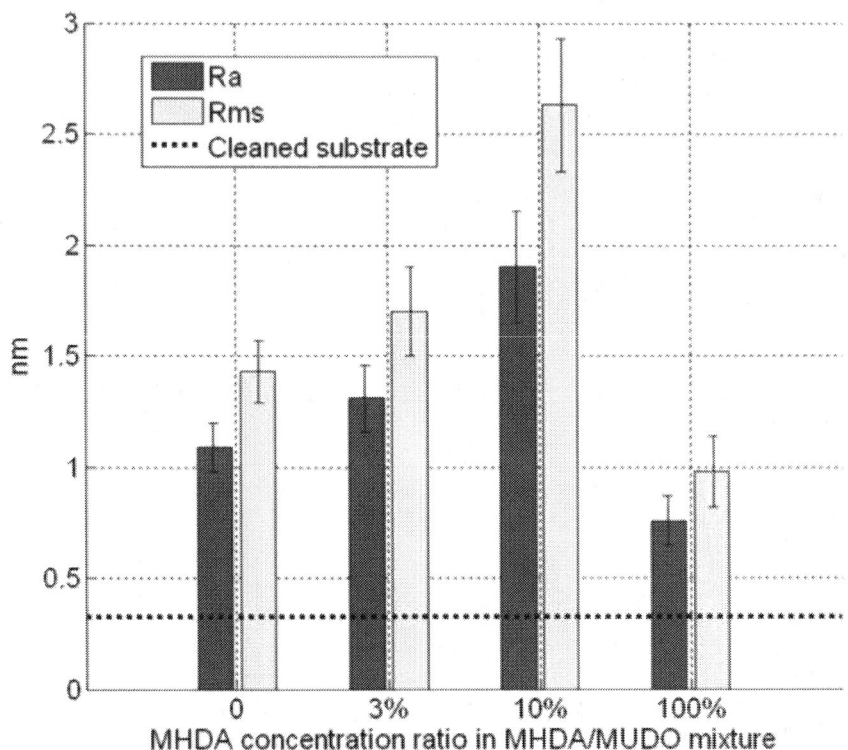

Figure 4. GaAs surface roughness after protein incubation on different chemical interfaces.

2.4. Protein Covalent Binding or Physisorption?

Figure 5 shows surface modifications after the protein deposition step, although the cleaned bare and 0% MHDA functionalized surfaces do not present functional chemical groups able to fix protein. To highlight the binding properties between the protein and each chemically modified surface, we tried to scratch the protein and thiolate layers by applying a controlled contact force with the tip on the surface. As explained in the experimental section, various AFM tips were used to apply medium or strong scraping tests. The applied force on the sample was adjusted by varying the contact setpoint and the same settings were used for each sample. A first set of tests was performed by applying a moderate force on the surface (scraping zone of 500 nm by 500 nm) thanks to a medium stiff tip (NPS10-B, k = 0.12 N/m). Results are reported in Figure 5.

We observed that:

(i) The 0% MHDA and bare substrate surface (not presented) have the same typical physisorption behavior when incubated with the protein solution. The gallium arsenide surface is known to be attractive for protein adsorption but in our experiments, the GaAs side was relatively poorly covered after protein incubation, probably due to the efficiency of OG washing, as shown previously by Ding et al. [37]. On AFM images obtained on naked GaAs surfaces (data not shown), the step between the protein layer and the substrate was almost invisible and the rolls were very small. The GaAs naked surface does not seem to adsorb a lot of RSA protein.

(ii) On MUDO functionalized GaAs surfaces, the deposited proteins were easily scratched by this test and a square zone 500 nm wide and 3.5 nm thick appeared. On the sides of this square, we observed rolls corresponding to a heap of proteins displaced by the tip. The visible rolls prove that proteins are present on MUDO surfaces, but only weakly adsorbed since they are easily removed from the surface by a medium scraping test.

(iii) On the contrary, the 100% MHDA surface is not fundamentally affected by this scraping test. In the scraped zone, the surface appears "compacted" and the molecules appear to be stretched but not removed under the force applied by the tip. As the consequences of the scraping were limited to the apparent spreading (without pulling off) of molecules, it tends to prove that proteins are strongly attached to the GaAs substrate.

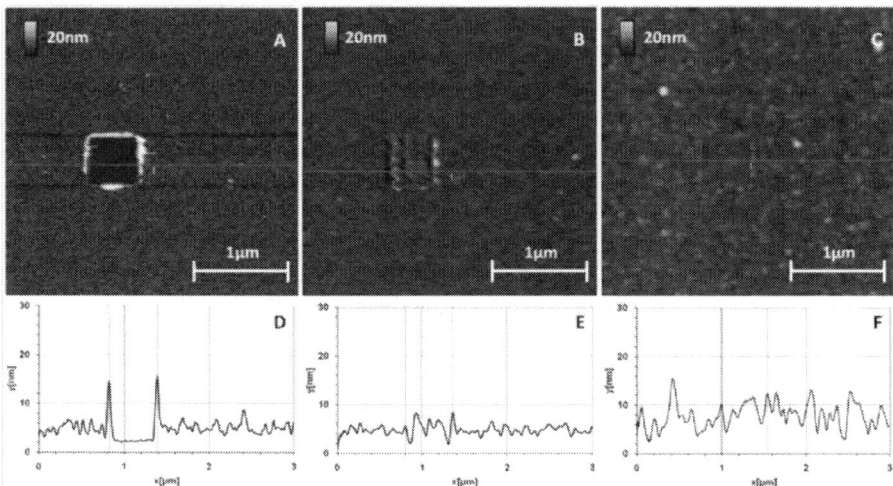

Figure 5. Atomic force microscopy (AFM) images of medium scrapings (stiff tip k = 0.12 N/m) applied on GaAs bio-functionalized surfaces and corresponding sections: 0% MHDA (**A,D**); 100% MHDA (**B,E**); and 10% MHDA (**C,F**).

The mixed layer containing 10% MHDA behaved similarly to 100% MHDA: the surface on the scraped zone appeared "compacted" with molecules that stayed on the surface while applying moderate force. Surprisingly, only 10% MHDA thus seems sufficient to toggle to a strongly bound protein layer. This result highlights the covalent binding that exists between the substrate and the protein, when MHDA molecules constitute the chemical interface.

On the section of Figure 5D, corresponding to the 0% MHDA surface, we show the height of rolls due to the scraping. The considerable height (10–15 nm) is evidence of the material having been removed from the scraped zone. At the bottom of this 500 nm wide hole, we observed a very flat surface corresponding to the GaAs substrate, and we measured a protein layer thickness of 3.5 nm (see explanation below for strong scraping and Figure 6).

The protein adsorption observed on the 0% MHDA layer is surprising because the hydroxyl tail group of MUDO is known to limit non-specific adsorption compared to other tail groups, and the OG washing should have completely removed the protein on the bare substrate. Recent articles [18,29,38,39] provide a possible explanation for this phenomenon. The length of the MUDO alkane chain, composed of 11 carbon atoms, is relatively short and it seems that a chain of 15 or 16 carbon atoms is the minimum chain length required to obtain a high degree of self-organization. This disorganization in the MUDO layer could then induce a sort of "porous" and loose chemical layer, in which aliphatic chains could interact with proteins through hydrophobic interactions. The addition of MHDA molecules (16 carbon atoms) in the functionalization process seems to facilitate the formation of a more densely packed protein layer, because no pulling-off at all was observed on mixed and 100% MHDA layers, as can be seen in Figure 5B,C.

2.5. Thickness of the Combined Thiol/Protein Layer

Additional scraping experiments were performed in order to totally remove the grafted protein layer and the thiol chemical interface. Similar experiments as in the previous section were performed, on a square area of 1 μm wide using the stiffest tip, an NPS10-A (k = 0.58 N/m). The force applied was considerably greater with this approach. The images and corresponding sections are presented in Figure 6.

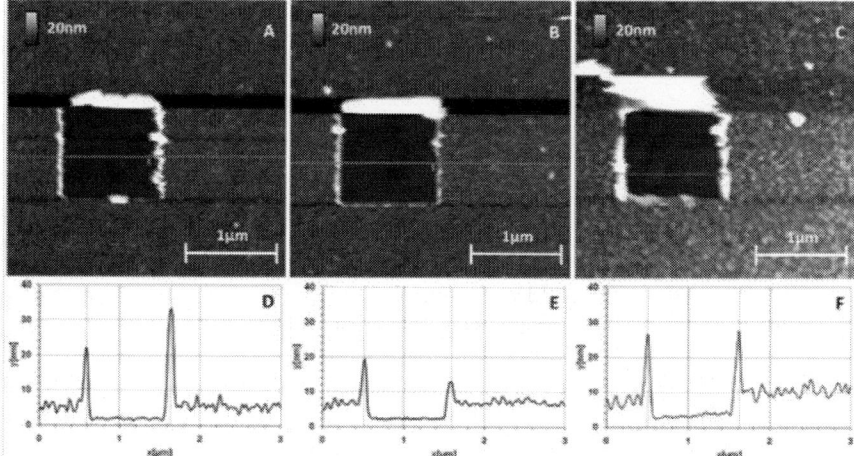

Figure 6. AFM images after strong scraping applied to GaAs biofunctionalized surfaces. Images and corresponding sections: 0% MHDA (**A,D**); 100% MHDA (**B,E**); and 10% MHDA (**C,F**).

For each surface, the thiolate/protein layers were scratched and we observed a well-defined square of 1 μm wide corresponding to the scraped zone. For bare substrate (not presented) and 0% MHDA (Figure 6A), the images are identical to these obtained by the first medium-scraping test, supporting the idea that the protein layer is just adsorbed on the surface and is not strongly fixed. Applying a high force made it possible to remove materials from the surfaces composed of 100% MHDA (Figure 6B) and mixed interfaces (3% MHDA (data not shown), 10% MHDA (Figure 6C)). Large rolls were formed around the scraped zone, proving the displacement of proteins and probably of thiolate molecules. The volume of this roll makes it possible to establish a comparison of the quantity of scraped molecules. Cleaned bare substrate (not presented), 0% MHDA (Figure 6A), 100% MHDA (Figure 6B), 3% MHDA (not presented) and 10% MHDA (Figure 6C) show, in this order, the smallest to largest roll of scratched molecules. The corresponding sections of these images are presented in Figure 6D–F. In the scraping zone, we observe similar flat surfaces. In order to analyze the roughness of these flat surfaces, we performed cross-sections inside and outside the scraped zone. The corresponding RMS values are reported in Figure 7.

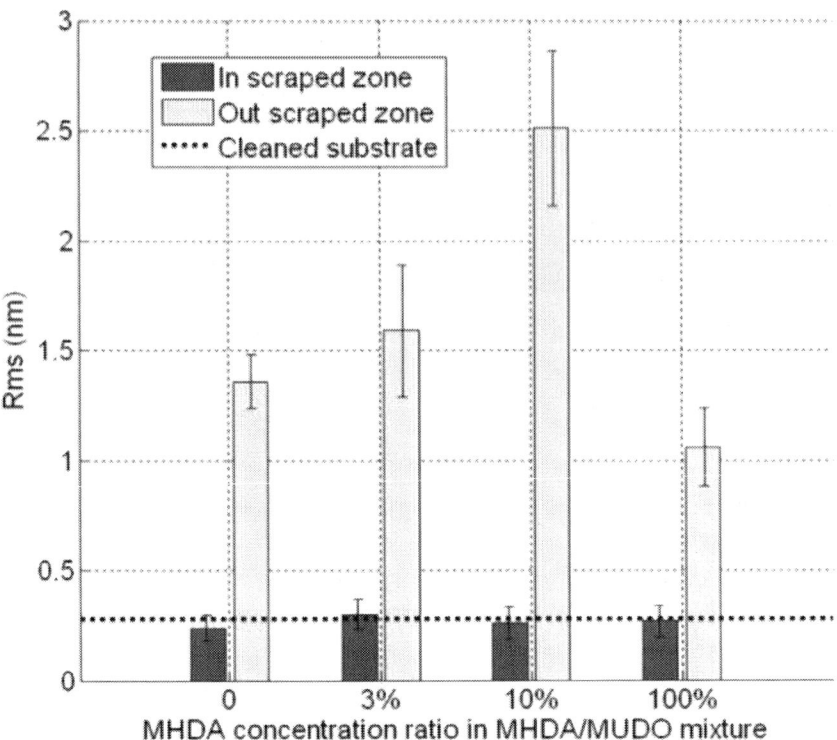

Figure 7. RMS values of protein modified GaAs surfaces, inside and outside the scraped zones, for each chemically modified GaAs interface.

As expected, outside the scraped zone, the RMS values are similar to these obtained in Figure 4. This result demonstrates the reproducibility of our GaAs biofunctionalization process. In the scraped zones, RMS values are approximately equal to 0.3 nm, which corresponds to the initial roughness of unfunctionalized substrate (dotted line). This result would indicate that, with the high force applied here, we scratched both thiolate and protein layers. Similarly, we measured the thickness of the bio-interface (Figure 8). The 0% MHDA profile (Figure 6D) is similar to Figure 5D and the average thickness of the layer is 3.5 nm (Figure 6D). This again highlights the reproducibility of our surface biofunctionalization process. The 100% MHDA sample presents a thickness of 4.4 nm (Figure 6E); mixed surfaces have a thickness of 4 nm and 5.8 nm for 3% MHDA and 10% MHDA (Figure 6F) respectively.

These thickness values could correspond to the superposition of the two layers: a 1.5 nm thin thiol layer [34] and the Rat Serum Albumin (RSA) protein layer [35]. The highest value, obtained at 10% MHDA, is in agreement with previous experiments, because the protein molecules do not tend to flatten on the surface due to the proximity of other neighboring RSA proteins. These results are in line with previous observations, namely:

The 0% MHDA surface allows protein adsorption.

The 100% MHDA surface presents a number of docking sites in excess of those required to obtain a protein monolayer. Indeed, on highly dense MHDA covered surfaces, proteins could graft to the surface, by engaging multiple free primary amine groups. The consequence of this could be the flattening of the protein on the surface.

The 3% MHDA surface is certainly limited in terms of number of docking sites, thereby reducing the density of protein coverage on this surface.

The 10% MHDA surface appears to be the best candidate, since this surface allows grafting of a dense, homogeneous and stable layer of proteins, and the number of docking sites appears to be well adapted for a biosensor interface.

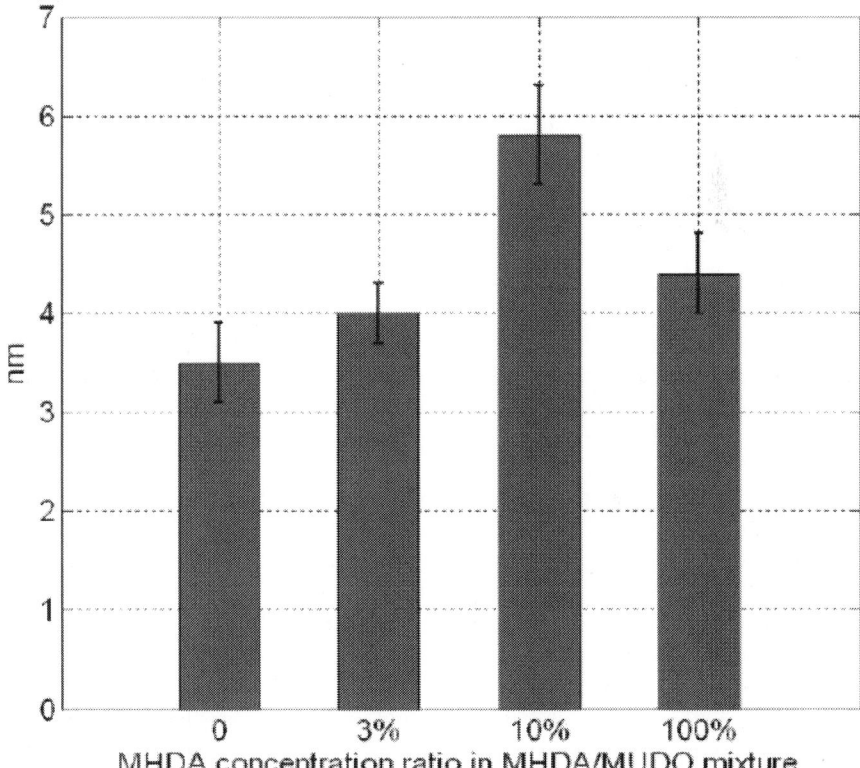

Figure 8. Thickness of the combined thiolate and protein layer, measured for each chemical interface.

2.6. *In Situ MALDI-TOF MS Analysis on GaAs Surfaces Modified by Proteins: Identification and Quantitative Distribution*

In parallel to AFM characterization of protein modified GaAs surfaces, we performed mass spectrometry analysis, in order to analyze the potential of the proteins present on GaAs to be ionized and identified. To this end, we split each GaAs sample into two parts to perform, firstly, AFM characterization, and secondly, mass spectrometry analysis. With these measurements, we aimed to investigate the correlation between the density of motifs observed by AFM and the MS signal intensity of the protein. The MS results yielded qualitative information regarding the protein layer coverage on the different chemically modified GaAs surfaces. Moreover, thanks to automatic measurements at different places on the whole surface, we were able to perform quantitative comparisons between each sample. This AFM/MALDI-MS coupled analysis is truly original, since, without the need for labeling, it provides complementary qualitative information beyond that obtained by photoluminescence [5,40] or labeled techniques like fluorescence [32,37,40].

The surface preparation, consisting of spraying TCEP, trypsin and HCCA matrix on the GaAs modified surfaces, was done on the whole surface, corresponding to a square area of 10 mm width. Afterwards, we screened the whole surface pitch by pitch with MS measurements to verify the homogeneity of protein coverage. Each pitch of this test is defined by a 120 μm wide square, composed of 2000 shots randomly distributed at steps of 20 μm. Among these measurements, we chose one position on each chip that showed the highest signal intensity. The protein generated trypsin fragments were matched to a database sequence. The protein having the closest peptide sequence was defined as the identified protein. A score, based on Mascot algorithm, was then attributed. This algorithm evaluates the probability that the identified protein is not a random match, i.e. that the identified protein is unambiguously the right one. In the same way, a Mascot score was attributed for MS^2 analysis, to compare detected and theoretical fragmentations of a specific peptide (here the 1960 Da peptide). Table 1 summarizes these results.

Table 1. MS and MS^2 results obtained on protein modified GaAs substrates.

MHDA concentration	MS			MS^2 (peptide 1960 Da)
	Identified protein	Matched peptides	Mascot score	Mascot score
Bare substrate	–	–	–	–
0%	RSA	8	85.8	16.42
3%	RSA	8	66.9	21.78
10%	RSA	9	87	59.92
100%	RSA	10	101	17.68

The RSA protein was identified on each surface, except on bare GaAs substrate. The unfunctionalized sample spectra do not exhibit peaks, proving that RSA protein was not adsorbed on the bare GaAs surface. For other MHDA/MUDO surfaces, the mascot scores were significant for all GaAs substrates, which confirms that the motifs observed by AFM truly correspond to RSA proteins. Among eight peptides matched on 0% MHDA and 3% MHDA samples, only one peptide differed between samples. This could partly explain the lowest score of RSA identification on the 3% MHDA sample. For 10% MHDA, the sequence covered was the combination of the two spectra detected on the 0% MHDA and 3% MHDA samples, therefore increasing the score. The addition of the 983.6 Da matched peptide to this sequence resulted in the 100% MHDA surface having the highest score. The mascot score obtained and the number of matched peptides with GaAs substrates were close to the values obtained on a gold reference chip with the same biofunctionalization protocol.

The MS^2 measurements were performed on each functionalized chip. Mascot scores were calculated for the 1960 Da peptide, which is a specific peptide of RSA protein. A score of 60 was obtained on the 10% MHDA sample, but the score was around 20 for the other functionalized GaAs chips. The score obtained at 10% MHDA was the same as on the gold reference chip, providing unambiguous proof of the presence of the peptide 1960 Da on the surface, and highlighting that it is possible to investigate peptides in situ on the GaAs surface. The MS^2 spectrum of this surface is presented in Figure 9.

If the MS intensity of peak 1960 Da (Figure 10) reflected the amount of this peptide present on the surface, we would get the highest MS^2 score for 0% MHDA, a medium MS^2 score for 10% MHDA, and low MS^2 scores for 3% and 100% MHDA. Our results (Table 2, fifth column) seem to follow this pattern, except for the 0% MHDA surface, which gave the lowest score. This low value indicates that there are many species that are not attributed to this 1960 Da peak, giving us reason to think that this chemically modified surface induces perturbations in either protein fragmentation or peptide desorption, or even both.

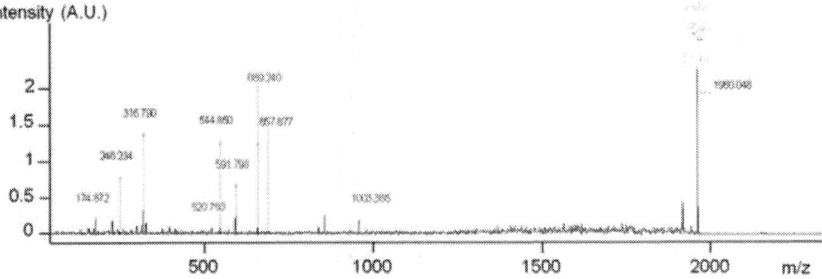

Figure 9. MS^2 analysis of the 1960 Da peak on GaAs (100) functionalized with 10% MHDA after Rat Serum Albumin (RSA) grafting.

Using the same experimental conditions, on a 2 mm by 2 mm defined area, we obtained semi-quantitative characterization of the protein interface on the GaAs surface, through the comparison of the matched peptide intensities. On each protein-modified GaAs surface, we collected a matrix of 17 by 17 MS spectra with the previous shooting conditions, and we summed the intensities of each matched peptide peak. The result is reported in Figure 10.

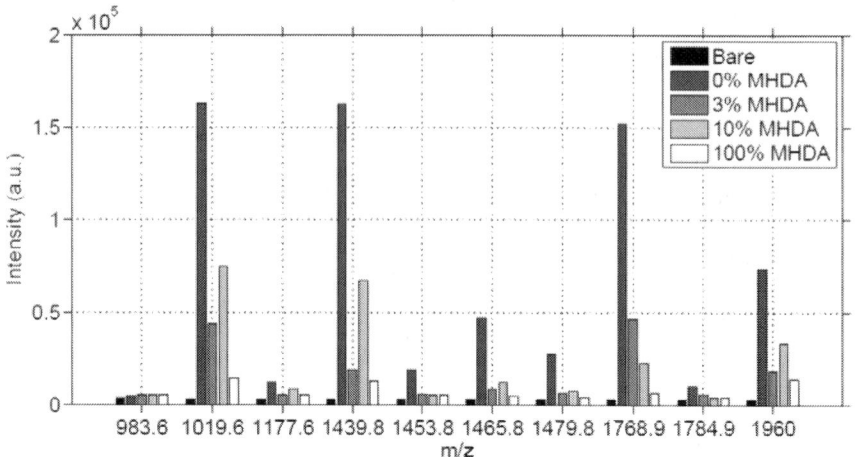

Figure 10. Sum of MS intensities for each matched peak, obtained on 2mm by 2mm unmodified (bare) and functionalized GaAs surfaces, after incubation in the RSA solution.

In Figure 10, for most of the MS peaks, we observed a high intensity for 0% MHDA, a medium intensity for 10% MHDA and low intensities for 3% and 100% MHDA. The bare substrate gives a noise reference and the low intensities of its peaks show once again that no RSA protein was adsorbed, proving the efficiency of the washing protocol. The highest intensity of the 0% MHDA surface is very surprising, meaning that this MUDO interface would authorize albumin non-specific interactions. Even after the OG washing step on this MUDO surface, proteins remained stuck on it. Probably the presence of hydroxyl terminal groups on MUDO enables the establishment of weak but numerous interactions with proteins.

The intensities of surfaces covered with MHDA molecules show a gradation, from lowest to highest intensity, with MHDA concentrations of 100%, 3% and 10%. This gradation is expected, although it differs slightly from the AFM analysis. To illustrate these results, we report in Figure 11 the intensity of the 1960 Da peak for each position on the 4 mm^2 tested surfaces.

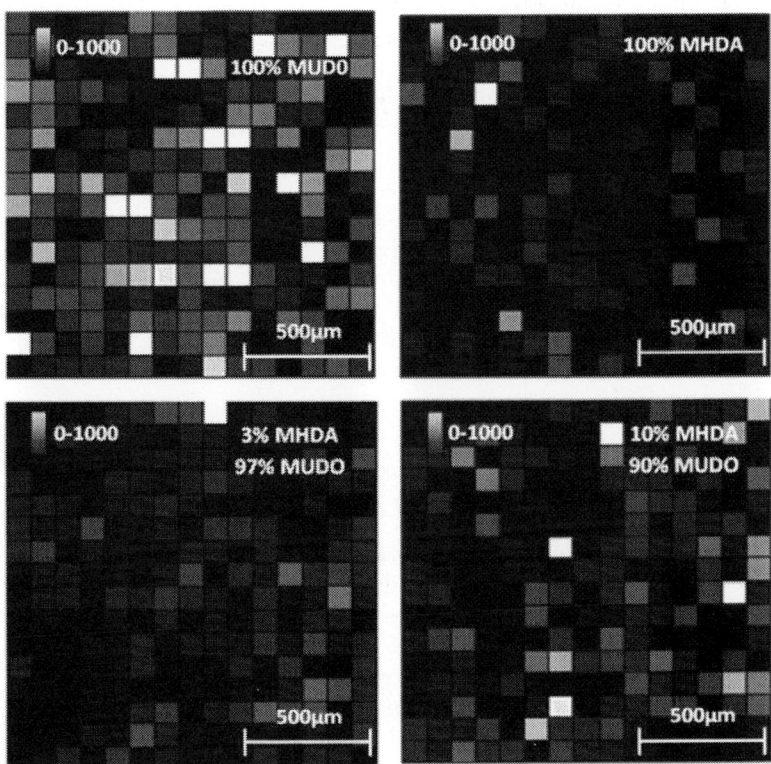

Figure 11. Intensity images (arbitrary unit) of the 1960 Da peak for each chemically functionalized GaAs surface after incubation in the RSA solution.

We observe that the 0% MHDA surface (Figure 11) apparently presents the highest amount of RSA proteins. This observation differs from the AFM images, showing small and scattered RSA motifs, in comparison with MHDA surfaces. How can this difference between AFM and MS results be explained? It could be purported that peptide ionization is easier on the 100% MUDO (0% MHDA) interface than on the MHDA surface, and that more protein molecules are desorbed by the laser from the 100% MUDO surface. Peptide desorption could be facilitated by the disorganized MUDO thiolate monolayer, and by the lower interactions existing between the RSA protein and the chemical interface. In fact, as was observed during AFM scraping measurements, the link between protein and chemical interface is weak for 100% MUDO and strong for MHDA interfaces. Griesser et al. [41] proved that peptide MALDI ion signals decrease as the surface-peptide binding affinity increases, and that peptide ionization is very sensitive to the nature of the chemical interaction between the surface and the peptide [42]. This phenomenon is particularly true for whole protein detection. Although trypsin digestion strongly reduces this limitation, it may be possible that in our case, the different chemical interfaces play an important role during this process. The weak interaction between peptides and the alkane chain and the non-presence of carboxylic acids could facilitate molecule desorption, and generate a higher intensity of signal response.

This theory seems to be correct, in light of the results obtained on the 100% MHDA surface. The AFM images show that the protein layer is not significantly less dense than on 3% MHDA, but an over-abundance of docking sites that multi-bond RSA peptides could explain the low MS signal. For 3% and 10% mixed chemical interfaces, the MS results are less surprising, and are consistent with AFM observations: the more MHDA in the chemical layer, the more intense the MS signal. To determine the optimal number of docking sites necessary to obtain a well-organized and dense protein layer, we recently performed complementary analyses of surfaces containing percentages of MHDA molecules greater than 10 percent. These results are given in the following section.

2.7. Density of Carboxylic Docking Sites and Consequences on the Reconstituted RSA Protein Layer on (100) GaAs

In the previous results, we showed that, on the one hand, a concentration of less than 10% MHDA is not sufficient to graft a dense and homogeneous layer of proteins; and on the other hand, that a layer composed of 100% MHDA molecules induced multi-bonding of the proteins to the surface, which somehow flattened the proteins and reduced laser desorption of this protein layer. We investigated the behavior of the protein layer on chemical interfaces composed of more than 10% MHDA.

We applied the same AFM and MS characterization methods as previously described to chemical layers obtained from a solution containing respectively 20% and 50% MHDA molecules. The AFM and MS images are reported in Figure 12.

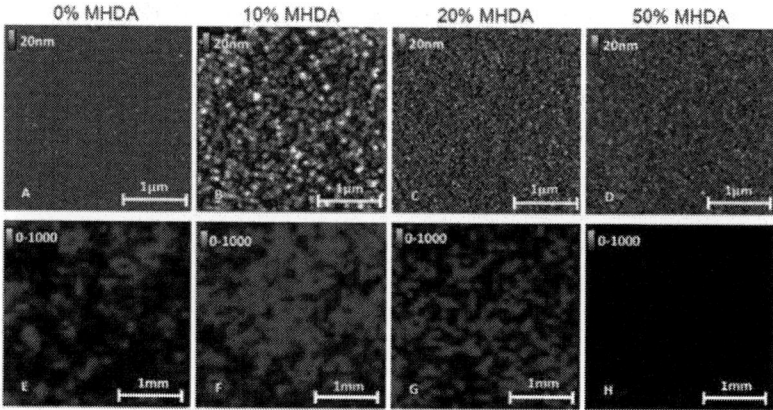

Figure 12. AFM images (**A–D**) and MS imaging results (**E–H**) for 0%–50% MHDA concentration: 0% MHDA (**A,E**); 10% MHDA (**B,F**); 20% MHDA (**C,G**); 50% MHDA (**D,H**).

Images A, B, E and F in Figure 12 give the same results as those previously observed for 0% MHDA (A,E) and 10% MHDA surfaces (B,F). On the 0% MHDA surface, we observed sparse patterns corresponding to RSA protein, and only locally high MS signal intensity. The 10% MHDA surface shows the highest organized protein layer and a strong MS signal. For highest values of MHDA (C,D,G,H), we observe a drastic decrease in the pattern sizes. The AFM images of these surfaces show the lowest patterns, whereas the MS signal intensity dramatically dropped and was even absent for 20% and 50% MHDA respectively. These results are in accordance with the hypothesis of multi-bonded proteins on such highly MHDA-concentrated surfaces. This observation suggests that a concentration of 10% MHDA apparently offers an optimal number of docking sites to graft a protein monolayer homogeneously and correctly.

3. EXPERIMENTAL SECTION

3.1. GaAs Samples

All GaAs samples were cut from a unique wafer provided by Azelis Electronics. This wafer was (100) oriented, semi-insulating and undoped. Only one side was polished, which avoids confusion during handling. We divided the wafer into several samples according to the advancement of the protocol using the (110) cleavage planes of the material to confirm that the initial surface was identical and that AFM and MALDI-TOF MS measurements were performed on the same original sample.

3.2. Functionalization of GaAs Samples

Samples were functionalized using a protocol previously described and characterized [28]. The first step consists of cleaning the surface using successively Decon®, acetone, sulfochromic acid (98% H_2SO_4, 2% $K_2Cr_2O_7$) and ethanol. Samples were rinsed with deionized water and were dried with nitrogen gas between each step. Then, to deoxidize the GaAs surface, a $HCl:H_2O$ (1:1) solution was used. The reaction is automatically stopped (after 2 min at ambient temperature) because the acid solution only etches the oxide present on the surface revealing Ga and As atoms, which are necessary for thiolate functionalization (Ga–S and As–S bonds)[4]. Next, surfaces were functionalized by a self- assembled monolayer (SAM) of thiols. Two kinds of thiols were tested (Sigma-Aldrich®): mercaptohexadecanoic acid ($HS(CH_2)_{15}CO_2H$, called MHDA) and 11-mercapto-1-undecanol ($HS(CH_2)_{11}OH$, called MUDO). Thiolate functionalization was achieved by immersion in solutions composed of 1 mM MHDA and/or MUDO diluted in ethanol, and incubation was carried out overnight at room temperature. Six solutions composed of X% of MHDA and Y% of MUDO, noted X% MHDA, were tested with X = {0, 3, 10, 20, 50, 100} and Y = 100 − X. After functionalization, protein grafting was performed. This protocol consists of:

The activation of carboxyl groups presented by MHDA molecules with N-hydroxysuccinimide (NHS) and N-(3-dimethylaminopropyl)-N-ethylcarbodiimide (EDC). (NHS coupling kit, Biacore®, 30 min), Rat serum albumin protein (RSA) grafting in an ultrasonic bath, prepared at 40 μg/mL in an acetate buffer solution at pH 5.2. (Sigma-Aldrich®, 20 min). Before any grafting of protein to a functionalized surface, the conditions for a better grafting have to be optimized. For that, we used to screen a range of pH values, usually from 4 to 7, for each protein, in order to determine the pH that allows the highest quantity of proteins to bind to the surface. Indeed, on MUDO/MHDA, after activation of COOH by EDC-NHS, the better pH for protein grafting is often a pH close to the isoelectric point of the protein. This pH is close to pH 5 for RSA protein.

Non-specific adsorption removed using Octyl β-d-glucopyranoside (OG) diluted in phosphate buffer saline solution (40 mM, PBS 1x, pH 7). (Sigma-Aldrich®, 10 min).

Passivation of GaAs surface with Ethanolamine (NHS coupling kit, Biacore®, 30 min).

3.3. Characterization Methods

In order to prove the chemical functionalization of the GaAs samples by the mixture of thiolates, we performed fluorescence characterization of the surface, using an NHS-modified probe, Cyanin5-mono NHS ester (from Amersham Biosciences), which is able to bind specifically to COOH chemical groups present on the functionalized GaAs surface. Using this probe, we sought to prove: 1) the presence of COOH groups on the GaAs surface after functionalization, and 2) that specific binding of the protein on the functionalized GaAs surface occurs after the EDC-NHS activation step of the thiolate layer. For this characterization, we used a Confocal fluorescence sensor (Fluo Sens DD, from ESE Embedded Systems Engineering, GmbH, Germany) that enables scanning of the sample at a desired and constant distance, while registering the fluorescence emitted by the sample at 680 nm. This sensor was used in E2D2 mode, which means with excitation and emission wavelengths of 625 and 680 nm respectively. To achieve this characterization, after deoxidation and thiolate functionalization of the GaAs sample (either by 10% MHDA/90% MUDO or by 0% MHDA/100% MUDO), we incubated Cy5-NHS (at 10 μg/mL, in distillated water) for 30 min, rinsed extensively with water, then measured the fluorescence intensity of the surface. This approach gives a graph in arbitrary units (A.U.) corresponding to the intensity of fluorescence emission at 680 nm, as a function of the scanned sample area in cm.

An original combination of characterization methods, namely atomic force microscopy (AFM) and MALDI-TOF mass spectrometry (MS), was used to highlight respectively topographical modifications after protein immobilization, and to prove the potential for in situ protein detection on the GaAs surface. The AFM is a Veeco® Multimode 2 with a nanoscope IIIa controller, used in contact-air mode. The Veeco® silicon nitride NPS10 tips and appropriate scan settings were used to obtain the best scanning conditions for each test and were kept constant from one sample to another. Cantilever stiffness was adapted: we used an NPS10-D tip (low stiffness, k = 0.06 N/m) for classical imaging, a NPS10-B (k = 0.12 N/m) for medium scraping and a NPS10-A (greatest stiffness, k = 0.58 N/m) for strong scraping. In this paper, we present only images of 3 µm width, but we also performed smaller and larger scans to respectively obtain better resolution and verify the homogeneity of the layer on a large scale. Measurements were repeated in several places in order to validate images for each surface. The MALDI-TOF mass spectrometer is a Bruker Daltonics® UltrafleXtreme. A home-made MALDI target was used to analyze samples in the chamber. The chips were processed with the automatic imageprep standard protocol. This procedure includes three steps: TCEP reduction (Tris(2-carboxyethylphosphine) 20 mM in NH_4HCO_3 buffer, 10 min at 37 °C), trypsin digestion (Trypsin Gold Mass Spectrometry Grade, Promega®, 10 ng/µL, 37 °C) and HCCA matrix deposition (α-cyano-4-hydroxycinnamic acid, Bruker Daltonics®, 1.5 mg/mL in 50/50 acetonitrile/TFA 0.25%). The automatic mode, FlexImaging, was used with constant parameters (number of shots 2000, random walk 20 µm and pitch 120 µm) to allow comparison of each sample. The data from mass spectrometry images were exported and treated with home-made software. Additional tandem mass spectrometry (MS^2) analysis, consisting in a second fragmentation of one specific peptide (here, the 1960 Da peptide) was performed to identify the protein unambiguously. The research for identification was performed with Mascot Matrix Science in the SwissProt database. Table 2 summarizes the different chemical interfaces tested on GaAs surfaces with or without RSA grafting.

Table 2. Different chemical interfaces tested on the GaAs surface with or without RSA grafting.

N	Thiol SAM	RSA	Characterization methods
1	without	without	AFM
2	10% MHDA/90% MUDO	without	
3	without	with	
4	0% MHDA/100% MUDO	with	
5	3% MHDA/97% MUDO	with	
6	10% MHDA/90% MUDO	with	AFM + MS/MS^2
7	20% MHDA/80% MUDO	with	
8	50% MHDA/50% MUDO	with	
9	100% MHDA/0% MUDO	with	

4. CONCLUSIONS

RSA protein grafting on (100) Gallium Arsenide substrate functionalized by a mixture of thiols was carried out and proved using an original combination of methods, namely atomic force microscopy and mass spectrometry. These characterization methods provided complementary information about the binding type and composition of the protein layer according to the chemical interface. Moreover, through fluorescence characterization of the functionalized surface, we proved that MHDA thiolates were present and reactive on the GaAs surface, since they were able to specifically immobilize NHS ester molecules. In addition, MS investigation of the protein layer proved the ability to analyze in situ the protein present on the surface. The 10% MHDA chemical interface creates specific covalent bonds between protein and the GaAs (100) surface, and presents an appropriate number of docking sites for grafting of a dense protein layer. Coupled with recent advances in the development of the resonant piezoelectric sensor and in the conception of a specific test bench, these results on biointerface optimization will enable us to propose a selective and efficient biosensor, presenting an optimized and highly controlled biointerface.

ACKNOWLEDGMENTS

We thank Wilfrid Boireau, Fabien Remy-Martin and Alain Rouleau for fruitful discussions concerning the protein binding and their contributions for comparative SPR studies. We greatly acknowledge the region Franche-Comte for financial support.

REFERENCES

1. Homola, J. Present and future of surface plasmon resonance biosensors. Anal. Bioanal. Chem. **2003**, 377, 528–539.
2. Länge, K.; Rapp, B.; Rapp, M. Surface acoustic wave biosensors: A review. Anal. Bioanal. Chem. **2008**, 391, 1509–1519.
3. Bienaime, A.; Liu, L.; Elie-Caille, C.; Leblois, T. Design and microfabrication of a lateral excited gallium arsenide biosensor. Eur. Phys. J. Appl. Phys. **2012**, 57, 21003:1–21003:11.
4. Bienaime, A.; Elie-Caille, C.; Leblois, T. Micro structuration of GaAs surface by wet etching: towards a specific surface behavior. J. Nanosci. Nanotechnol. **2012**, 12, 1–9.

5. Duplan, V.; Frost, E.; Dubowski, J.J. A photoluminescence-based quantum semiconductor biosensor for rapid in situ detection of Escherichia coli. Sens. Actuators Chem. **2011**, 160, 46–51.

6. Onodera, K.; Hirano-Iwata, A.; Miyamoto, K.-I.; Kimura, Y.; Kataoka, M.; Shinohara, Y.; Niwano, M. Label-free detection of protein-protein interactions at the GaAs/water interface through surface infrared spectroscopy: Discrimination between specific and nonspecific interactions by using secondary structure analysis. Langmuir **2007**, 23, 12287–12292.

7. Addadi, L.; Rubin, N.; Scheffer, L.; Ziblat, R. Two and three-dimensional pattern recognition of organized surfaces by specific antibodies. Acc. Chem. Res. **2008**, 41, 254–264.

8. Goede, K.; Busch, P.; Grundmann, M. Binding specificity of a peptide on semiconductor surfaces. Nano Lett. **2004**, 4, 2115–2120.

9. Schnirman, A.A.; Zahavi, E.; Yeger, H.; Rosenfeld, R.; Benhar, I.; Reiter, Y.; Sivan, U. Antibody molecules discriminate between crystalline facets of a gallium arsenide semiconductor. Nano Lett. **2006**, 6, 1870–1874.

10. Whaley, S.R.; English, D.S.; Hu, E.L.; Barbara, P.F.; Belcher, A.M. Selection of peptides with semiconductor binding specificity for directed nanocrystal assembly. Nature **2000**, 405, 665–668.

11. Willett, R.L.; Baldwin, K.W.; West, K.W.; Pfeiffer, L.N. Differential adhesion of amino acids to inorganic surfaces. Proc. Natl. Acad. Sci. USA. **2005**, 102, 7817–7822.

12. Estephan, E.; Bajoni, D.; Saab, M.-B.; Cloitre, T.; Aulombard, R.; Larroque, C.; Andreani, L.C.; Liscidini, M.; Malvezzi, A.M.; Gergely, C. Sensing by means of nonlinear optics with functionalized GaAs/AlGaAs photonic crystals. Langmuir **2010**, 26, 10373–10379.

13. Cho, Y.; Ivanisevic, A. In vitro assessment of the biocompatibility of chemically modified GaAs surfaces. NanoBioTechnology **2006**, 2, 51–59.

14. Arya, S.K.; Solanki, P.R.; Datta, M.; Malhotra, B.D. Recent advances in self-assembled monolayers based biomolecular electronic devices. Biosens. Bioelectron. **2009**, 24, 2810–2817.

15. Stutzmann, M.; Garrido, J.A.; Eickhoff, M.; Brandt, M.S. Direct biofunctionalization of semiconductors: A survey. Phys. Status Solidi **2006**, 203, 3424–3437.

16. Lunt, S.R.; Ryba, G.N.; Santangelo, P.G.; Lewis, N.S. Chemical studies of the passivation of GaAs surface recombination using sulfides and thiols. J. Appl. Phys. **1991**, 70, 7449–7467.

17. Nakagawa, O.S.; Ashok, S.; Sheen, C.W.; Mårtensson, J.; Allara, D.L. GaAs interfaces with octadecyl thiol self-assembled monolayer: Structural and electrical properties. Jpn. J. Appl. Phys. **1991**, 30, 3759–3562.

18. McGuiness, C.L.; Diehl, G.A.; Blasini, D.; Smilgies, D.-M.; Zhu, M.; Samarth, N.; Weidner, T.; Ballav, N.; Zharnikov, M.; Allara, D.L. Molecular self-assembly at bare semiconductor surfaces: Cooperative substrate-molecule effects in octadecanethiolate monolayer assemblies on GaAs(111), (110), and (100). ACS Nano. **2010**, 4, 3447–3465.

19. Seker, F.; Meeker, K.; Kuech, T.F.; Ellis, A.B. Surface chemistry of prototypical bulk II-VI and III-V semiconductors and implications for chemical sensing. Chem. Rev. **2000**, 100, 2505–2536.

20. Wink, T.; van Zuilen, S.; Bult, A.; van Bennekom, W. Self-assembled monolayers for biosensors. Analyst **1997**, 122, 43R–50R.

21. Bent, S.F. Heads or tails: Which is more important in molecular self-assembly? ACS Nano. **2007**, 1, 10–12.

22. Dubowski, J.J.; Voznyy, O.; Marshall, G.M. Molecular self-assembly and passivation of GaAs (001) with alkanethiol monolayers: A view towards bio-functionalization. Appl. Surf. Sci. **2010**, 256, 5714–572.

23. Adlkofer, K.; Shaporenko, A.; Zharnikov, M.; Grunze, M.; Ulman, A.; Tanaka, M. Chemical engineering of gallium arsenide surfaces with 4-Methyl-4-mercaptobiphenyl and 4′-Hydroxy-4-mercaptobiphenyl monolayers. J. Phys. Chem. **2003**, 107, 11737–11741.

24. Boireau, W.; Rouleau, A.; Lucchi, G.; Ducoroy, P. Revisited BIA-MS combination: Entire "on-a-chip" processing leading to the proteins identification at low femtomole to sub-femtomole levels. Biosens. Bioelectron. **2009**, 24, 1121–1127.

25. Briand, E.; Salmain, M.; Compère, C.; Pradier, C.-M. Immobilization of protein A on SAMs for the elaboration of immunosensors. Coll. Surf. Biointerfaces **2006**, 53, 215–224. [Google Scholar] [CrossRef]

26. Briand, E.; Salmain, M.; Herry, J.-M.; Perrot, H.; Compère, C.; Pradier, C.-M. Building of an immunosensor: How can the composition and structure of the thiol attachment layer affect the immunosensor efficiency? Biosens. Bioelectron. **2006**, 22, 440–448.

27. Frederix, F.; Bonroy, K.; Laureyn, W.; Reekmans, G.; Campitelli, A.; Dehaen, W.; Maes, G. Enhanced performance of an affinity biosensor interface based on mixed self-assembled monolayers of thiols on gold. Langmuir **2003**, 19, 4351–4357.

28. Bienaime, A.; Leblois, T.; Lucchi, G.; Blondeau-Patissier, V.; Elie-Caille, C. Reconstitution of a protein monolayer on thiolates functionalized GaAs surface. Int. J. Nanosci. **2012**, 11, 1240018:1–1240018:5.

29. Ding, X.; Dubowski, J.J. Surface passivation of (001) GaAs with self-assembled monolayers of long-chain thiols. Proc. SPIE **2005**, 5713, 545–551.

30. Ding, X.; Moumanis, K.; Dubowski, J.J.; Tay, L.; Rowell, N.L. Fourier-transform infrared and photoluminescence spectroscopies of self-assembled monolayers of long-chain thiols on (001) GaAs. J. Appl. Phys. **2006**, 99, 054701:1–054701:6.

31. Abdelghani, A. Atomic force microscopy on bare and thiol monolayer covered gallium arsenide. Mater. Lett. **2001**, 50, 73–77.

32. Duplan, V.; Miron, Y.; Frost, E.; Grandbois, M.; Dubowski, J.J. Specific immobilization of influenza A virus on GaAs (001) surface. J. Biomed. Opt. **2009**, 14, 054042:1–054042:6.

33. Frolov, L.; Rosenwaks, Y.; Richter, S.; Carmeli, C.; Carmeli, I. Photoelectric junctions between GaAs and photosynthetic reaction center protein. J. Phys. Chem. **2008**, 112, 13426–13430.

34. Zhou, C.; Walker, A.V. UV photooxidation and Photopatterning of alkanethiolate self-assembled monolayers (SAMs) on GaAs (001). Langmuir **2007**, 23, 8876–8881.

35. Peters, T.; Anfinsen, C.B.; Edsall, J.T.; Richards, F.M. Serum albumin. In advances in protein chemistry. Adv. Protein Chem. **1985**, 37, 161–245.

36. Voznyy, O.; Dubowski, J.J. Structure of thiol self-assembled monolayers commensurate with the GaAs (001) surface. Langmuir **2008**, 24, 13299–13305.

37. Ding, X.; Moumanis, K.; Dubowski, J.J.; Frost, E.H. A study of binding biotinylated nano-beads to the surface of (001) GaAs. Proc. SPIE **2006**, 6106, 61061L1–61061L7.

38. Jun, Y.; Zhu, X.Y.; Hsu, J.W.P. Formation of alkanethiol and alkanedithiol monolayers on GaAs(001). Langmuir **2006**, 22, 3627–3632.

39. McGuiness, C.L.; Blasini, D.; Masejewski, J.P.; Uppili, S.; Cabarcos, O.M.; Smilgies, D.; Allara, D.L. Molecular self-assembly at bare semiconductor surfaces: Characterization of a homologous series of n-alkanethiolate monolayers on GaAs(001). ACS Nano. **2007**, 1, 30–49.

40. Ding, X.; Moumanis, K.; Dubowski, J.J.; Frost, E.H.; Escher, E. Immobilization of avidin on (001) GaAs surface. Appl. Phys. Mater. Sci. Process. **2006**, 83, 357–360.

41. Griesser, H.J.; Kingshott, P.; McArthur, S.L.; McLean, K.M.; Kinsel, G.R.; Timmons, R.B. Surface-MALDI mass spectrometry in biomaterials research. Biomaterials **2004**, 25, 4861–4875.

42. Walker, A.K.; Qiu, H.; Wu, Y.; Timmons, R.B.; Kinsel, G.R. Studies of peptide binding to allyl amine and vinyl acetic acid-modified polymers using matrix-assisted laser desorption/ionization mass apectrometry. Anal. Biochem. **1999**, 271, 123–130.

CHAPTER 11

Neuron Biomechanics Probed by Atomic Force Microscopy

Elise Spedden and Cristian Staii *

Department of Physics and Astronomy and Center for Nanoscopic Physics, Tufts University, 4 Colby Street, Medford, MA 02155, USA

ABSTRACT

Mechanical interactions play a key role in many processes associated with neuronal growth and development. Over the last few years there has been significant progress in our understanding of the role played by the substrate stiffness in neuronal growth, of the cell-substrate adhesion forces, of the generation of traction forces during axonal elongation, and of the relationships between the neuron soma elastic properties and its health. The particular capabilities of the Atomic Force Microscope (AFM), such as high spatial resolution, high degree of control over the magnitude and orientation of the applied forces, minimal sample damage, and the ability to image and interact with cells in physiologically relevant conditions make this technique particularly suitable for measuring mechanical properties of living neuronal cells. This article reviews recent advances on using the AFM for studying neuronal biomechanics, provides an overview about the state-of-the-art measurements, and suggests directions for future applications.

Keywords: Atomic Force Microscopy; neurons; cellular elasticity; cellular biomechanics; cytoskeletal dynamics

1. INTRODUCTION

Neurons are highly specialized cells primarily responsible for transmitting information through chemical and electrical signaling in both the central and peripheral nervous system. They are composed of a cell body (soma), as well as dendrites and axons (neurites) which extend from the soma. The structural and mechanical properties of living neurons are essential components that govern many neuronal growth and regeneration processes including axonal extension, generation of traction forces and interactions between neurons and the surrounding environment such as the growth substrate, extracellular matrix, glial cells or other neurons [1]. The structural and mechanical properties of neurons are influenced by cytoskeletal components (microtubules and actin filaments), the cytoplasm and cell nucleus, and by the coupling between neurons and the extracellular matrix, which is involved in cell-substrate adhesion, and provides the anchoring needed for processes such as axonal extension and cell migration [2]. Recent advances in micro-spectroscopy techniques have allowed for the study of the structural and mechanical properties of living cells at increasing resolution. The Atomic Force Microscope (AFM) is at the forefront of this exploration. The AFM is a unique tool which enables researchers to obtain high resolution topographical data, control forces applied to cells, measure cellular elastic properties, and monitor variations in elastic modulus across living cells. Controlled forces have also been used to precisely manipulate the cells under study.

Excellent review papers are available concerning a wide array of topics related to the brain and nervous system, as well as AFM study of many types of living cells. These range from broader reviews of nervous system mechanics, mechanosensitivty and growth [1,3], to more specific reviews of AFM applications in biology [4–10]. Here, we present a focused and comprehensive review of the use of the AFM for the direct study of individual neuronal cells. We cover recent advancements in AFM-based techniques for neuronal measurement and summarize their results. We also describe in detail AFM measurements of neuronal mechanics, and how the measured elastic properties vary across different neuronal cell types, and between different areas of the same cell.

2. CELLULAR TOPOGRAPHY AND ELASTIC PROPERTIES MEASURED BY AFM

AFM measurement relies on the motions of a sharp tip positioned at the end of a flexible cantilever, which interacts with the sample via surface forces such as Van der Waals forces, electrostatic forces and capillary forces [5,11]. The cantilever behaves like a spring, so that any forces acting on the AFM tip cause the cantilever to deflect. These interactions are detected by the positioning of a laser reflected off of the back of the cantilever onto a 4-quadrant photo sensor.

2.1. Modes of AFM Measurement

Several modes of topographical measurement are possible. Here we focus on contact mode and tapping mode, which are the two most commonly used for measuring cells. In basic contact mode scanning, topography is determined by a feedback mechanism which maintains a constant cantilever deflection against the sample surface. This is done by moving the z position of the cantilever up and down in response to changes in deflection (Figure 1a). This maintains a constant force on the surface during scanning. Another widely used imaging mode is intermittent contact (AC), or "tapping" mode, where the cantilever is driven to oscillate near its resonant frequency, and surface forces are measured by the damping of these oscillations [5]. In this mode the cantilever z position is adjusted to maintain a constant set-point oscillation amplitude of the cantilever, and its vertical position at each point forms the topographical image of the sample. For both contact and tapping modes, adjustments in the z position of the cantilever yield the sample height, so the actual values for deflection or oscillation amplitude are not important for acquiring the sample topography. For elastic modulus measurements, however, both the actual cantilever deflection distance, and the cantilever spring constant are needed to determine cantilever force and indentation values, as described below.

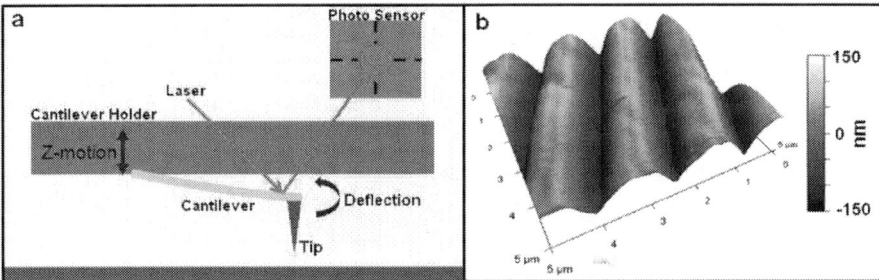

Figure 1. (a) Schematic of Atomic Force Microscope (AFM) cantilever holder and laser detector setup. Before calibration, tip deflection is measured as a voltage value generated by a difference in light falling on the upper vs. lower quadrants of the photo sensor. In order to measure cantilever deflection as a distance, the cantilever must be calibrated. This is typically done by performing a force curve on an "infinitely" hard surface (zero indentation), and equating the decrease in cantilever z height after contact with the increase in tip deflection [5]; **(b)** AFM topography image of ridges on a micropatterned silicon substrate shown as an example of a representative AFM image. Image taken by Elise Spedden in Staii lab at Tufts University.

Determining the elastic modulus of a cell via AFM typically involves taking force vs. indentation curves on the sample (Figure 2a). A force curve will move the z position of the cantilever down until the cantilever is deflected a set amount (the trigger value) and then retracted. The elastic modulus of the sample can be determined by fitting the force vs. indentation curve above the contact point [6,9]. Indentation is obtained by subtracting cantilever upward deflection from the downward z-motion of the AFM head, and force is obtained from the deflection value by treating the cantilever as a spring obeying Hooke's law. These force vs. indentation curves can be measured on specific locations on a living or fixed sample, or they can be taken in a grid pattern across the sample surface (Figure 2b–d). This allows maps of elastic modulus values to be obtained across a sample [12]. This technique is sometimes referred to as force-volume mode as these maps also contain simultaneous topographical information over the cells which can be used to model the cells in 3 dimensions (see Figure 3).

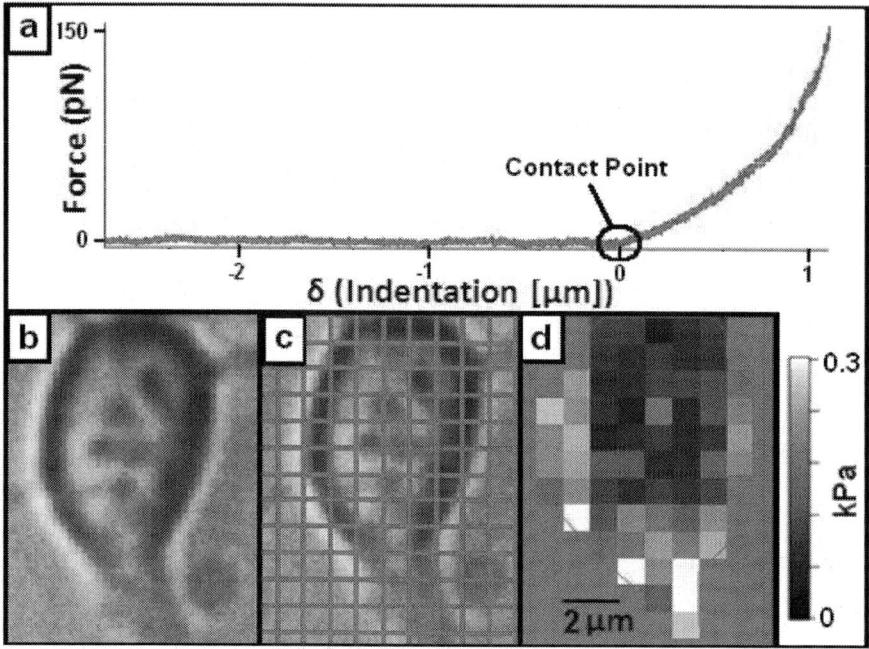

Figure 2. (a) Typical AFM force vs. indentation curve. The contact point is defined as the point at which the force vs. indentation slope begins to increase above noise level. The flat line indicates the approach portion, where the cantilever is moving down but it is not yet in contact with the surface. The curved portion after the contact point indicates how the total force of the tip on the sample changes as the cantilever indents the surface; **(b)** Optical image of living cortical neuron; **(c)** Neuron shown in **(a)** with superimposed 1 × 1 micron grid; **(d)** Elastic modulus map of neuron from **(c)**, where force vs. indentation curves are taken at each point on the grid and elastic modulus values are calculated from the indentation curves (see text). These images illustrate the elasticity mapping procedure where optical and AFM measurements are taken simultaneously. Images taken by Elise Spedden, in Staii lab at Tufts University.

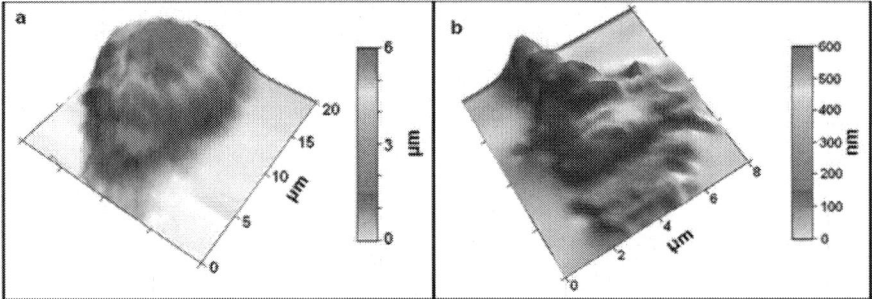

Figure 3. (**a**) Example of a 3D AFM topography image of a live cortical neuron soma; (**b**) 3D AFM topography image of a live neuron growth cone. Both images were taken in force-volume mode, thus rendering lower spatial resolution than high-resolution AFM topography in tapping mode. However, force-volume mode images contain both topographical and elasticity information as explained in the main text. Images taken by Elise Spedden in Staii lab at Tufts University.

2.2. The Hertz Model Applied to AFM

The elastic modulus of a material (Young's modulus) is defined as the ratio between the uniaxial stress (force) applied to the material and the uniaxial strain (deformation) that it undergoes. The Hertz model, as applied to AFM relates the total cantilever force and maximum indentation into the sample to the sample elastic modulus [6,9]. The specific relationship varies for different cantilever tip geometries. Most common for AFM applications are the models for the sphere and cone shaped tips [9]. There are also other tip shapes, such as 3 and 4-sided pyramids, although those tip shapes are often approximated as conical [10]. Conical or pyramidal shaped tips are best for high resolution topographical scanning, as well as for mapping elastic modulus values with high spatial resolution. Spherical tips have the advantage of distributing the applied forces over larger areas. This is useful for limiting any potential damage to the cell (which is an important concern if high forces are used), and for obtaining averaged elastic modulus values of a whole cell soma or entire tissue areas. The Hertz model as applied to AFM requires knowledge of the tip geometry. In the case of the spherical tip, this means the radius of sphere used (R). In the case of conical indenters the relevant quantity is the half-angle of the conical indenter (α). These values relate to how the tip-sample contact area and forces are expected to change during indentation. The Hertz model as applied to AFM for a spherical indenter is given in Equation 1, and for a conical indenter in Equation 2 [9].

$$F = \frac{4}{3} \frac{E}{(1-v^2)} \sqrt{R} \delta^{\frac{3}{2}}$$

(1)

$$F = \frac{E}{(1-v^2)} \frac{2\tan(\alpha)}{\pi} \delta^2$$

(2)

In these equations F is the force, E is the elastic modulus, and δ is the indentation depth. Also important in these equations is a value known as the Poisson ratio (v). This value relates to the samples ability to "bulge out" when compressed. For example, in the case of a rubber ball, the ball expands laterally as it is being compressed from above, maintaining a constant volume. This redistributes some of the material strain in the lateral direction. Cork, on the other hand, typically compresses without bulging. In general, it is important to know for any material how much of the stress on the system goes into uniaxial strain in the direction of pressure, and how much is distributed into lateral strains. This relationship between strains parallel and perpendicular to the stress on a system is called the Poisson ratio of the material. The bulging rubber ball exemplifies a Poisson ratio close to 0.5, whereas the fully compressible cork would be closer to 0. For living cells the value is typically chosen as being between 0.3 and 0.5 representing a mostly or fully incompressible system. While most cell types have not been measured for their Poisson ratio, a value of 0.5 is often chosen, assuming that typical living cells are fully incompressible [4]. However, several types of cells that have been measured yield lower values for their Poisson ratio. For example, a value of 0.38 has been measured for chondrocytes [13].

While the Hertz model is currently by far the most common method of elastic modulus evaluation of force curves from living cells, the model makes several assumptions. These assumptions result in limitations both on what can be measured using this method and how to compare the results with other measurements obtained using different models.

First, the Hertz model assumes that the material is behaving elastically and will return to its initial state after the force is removed. To an extent, the validity of this assumption for any given sample can be checked by the shape and repeatability of the force curves acquired. Though cells do not behave exactly like perfect elastic materials, the better this condition is met, the more valid the model for the sample in question. Secondly, the Hertz model assumes that the materials are homogeneous and isotropic. This is typically true for the AFM indenters, but it is only an approximation for living cells. In the measurement of living cells multiple layers of inhomogeneous material are being indented together. When the traditional AFM adaptation of the Hertz model is applied to this situation, the "true" elastic modulus values of all materials being intended contribute to the final value measured. This can give a reasonable and consistent measurement of the cell's elastic response to external forces, but does not give specific information about the individual components being measured. Models have been proposed which isolate elastic modulus information from various portions of a force vs. indentation curve to accommodate the multiple-layers encountered by the AFM tip [14,15]. This type of model may be preferred by researchers interested in isolating the individual contributions within multi-layer systems to AFM-acquired force vs. indentation curves. A third consideration in this model is the assumption of measurement on a flat surface. As such, researchers must expect additional error in the application of the typical AFM Hertz model for curves near the very edges of a highly curved sample (such as a living soma). These regions also may experience problems such as lateral movement of the sample during curve acquisition. Special care must be taken in considering the validity of curves taken in these areas [12].

3. AFM TOPOGRAPHY OF NEURONS

AFM topography is used to image fixed and living neurons at high resolution [16–27]. Many topographical studies fix the neurons prior to measurement [16,17,25,26]. This can combat damage to the cell structure or removal of the cell from the surface due to forces exerted by the AFM tip. AFM topography of fixed hippocampal neurons has provided insight into 3-dimensional (3D) cell structure and location of the cell nucleus, mitochondria, and filaments [25]. This early study was among the first to visualize high resolution neuronal structure in 3D. Fixed growth cones of neurons have been imaged in detail [16] and AFM topography of fixed hippocampal somas, neurites and growth cones can also be utilized to track growth and detailed changes in morphology along patterned substrates [17,26]. The AFM has been used to perform high resolution analysis of axon morphology and height-to-width ratio over time on micropatterned substrates [26]. Topography of fixed growth cones at various stages of growth has presented a decreasing height-to-width ratio as axon development progressed, as well as substantial increase in growth cone flattening over the adhesion molecule L1 as compared to Poly-d-Lysine (PDL). This increase in flattening produced a larger growth cone area and was correlated with highly dynamic outgrowth. This result indicates that an increase in sensing and progression accompanies an increase in growth cone spreading [26].

AFM topography of both fixed and living cells has been obtained for Dorsal Root Ganglia (DRG) neurons, embryonic stem cell-derived neurons, and chick embryo spinal cord neurons [18–23,27]. Nanoscale structural details in fixed growth cones, including low-height regions of the growth cone with an area of 0.01–3.5 μm^2 and ranging from 2 to 180 nm deep were discovered through AFM topography. These measurements show the significance of the structural properties of actin and tubulin in the growth cone, as the low regions were present primarily in areas devoid of actin or tubulin concentrations. These structures were then confirmed by measurements on living DRGs [21], and later, similar structures were found in the growth cones of living and fixed embryonic stem cell derived neurons [22]. Fluorescence has been combined with AFM to track morphological changes and the destruction of the DRG cytoskeleton upon chemical modification [20]. The effects of acrolein on neuronal morphology have also been studied with the AFM. The study showed that increased presence of acrolein (an aldehyde produced during traumatic brain injury) is a likely contributor to secondary injury after an initial traumatic injury as it produced significant degradation of cytoskeletal structures and a major decrease in cell viability [20]. In a different set of experiments, the AFM tip has been used as an intentional source of neuronal damage prior to imaging. In these studies, living DRG somas and axons were intentionally sliced with the AFM and the cell volume and morphology tracked over time after damage [18]. This use of the AFM yielded unprecedented precision in the study of neuronal injury. Microscale slices in the neuronal soma membrane produced cell death and elimination of cytoplasm, while identical injury to the membrane of neurites did not lead to their destruction. This indicates a higher susceptibility of the soma to injury of this type than of neurites. This study also used AFM to resolve new dynamic architectures in the growth cone at a scale inaccessible from traditional optical microscopy [18].

Live hippocampal neurons are not typically imaged using traditional tapping or contact mode in fluid due to their extremely soft and malleable nature. Magnetic AC (MAC) mode, a modification of tapping mode where a specialized AFM cantilever is driven at its tip via an external magnetic field, has been used to image the topography of living hippocampal neurons [24]. Driving the cantilever at the tip rather than the base minimizes changes in the tip drive amplitude and phase due to the surrounding fluid and allows for a more precise measurement of changes in cantilever oscillation due to surface interactions. This also allows clearer images to be taken in fluid at a lower scanning force. This low-force imaging allowed the researchers to monitor cellular damage. Neural spine damage was observed under moderate force, and MAC imaging was used to track neuronal regeneration. Neurons were also chemically modified with Aβ25–35 to determine its effects on growing neurons. This chemical is thought to be involved in the initiation of Alzheimer's disease. Changes induced by this modification were studied via MAC mode topography scanning and included the gradual contraction and obliteration of the growth cone of previously growing axons [24].

AFM topography can be used to monitor changes in living neurons over time, and is a valuable tool for high resolution detection and monitoring of neuronal morphology. Limitations of this technique include the necessity for neurons to be well adhered to the sample, to be confined to 2-dimensional substrates, and to change their morphology over time scales which are slower than the typical time necessary to acquire the AFM image (1–5 min). These constrains are particularly important when imaging active growth cones [23].

4. MECHANICAL PROPERTIES OF NEURONS

The AFM is an ideal tool for micromechanical measurements. AFM has been used to determine elastic modulus values on live and fixed neurons, as well as slices and explants from the brain (see Table 1). Techniques for micromechanical measurements vary. Spherical AFM tips are common for tissue or bulk soma measurements [28–32]. Smaller cone or pyramidal AFM tips are typically used for measurements on particular points across the cell or for mapping the elastic modulus over entire regions [12,27,33,34].

Table 1. Elastic modulus values (E values) for neurons measured via AFM.

Fixed/living	Area measured	Animal	Type of neuron	E values	Citation
Fixed	Soma	Rat	Primary spinal cord neurons	<25–40 kPa	Jiang [33]
Fixed	Neurite	Rat	Primary spinal cord neurons	<7.5 kPa	Jiang [33]
Fixed	Growth Cone	Aplysia	Bag cell neurons	45–225 kPa	Xiong [35]
Living	Tissue	Rat	Hipocampal slices	52–308 Pa	Elkin [28]
Living	Tissue	Rat	Cortex explants	305 ± 25 Pa	Norman [29]
Living	Soma	Mouse	Hippocampal neurons	480–970 Pa	Lu [30]
Living	Soma	Guinea Pig	Retinal neurons	650–1590 Pa	Lu [30]
Living	Soma	Chick	Dorsal root ganglia neurons	10–140 kPa	Mustata [34]
Living	Soma	Chick	Dorsal root ganglia neurons	1–8 kPa	Spedden [12]
Living	Soma	Mouse	P19-derrived neurons	200–2000 Pa	Spedden [12]
Living	Soma	Mouse	P19-derrived neurons	230 ± 180 Pa	Spedden [31]
Living	Soma	Rat	Cortical neurons	30–200 Pa	Bernick [32]
Living	Soma	Rat	Cortical neurons	100–200 Pa	Spedden [31]
Living	Soma	Rat	Cortical neurons	80–500 Pa	Spedden [12]
Living	Growth Cone	Aplysia	Bag cell neurons	3–40 kPa	Xiong [35]
Living	Growth Cone	Mouse	Dorsal root ganglia neurons	0.4–33 kPa	Martin [27]

4.1. Fixed Neurons

Fixing neuronal cells prior to elastic modulus measurement helps circumvent many potential issues that accompany live measurement such as problems with cell-substrate adhesion, cell motility and the maintenance of neuronal viability during measurement. Fixation also may allow for the imaging of structures which are otherwise too dynamic to capture on the timescales required for AFM scanning. On fixed neuronal cells mechanical measurements have been performed on soma and growth cone regions. Fixed neurons typically yield elastic modulus values on the order of 10 to hundreds of kPa [33,35]. These values depend on both the type of neuron, and region of cell measured. Fixed primary spinal cord neurons measured at various locations on the soma with a pyramidal tip yield elastic modulus values of the order of 10 kPa [33]. This study aimed to determine if the elastic modulus of neurons responds to changes in substrate stiffness, as other cell types, such as fibroblasts [36], have been shown to do. It was found that fixed neurons do show a dependence of the cell elastic modulus on the plating substrate; plating on stiffer polyacrylamide gels yields stiffer elastic modulus values over the fixed cell soma [33]. These results have led to the hypothesis that differences in the production of extracellular matrix between cells on different substrates may account for some of the differences in elastic modulus. Other cell types which exhibit this trend, such as fibroblasts, exhibit changes in cell spreading and stress fiber formations which contribute to the observed differences in the elastic modulus [36]. The above results suggest that a similar process may exist in neurons. Growth cones of fixed Aplysia bag cell neurons have also been examined via micromechanical measurements. Different growth cone regions yield varying elastic modulus values in the range of 45–225 kPa [35]. Isolated dried and re-hydrated peripheral nerve fibers yield relatively high values, with myelinated and demyelinated fibers both averaging around 800–900 kPa [37].

4.2. Living Neurons

Fixation of cells and tissue has been shown to consistently increase the elastic modulus of the sample being measured [9]. As such, measurements on living cells and tissue are often of greater biological significance. While measurement of the elastic modulus of fixed cells may provide useful data on relative stiffness, measuring living cells allows for a better understanding of how the cell might respond to forces within its native environment. For example, chick DRG neurons exhibit preferential outgrowth on substrates with elastic modulus that match the corresponding average elastic modulus of living DRG somas [12,38]. This type of mechanical coupling can only be observed through the measurement of living cells. These types of measurements, however, present additional difficulties over measurements taken on fixed cells. Cell health must be maintained throughout the experiment, which often restricts the timeframe available for the measurements, as well as the maximum forces that can be used. Cell adhesion to the substrate can also be an issue, as cells that are not well attached may slip around under the cantilever, preventing useful measurement. Many techniques are available to circumvent such issues, including the use of heated AFM chambers for measurement [12]. Several studies, described below, have shown that physiologically relevant values for elastic modulus are obtained when cell health and attachment are successfully maintained during the AFM measurements. These techniques are extremely valuable for obtaining information on the neuron structure and elastic properties in response to external stimuli and controlled external forces. The necessary requirement that cells be well adhered to a 2-dimensional substrate, does impose limitations on the technique, such as the capability of measuring the cell structure and responses to controlled external forces in vivo, in the native environment of the cell. Despite these limitations, however, the AFM measurements could give valuable insight into how neuronal cells sense external forces and react to mechanical stresses in their natural environment.

4.2.1. Elastic Modulus Measurements

Several types of living neurons and neuronal tissues have been measured with the AFM. Living rat hippocampal slices have been measured by large (25 μm) sphere-tipped AFM cantilevers. It was found that elastic modulus values fall between 50 and 300 Pa, and are dependent on the region of the cell being measured [28]. For example, measurements in CA3 regions (Cornu Ammonis area 3 of the hippocampus) yielded stiffer average measurements (230 ± 150 Pa, and 300 ± 180 Pa) than those taken in CA1 regions (170 ± 50 Pa, and 200 ± 130 Pa). Living explants of the fetal rat cortex yield similar values as the hippocampal slices, with an average of 300 ± 25 Pa measured with a 2.5 μm sphere-tipped probe [29].

Elastic modulus values measured via AFM for living neuron somas are available for various cell types. Hippocampal and retinal neurons were measured at various indentation frequencies. An increase in measured elastic modulus value is seen with an increase in indentation frequency. These experiments yielded elastic modulus values around 650 Pa at 30 Hz, and 1590 Pa at 200 Hz [30]. AFM measurements of neuronal cells are typically taken at substantially lower indentation frequencies to limit viscoelastic effects. One study showed that cells respond with an almost ideal elastic response (little to no loading rate dependence) at very low force (30 pN or less), and that at larger forces the measured elastic modulus depends on the loading rate [39]. For larger forces, the dependence of the measured elastic modulus on the loading rate can be minimized by choosing weaker cantilevers, smaller maximum force values, and by minimizing the cantilever loading rates used.

Living DRG, p19 (embryonic mouse teratocarcinoma stem cell) derived, and cortical neurons have all been measured by both individual force curves on varying regions of the cell soma, and by high resolution elastic modulus mapping [12,31,34]. The AFM-based mapping of the cell elastic modulus allows obtaining both height and elastic modulus information with sub-micron resolution. This can be used to measure the distribution of elastic modulus values across a cell (Figure 4).

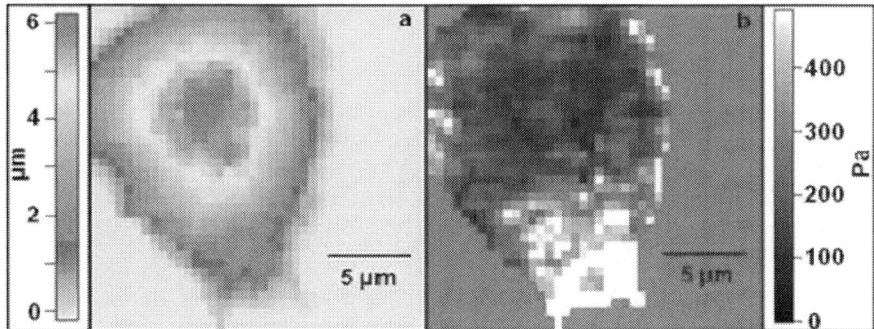

Figure 4. (a) Topography of live neuron soma obtained during elastic modulus mapping; (b) High resolution elastic modulus map of cell shown in (a). The cell displays regions of high elastic modulus localized at the lower end (bright areas in the figure). Images taken by Elise Spedden in Staii lab at Tufts University.

Live DRG neurons measured by individual force curves are reported to yield elastic modulus values averaging around 60 kPa [34]. Elastic modulus mapping of living DRG's, however, has yielded averages of around 1 kPa, with individual points on DRG somas ranging between 0.1 and 8 kPa [12]. These values are higher on average than those measured for cells originating from the cortex. This may result from mechanical coupling between cell types and their native growth environment. While cortical and hippocampal neurons must navigate and make connections within tissue measured on the order of hundreds of Pa [28,29], DRG neurons are located along the spinal column connecting the sensory nerves to the central nervous system. This presents a stiffer and more varied environment for growth, as elastic modulus values of the tissue outside the brain typically lie in the kPa range, with spinal cord elastic modulus values falling in the tens of kPa [40]. P19 derived neurons are neuron-type cells differentiated from the mouse p19 stem-cell line. These cells have been measured via sphere-tip AFM indentation yielding average elastic modulus values around 230 Pa [31]. P19 derived neurons have also been measured via AFM elastic modulus mapping, yielding soma average values of around 400 Pa, with individual points on the elastic modulus maps ranging between 100 and 2000 Pa [12]. Indeed, since P19 neurons are typically considered a model system for cortical neurons, they are expected to have similar mechanical properties to those obtained on cortical region neurons [41].

Cortical neuron somas have also been measured by a variety of AFM indentation techniques. Bulk elastic modulus of the cortical neuron soma was measured via repeated oscillations of 45 μm spherical tip probes against the cell body. Rather than computing the elastic modulus using a typical force vs. indentation model, such as the commonly applied Hertz model, this study recorded and analyzed repeated oscillations with the aid of a finite element framework simulating experimental testing conditions. The study models cortical soma response to large deformations in a large-strain kinematics framework at various deformation rates. The average value of elastic modulus obtained for the cortical neuron soma was 78 Pa, with a range of 30–200 Pa [32]. These values are consistent with the corresponding values measured through traditional Hertz model analysis of force curves. Cortical neurons measured using individual force vs. indentation curves using spherical AFM tips yield soma averages around 100–200 Pa [31]. High resolution elastic modulus mapping of cortical neurons also yields similar values for average soma elastic modulus of around 200 Pa [12]. The observed consistency between different types of measurements, as well as the similarity between the elastic modulus values obtained on individual neurons and the values obtained on bulk tissue measurements of the corresponding cortical region, demonstrate a match between individual cellular elastic modulus values and those for the environment in which they grow [28,29].

AFM topography and elastic modulus measurements have also been performed on living neuronal growth cones. The Aplysia bag cell neuron has a remarkably large and stable growth cone structure, lending itself to detailed topographic and nanomechanical analysis via AFM. AFM topographical imaging allowed one study to locate the P domain lamellipoda, P domain filipodia bundles, T zone ruffles, and C domain of the living growth cone. Individual force vs. indentation measurements were taken within each of these regions yielding average elastic modulus values of 16.7, 29.8, 3.5 and 0.5 kPa respectively [35]. It was hypothesized that differences in mechanical stiffness between regions measured, particularly the high values found for the P and T domains, could be important for support and traction generation during growth cone advancement and steering [35]. DRG topography and elastic modulus values have been studied before and after induced sciatic nerve injury, showing regenerative growth post-injury with a lower elastic modulus growth cone (3–13 kPa) than that seen in healthy uninjured growth cones (16–33 kPa). A change in the relative elastic modulus values across measured growth cone regions is also observed [27]. In this study AFM elastic modulus measurements provide unique insight into the post-injury changes of the growth cone, highlighting the differences in elastic modulus of the growth cones between injured and control neurons, along with a drop in actin and increase in βIII-tubulin content [27]. This represents one of several studies that use AFM topography or elastic modulus mapping to track changes to neurons and their structural components over time. Another novel use of this technique was the mapping and tracking of neuronal soma elastic modulus during axonal elongation events. The soma of cortical neurons adjacent to an axon was found to stiffen reversibly during active extension of the axon, with dramatic local stiffening resulting in the average elastic modulus over the whole soma increasing by 30%–180% [12]. AFM provides the unique ability to study such changes in the neuronal cytoskeleton without the need for potentially disruptive fluorescent modification. The same AFM force vs. indentation method used to measure living and fixed neurons as described above has also been used to characterize the elastic modulus and topography of various substrates used in neuronal growth experiments. For example, this approach is often used to characterize gels such as polyacrylamide and silk-based surfaces [1,29,33,42].

4.2.2. Combined AFM/Fluorescence Measurements

AFM elastic modulus measurements have been used to monitor live neuronal structures over time. Components within the neuronal soma have been tracked over time using combined AFM mechanical measurements and fluorescence microscopy. For cortical neuron somas maintained at 37 °C, relative elastic modulus values have been used as an indicator for tubulin density. Higher elastic modulus regions typically correlate to dense tubulin aggregations (see Figure 5). Actin density in cortical neurons at 37 °C, however, does not correlate strongly to measured elastic modulus [12]. This technique allows for the tracking of tubulin density within living, unstained cells. The ability to obtain relationships between density of cytoskeletal components and local elastic modulus values at high resolution is currently unique to the AFM. Elastic modulus values can also be tracked for neuronal cells during chemical [12,34] or protein [43] modification. Types of modification explored include Taxol, a stabilizer of microtubules, Nocodazole, an inhibitor of microtubule polymerization, and Blebbistatin, an inhibitor of myosin II contraction of the actin cytoskeleton [12]. This type of measurement offers new insights into the effects that chemical modifiers have on the structure and stiffness of cytoskeletal components of living cells over time. For example, application of a 10 μM dose of Taxol to living neurons has been shown to increase soma elastic modulus by 30% or more, with particular increases in elastic modulus near the tubulin-dense region of the soma close to the axon [12].

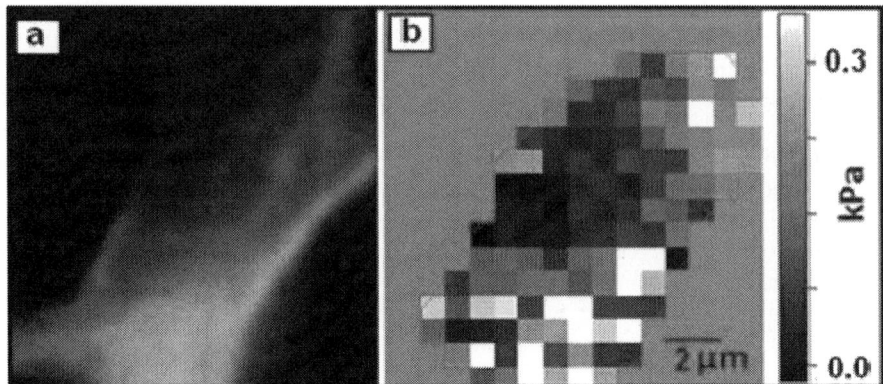

Figure 5. (a) Fluorescent image of a living neuron soma stained for tubulin density; (b) Elastic modulus map of cell shown in (a). The images have the same scale bar shown in (b). There is a significant overlap between areas of high actin concentration (high fluorescence intensity) and the regions of high stiffness. Images taken by Elise Spedden in Staii lab at Tufts University.

5. OTHER TYPES OF AFM-BASED MEASUREMENTS

In addition to the modes discussed in the previous sections, AFM probes can be used in a several other ways to study neurons and their growth. The ability to apply controlled forces using an AFM cantilever can be used to study the mechanosensitive properties of neuron-like cells. By applying controlled mechanical forces via the AFM tip to the growth cones of living NG108-15 and PC12 cells one study determined that mechanical stimulation above a threshold limit resulted in retraction and re-exploration of the stimulated neurite, indicating a mechanosensitive growth-cone response [44]. AFM allowed precise measurement of this stimulation threshold indicating a possible method through which growth cones may sense and react to the stiffness of the growth substrate. AFM can also be used to characterize and/or manipulate the growth environment of neurons. Application of controlled forces to a substrate during AFM scanning can be used to remove surface-bound molecules that inhibit neuronal growth and to promote the attachment of extracellular matrix proteins or adhesion factors in controlled geometries. This technique has been used to position and guide the growth of neuronal cells, and to study the dynamics of axonal extension in controlled geometries [45,46].

Modified AFM probes have been used to track the distribution of AMPA-type glutamate receptors on live hippocampal neurons. Tips were functionalized with antibodies and force vs. indentation maps were taken. This technique yields the traditional height and elastic modulus data in addition to strength of binding between the tip and surface during tip retraction. The binding strength was used to monitor antibody-receptor interactions and track receptor density. The results show that receptor density tends to be higher in areas of high local elastic modulus, and that changes in the distribution of the receptors are accompanied by changes in elastic modulus. Similarly, AFM tips coated with nerve growth factors (NGF) have been employed to image surface distribution of NGF receptors across living PC12 cells. This was accomplished through analysis of tip-neuron binding forces during force map acquisition [47]. Finally, we mention that recent experiments have used AFM probes modified with anti-α7 subunit nAChR antibody to measure the ligand-binding properties of α7-containing nicotinic acetylcholine receptors expressed on living neurons. Frequency of adhesion and detachment forces was used to monitor decreases in receptor interaction after exposure to nicotine [48].

6. FUTURE DIRECTIONS

6.1. Elastic Modulus and Neuronal Growth

The AFM has proven to be a very effective tool for studying topographic and mechanical features of living and fixed neuronal cells The high-resolution capability of the AFM to obtain physiologically relevant elastic modulus data on living cells may have important future applications. For example, continued exploration of changes in topography and elastic modulus of diseased and dysfunctional neurons may help us gain additional insight into how different neuronal defects contribute to various neurological diseases. As recent studies have shown that certain types of neurons exhibit preferential growth on substrates close to their own average elastic modulus [12,38], expanded data on live neuron elastic modulus under varying conditions might be used to generate better growth substrates for different types of neurons, with applications in regenerative medicine.

6.2. Receptor Tracking, Ion Tracking and Membrane Electrostatics

AFM has also been used to track changes in ion concentration along living cells. AFM tips coated with ion-selective polymer have been used to locate areas of high potassium concentrations on the surfaces of MDCK-F1 cells. Adhesion of the modified tips to the surface during force curves was recorded, with higher values of adhesion indicating higher concentrations of potassium ions. These areas of high concentration were used as indicators of the presence of active potassium channels. Relative regional ion channel activity can then be tracked by tracking this adhesion between different curves over time [49]. Such measurements have not yet been performed on neurons; however, similar imaging techniques might be employed to track signaling activity in axons.

This type of work is part of the emerging field of using modified AFM probes to assess binding properties between chemical or antibody modifiers and specific proteins, channels, or receptors along the surface of living neurons

[47,48]. Modifications of this type might be used to selectively monitor the activity of such receptors and sites under various chemical and environmental conditions with high spatial resolution, which is currently unavailable by other methods. This type of research is at the forefront of AFM measurement on living cells and is likely to continue as an active field of research in the future.

AFM can also be a powerful tool for analyzing membrane electrostatics [50–52]. Electrostatic parameters of living cell membranes such as surface charge density, surface potential, and transmembrane potential could be mapped via AFM with nanometer-scale resolution [50–52]. Electric Force Microscopy (EFM), for example, has been used to measure surface charge density [50] and membrane dipole potential [51] of lipid bilayers in fluid, and AFM-based electromechanical imaging of collagen fibrils have also been performed [52]. Similar techniques might be adapted to image or control the electrical activity, particularly action potentials, of living neurons cultured on various growth substrates. This technology will be beneficial to neuroscience by allowing researchers to query simple circuits in vitro to answer focused questions about neuronal growth and formation of functional connections between neurons.

7. CONCLUSIONS

During the past few years an increasing number of studies have used the versatility and high spatial resolution of the AFM to probe topographical features and biomechanical properties of neurons. While neuronal cells are some of the weakest in the body, AFM measurement can be sufficiently delicate to preserve the life and health of various types of neuronal cells. This allows for measurement of neuronal structure during growth or regeneration, and the tracking of cytoskeletal components through elastic modulus mapping. The ability to combine AFM measurements with bright field and fluorescence microscopy, as well as to perform these measurements in physiological conditions are valuable to our understanding of the mechanical structure of neuronal cells and how that depends on cytoskeletal components and their dynamics. AFM tip functionalization and novel scanning probe techniques open up new possibilities for tracking non-topographical or mechanical information on living neurons at high resolution.

ACKNOWLEDGMENTS

The authors thank David Kaplan and the Kaplan group for useful discussion. We also thank Steve Moss's laboratory at Tufts Center for Neuroscience for providing embryonic rat brain tissues. The authors gratefully acknowledge financial support from National Science Foundation (NSF-CBET 1067093) and Tufts University.

REFERENCES

1. Franze, K.; Guck, J. The biophysics of neuronal growth. Rep. Prog. Phys. **2010**, 73.
2. Letourneau, P.C.; Condic, M.L.; Snow, D.M. Interactions of developing neurons with the extracellular matrix. J. Neurosci **1994**, 14, 915–928.
3. Franze, K.; Janmey, P.A.; Guck, J. Mechanics in neuronal development and repair. Annu. Rev. Biomed. Eng **2013**, 15, 227–251.
4. Sokolov, I. Atomic force microscopy in cancer cell research. In Cancer Nanotechnology: Nanomaterials for Cancer Diagnostics and Therapy; Nalwa, H.S., Webster, T.J., Eds.; American Scientific Publishers: Valencia, CA, USA, 2007; Volume 1, pp. 1–17.
5. García, R.; Pérez, R. Dynamic atomic force microscopy methods. Surf. Sci. Rep **2002**, 47, 197–301.
6. Butt, H.-J.; Cappella, B.; Kappl, M. Force measurements with the atomic force microscope: Technique, interpretation and applications. Surf. Sci. Rep **2005**, 59, 1–152.
7. Franze, K. Atomic force microscopy and its contribution to understanding the development of the nervous system. Curr. Opin. Genet. Dev **2011**, 21, 530–537.
8. Fisher, T.E.; Carrion-Vazquez, M.; Oberhauser, A.F.; Li, H.; Marszalek, P.E.; Fernandez, J.M. Single molecular force spectroscopy of modular proteins in the nervous system. Neuron **2000**, 27, 435–446.
9. Kuznetsova, T.G.; Starodubtseva, M.N.; Yegorenkov, N.I.; Chizhik, S.A.; Zhdanov, R.I. Atomic force microscopy probing of cell elasticity. Micron **2007**, 38, 824–833.
10. Tao, N.J.; Lindsay, S.M.; Lees, S. Measuring the microelastic properties of biological material. Biophys. J **1992**, 63, 1165–1169. [Google Scholar]
11. Binnig, G.; Quate, C.F.; Gerber, C. Atomic force microscope. Phys. Rev. Lett **1986**, 56, 930–933.
12. Spedden, E.; White, J.D.; Naumova, E.N.; Kaplan, D.L.; Staii, C. Elasticity maps of living neurons measured by combined fluorescence and atomic force microscopy. Biophys. J **2012**, 103, 868–877.
13. Trickey, W.R.; Baaijens, F.P.; Laursen, T.A.; Alexopoulos, L.G.; Guilak, F. Determination of the Poisson's ratio of the cell: recovery properties of chondrocytes after release from complete micropipette aspiration. J. Biomech **2006**, 39, 78–87.
14. Kim, Y.; Kim, M.; Shin, J.H.; Kim, J. Characterization of cellular elastic modulus using structure based double layer model. Med. Biolog. Eng. Comput **2011**, 49, 453–462.

15. Iyer, S.; Gaikwad, R.M.; Subba-Rao, V.; Woodworth, C.D.; Sokolov, I. Atomic force microscopy detects differences in the surface brush of normal and cancerous cells. Nat. Nanotechnol **2009**, 4, 389–393.

16. Grzywa, E.L.; Lee, A.C.; Lee, G.U.; Suter, D.M. High-resolution analysis of neuronal growth cone morphology by comparative atomic force and optical microscopy. J. Neurobiolog **2006**, 66, 1529–1543.

17. Xing, S.; Liu, W.; Huang, Z.; Chen, L.; Sun, K.; Han, D.; Zhang, W.; Jiang, X. Development of neurons on micropatterns reveals that growth cone responds to a sharp change of concentration of laminin. Electrophoresis **2010**, 31, 3144–3151.

18. McNally, H.A.; Borgens, R.B. Three-dimensional imaging of living and dying neurons with atomic force microscopy. J. Neurocytol **2004**, 33, 251–258.

19. McNally, H.A.; Rajwa, B.; Sturgis, J.; Robinson, J.P. Comparative three-dimensional imaging of living neurons with confocal and atomic force microscopy. J. Neurosci. Method **2005**, 142, 177–184. [Google Scholar]

20. Liu-Snyder, P.; McNally, H.; Shi, R.; Borgens, R.B. Acrolein-mediated mechanisms of neuronal death. J. Neurosci. Res **2006**, 84, 209–218.

21. Laishram, J.; Kondra, S.; Avossa, D.; Migliorini, E.; Lazzarino, M.; Torre, V. A morphological analysis of growth cones of DRG neurons combining atomic force and confocal microscopy. J. Struct. Biol **2009**, 168, 366–377.

22. Ban, J.; Migliorini, E.; Di Foggia, V.; Lazzarino, M.; Ruaro, M.E.; Torre, V. Fragmentation as a mechanism for growth cone pruning and degeneration. Stem Cell. Dev **2011**, 20, 1031–1041.

23. Ricci, D.; Grattarola, M.; Tedesco, M. The growth cones of living neurons probed by the atomic force microscope. Method. Mol. Biol **2011**, 736, 243–257.

24. Yunxu, S.; Danying, L.; Yanfang, R.; Dong, H.; Wanyun, M. Three-dimensional structural changes in living hippocampal neurons imaged using magnetic AC mode atomic force microscopy. J. Electron Microsc **2006**, 55, 165–172.

25. Parpura, V.; Haydon, P.G.; Henderson, E. Three-dimensional imaging of living neurons and glia with the atomic force microscope. J. Cell Sci **1993**, 104, 427–432.

26. Messa, M.; Canale, C.; Marconi, E.; Cingolani, R.; Salerno, M.; Benfenati, F. Growth cone 3-D morphology is modified by distinct micropatterned adhesion substrates. IEEE Trans. Nanobiosci **2009**, 8, 161–168.

27. Martin, M.; Benzina, O.; Szabo, V.; Vegh, A.G.; Lucas, O.; Cloitre, T.; Scamps, F.; Gergely, C. Morphology and nanomechanics of sensory neurons growth cones following peripheral nerve injury. PLoS One **2013**, 8, e56286.

28. Elkin, B.S.; Azeloglu, E.U.; Costa, K.D.; Morrison, B., 3rd. Mechanical heterogeneity of the rat hippocampus measured by atomic force microscope indentation. Neurotrauma **2007**, 24, 812–822.

29. Norman, L.; Aranda-Espinoza, H. Cortical neuron outgrowth is insensitive to substrate stiffness. Cel. Mol. Bioeng **2010**, 3, 398–414.

30. Lu, Y.B.; Franze, K.; Seifert, G.; Steinhauser, C.; Kirchhoff, F.; Wolburg, H.; Guck, J.; Janmey, P.; Wei, E.Q.; Kas, J.; et al. Viscoelastic properties of individual glial cells and neurons in the CNS. Proc. Natl. Acad. Sci. USA **2006**, 103, 17759–17764.

31. Spedden, E.; White, J.D.; Kaplan, D.; Staii, C. Young's modulus of cortical and P19 derived neurons measured by atomic force microscopy. MRS Online Proc. Libr. **2012**, 1420.

32. Bernick, K.B.; Prevost, T.P.; Suresh, S.; Socrate, S. Biomechanics of single cortical neurons. Acta Biomater **2011**, 7, 1210–1219.

33. Jiang, F.X.; Lin, D.C.; Horkay, F.; Langrana, N.A. Probing mechanical adaptation of neurite outgrowth on a hydrogel material using atomic force microscopy. Ann. Biomed. Eng **2011**, 39, 706–713.

34. Mustata, M.; Ritchie, K.; McNally, H.A. Neuronal elasticity as measured by atomic force microscopy. J. Neurosci. Method **2010**, 186, 35–41.

35. Xiong, Y.; Lee, A.C.; Suter, D.M.; Lee, G.U. Topography and nanomechanics of live neuronal growth cones analyzed by atomic force microscopy. Biophys. J **2009**, 96, 5060–5072.

36. Solon, J.; Levental, I.; Sengupta, K.; Georges, P.C.; Janmey, P.A. Fibroblast adaptation and stiffness matching to soft elastic substrates. Biophys. J **2007**, 93, 4453–4461.

37. Heredia, A.; Bui, C.C.; Suter, U.; Young, P.; Schaffer, T.E. AFM combines functional and morphological analysis of peripheral myelinated and demyelinated nerve fibers. Neuroimage **2007**, 37, 1218–1226.

38. Koch, D.; Rosoff, W.J.; Jiang, J.; Geller, H.M.; Urbach, J.S. Strength in the periphery: Growth cone biomechanics and substrate rigidity response in peripheral and central nervous system neurons. Biophys. J **2012**, 102, 452–460.

39. Nawaz, S.; Sanchez, P.; Bodensiek, K.; Li, S.; Simons, M.; Schaap, I.A. Cell visco-elasticity measured with AFM and optical trapping at sub-micrometer deformations. PLoS One **2012**, 7, e45297.

40. Levental, I.; Georges, P.C.; Janmey, P.A. Soft biological materials and their impact on cell function. Soft Matter **2007**, 3, 299–306.

41. Bain, G.; Ray, W.J.; Yao, M.; Gottlieb, D.I. From embryonal carcinoma cells to neurons: The P19 pathway. BioEssays **1994**, 16, 343–348.

42. Hopkins, A.M.; De Laporte, L.; Tortelli, F.; Spedden, E.; Staii, C.; Atherton, T.J.; Hubbell, J.A.; Kaplan, D.L. Silk hydrogels as soft substrates for neural tissue engineering. Adv. Funct. Mater. **2013**.

43. Lulevich, V.; Zimmer, C.C.; Hong, H.S.; Jin, L.W.; Liu, G.Y. Single-cell mechanics provides a sensitive and quantitative means for probing amyloid-beta peptide and neuronal cell interactions. Proc. Natl. Acad. Sci. USA **2010**, 107, 13872–13877.

44. Franze, K.; Gerdelmann, J.; Weick, M.; Betz, T.; Pawlizak, S.; Lakadamyali, M.; Bayer, J.; Rillich, K.; Gogler, M.; Lu, Y.B.; et al. Neurite branch retraction is caused by a threshold-dependent mechanical impact. Biophys. J **2009**, 97, 1883–1890.

45. Staii, C.; Viesselmann, C.; Ballweg, J.; Shi, L.; Liu, G.-Y.; Williams, J.C.; Dent, E.W.; Coppersmith, S.N.; Eriksson, M.A. Positioning and guidance of neurons on Au surfaces by directed assembly of proteins using atomic force microscopy. Biomaterials **2009**, 30, 3397–3404.

46. Staii, C.; Viesselmann, C.; Ballweg, J.; Shi, L.; Liu, G.-Y.; Williams, J.C.; Dent, E.W.; Coppersmith, S.N.; Eriksson, M.A. Distance dependence of neuronal growth on nanopatterned gold surfaces. Langmuir **2011**, 27, 233–239.

47. Reddy, C.V.; Malinowska, K.; Menhart, N.; Wang, R. Identification of TrkA on living PC12 cells by atomic force microscopy. Biochim. Biophys. Acta **2004**, 1667, 15–25.

48. Clark, C.G.; Sun, Z.; Meininger, G.A.; Potts, J.T. Atomic force microscopy to characterize binding properties of alpha7-containing nicotinic acetylcholine receptors on neurokinin-1 receptor-expressing medullary respiratory neurons. Exp. Physiol **2013**, 98, 415–424.

49. Schar-Zammaretti, P.; Ziegler, U.; Forster, I.; Groscurth, P.; Spichiger-Keller, U.E. Potassium-selective atomic force microscopy on ion-releasing substrates and living cells. Anal. Chem **2002**, 74, 4269–4274.

50. Johnson, A.S.; Nehl, C.L.; Mason, M.G.; Hafner, J.H. Fluid electric force microscopy for charge density mapping in biological systems. Langmuir **2003**, 19, 10007–10010.

51. Yang, Y.; Mayer, K.M.; Hafner, J.H. Quantitative membrane electrostatics with the atomic force microscope. Biophys. J **2007**, 92, 1966–1974.

52. Kalinin, S.V.; Rodriguez, B.J.; Jesse, S.; Thundat, T.; Gruverman, A. Electromechanical imaging of biological systems with sub-10 nm resolution. Appl. Phys. Lett. **2005**, 87.

CHAPTER 12

Preparation of DOPC and DPPC Supported Planar Lipid Bilayers for Atomic Force Microscopy and Atomic Force Spectroscopy

Simon J. Attwood 1, Youngjik Choi 2 and Zoya Leonenko 1,2,3

[1] Department of Physics and Astronomy, University of Waterloo, Waterloo, ON N2L 3G1, Canada
[2] Department of Biology, University of Waterloo, Waterloo, ON N2L 3G1, Canada
[3] Waterloo Institute for Nanotechnology, University of Waterloo, Waterloo, ON N2L 3G1, Canada

ABSTRACT

Cell membranes are typically very complex, consisting of a multitude of different lipids and proteins. Supported lipid bilayers are widely used as model systems to study biological membranes. Atomic force microscopy and force spectroscopy techniques are nanoscale methods that are successfully used to study supported lipid bilayers. These methods, especially force spectroscopy, require the reliable preparation of supported lipid bilayers with extended coverage. The unreliability and a lack of a complete understanding of the vesicle fusion process though have held back progress in this promising field. We document here robust protocols for the formation of fluid phase DOPC and gel phase DPPC bilayers on mica. Insights into the most crucial experimental parameters and a comparison between DOPC and DPPC preparation are presented. Finally, we demonstrate force spectroscopy measurements on DOPC surfaces and measure rupture forces and bilayer depths that agree well with X-ray diffraction data. We also believe our approach to decomposing the force-distance curves into depth sub-components provides a more reliable method for characterising the depth of fluid phase lipid bilayers, particularly in comparison with typical image analysis approaches.

Keywords: DOPC; DPPC; AFM; force spectroscopy; supported lipid bilayer; vesicle fusion; breakthrough forces; force volume

1. INTRODUCTION

The surfaces of cell plasma membranes play a pivotal role in many biological processes including cell recognition, signalling, selective-ion transfer, adhesion and fusion [1]. The composition and lateral organisation of native membranes are complex, consisting for example of mixtures of phospholipids, glycolipids and various proteins. Such complexity makes the task of identifying the specific effects of membrane interactions with other molecules very difficult. Therapeutic drugs or protein molecules may target specific receptors, but also may interact non-specifically with the lipid membrane itself [2,3]. By simplifying the system, it is possible to systematically study the sub-components of cellular membranes and therefore gain valuable insights that would otherwise be obscured.

Atomic Force Microscopy is a very powerful technique that can be used to study not only the topographical changes but also a range of biomechanical properties. There has been a lot of interest recently in planar supported lipid bilayers (SLB) as model systems, comprised of either single or multiple component lipids, prepared either using vesicle fusion [4–12], or Langmuir–Blodgett or Langmuir–Schaefer deposition [13–15]. There are generally two approaches to studying these systems with the Atomic ForceMicroscope. Firstly, AFM imaging can be performed by scanning the AFMprobe across the surface of a lipid bilayer, which provides information on the topographical characteristics of the supported lipid bilayer, such as the lateral extent of domains, roughness and height of patches relative to the substrate. Then, after addition of an effector molecule of interest, the surface topography can be re-assessed. We can also find the timescale of the interaction by imaging the surface after incremental time steps and at each point assess the changes. Examples of these type of studies include lipid interactions with anesthetic halothane [16,17], ethanol [16], antibiotic azithromycin [18,19], immunodeficiency peptide [7], peptide gramicidin [20], amyloid beta [21–24], model peptide WALP23 [25].

The second approach is to apply force spectroscopy to assess the biomechanical changes due to some effector molecule. In this technique AFM probes are brought towards the supported lipid bilayer and a load increasingly applied until the bilayer ruptures and the probe senses the underlying hard substrate. Afterwards the probe is withdrawn and the cycle is repeated many times. The rupture events are manifested by a well-defined discontinuity in the force-distance approach curves, which can subsequently be analysed to determine the magnitude of the rupture force or break-through-force. The average or most probable rupture force has shown to be a fingerprint for the intrinsic properties of the bilayer. The effect of pH [26], ionic strength of medium [27], deposition pressure [28], temperature [29] and head/tail group composition [30] on membrane structure and function have all been studied. Furthermore, the effect of various proteins and drugs have also been studied, including Myelin based protein [31,32], cytochrome-c [31], bax protein [33], cholesterol [34–36], Synapsin I [37], general anesthetic halothane [17] and antibiotic azithromycin [19]. Atomic force spectroscopy (AFS) is commonly performed in a force volume mode in order to collect a statistically sound set of data. This requires defect-free supported planar lipid bilayers covering extended areas. In all of these studies an essential prerequisite is a well-developed protocol that can consistently be used to prepare lipid bilayers [38].

In the current work we detail two protocols that have been developed for preparing both fluid and gel phase planar bilayers on mica for use in AFM studies. Dioleoylphosphatidylcholine (DOPC) has a transition temperature of–16.5 °C [39] and therefore exists in the fluid like liquid crystalline state (L_β) at room temperature. Dipalmitoylphosphatidylcholine (DPPC) has a transition temperature of 41.3 °C [40] and therefore exists in the solid-like gel state (L_β) at room temperature. Lipids in different states can affect membrane functionality very differently and we therefore chose two lipids to represent the two main lipid phase classes. Furthermore, phospholipids containing the choline group moiety are the most abundant class in eukaryotic cells [41]. We also note that DOPC and DPPC are two of the most common model lipid systems studied.

We report defect-free bilayers that are ideally suited for AFM studies. We highlight the most significant experimental parameters, and introduce tests that can be used to confirm the presence of bilayer. We present optimization approaches to account for solution to solution differences that are difficult to control. We then highlight the major differences between fluid and gel phase lipids by presenting a protocol for the formation of DPPC bilayers. We illustrate through experiment the effect of parameters such as solution temperature, cooling rate, incubation time, concentration and ionic strength.

Without a thorough understanding of bilayer preparation, it is easy to produce misleading results. Even mature protocols are reported to take several months of practice before these model systems can be accurately and reliably reproduced [38]. We hope to highlight the relative importance of various experimental parameters, illustrate the significant differences between gel and fluid phase lipids, and present important tests for bilayer assessment. Debate still continues about the best way to prepare bilayer samples and the important experimental parameters as evident by every single laboratory using a different protocol. We hope that through this work, scientists new to the field can quickly and reliably produce model bilayer systems for their study.

Finally, we illustrate the effectiveness of the DOPC protocol by assessing the DOPC surface with force spectroscopy. Without a reliable, robust and defect-free bilayer surface, it is easy for the tip to get contaminated, producing inconsistent and misleading results. We demonstrate that rupture force and depth values can easily be obtained in good agreement with other AFM studies. We believe that obtaining the depth characteristics of a bilayer from the force spectroscopy measurements is a much more accurate and representative measure of the bilayer thickness. Our values agree well with previous X-ray diffraction studies.

2. RESULTS AND DISCUSSION

2.1. DOPC Bilayers from Vesicles in Water

The AFM is a very powerful tool for determining the topography of materials at the nanoscale and has the distinct advantage over other techniques in that it can be used in liquid over a range of salt concentrations and pH values. This makes it particularly amenable to the study of biological systems where physiological conditions are important. In the case of synthetic bilayers, which are important cell analogues, experiments must be carried out at least in pure water, as their assembly and stability is primarily driven by the hydrophobicity of these amphiphilic molecules.

A deficiency of the AFM is its inability to truly probe three dimensions. Topography maps lack a depth component into the sample and are really two dimensional surface maps. From a practical viewpoint, it is therefore very difficult to distinguish single bilayers that completely cover the surface from multilayers and/or bare mica, all of which would produce a completely featureless image. There are tricks however that can be employed to overcome some of these problems.

Protocols for the preparation of DOPC liposome solutions and sample preparation by vesicle fusion are described in detail in the experimental section. In order to produce and verify that a complete bilayer has been formed on the mica surface, we find it necessary to perform a time series of experiments. That is, we prepare several samples each incubated with the liposome solution for slightly different times. Thus, we are able to capture the sample at different stages during formation and, by washing, halt any further progress. As shown in Figure 1, this allows us to very precisely determine the exact experimental conditions for the specific liposome solution being used, which will produce complete coverage (we observe continuous bilayers for > 30 μm^2 areas). There are several methods for confirming whether we observe single bilayer patches spreading across a mica surface versus patches of bilayer forming on top of a complete first bilayer (or multilayer). Firstly, the phase signal, which is thought to reflect differences in mechanoelastic and surface chemical properties, can be used; if we see a large difference in phase (1–2 degrees for Agilent AFM) between the bilayer patches and the underlying surface, then this is a good indication that we are observing mica beneath the bilayer patches. However, if the tip gets contaminated, which is very likely whilst imaging the soft DOPC bilayers, then the contrast will not be as great and making a firm conclusion is difficult. We also know that if we incrementally decrease the time or concentration any further than as in the least covered surface, we will only see featureless samples, indicative of bare mica.

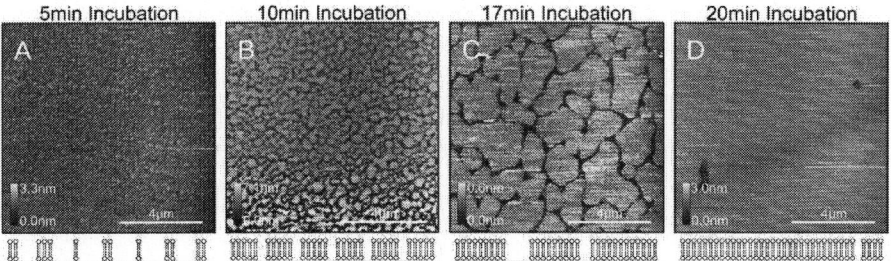

Figure 1. Time series of DOPC bilayer formation. Four separate samples of DOPC bilayers on mica were prepared that were incubated for (**A**) 5min; (**B**) 10min; (**C**) 17min and (**D**) 20min. Below each image is an illustration of the state of the lipid coverage across the mica surface. The time series experiment is a good way of determining that a complete single bilayer covers the mica surface. Without doing the time series experiment, it is very difficult to distinguish a complete single bilayer from multilayer or even bare mica. In addition, when faced with a DOPC sample with partial patchy coverage, it is a simple task to prepare a new sample with a slightly increased incubation time that will result in continuous coverage. All images were taken in pure water at room temperature.

It should be noted that it has previously been thought that the addition of calcium or other divalent cations was an essential step to forming bilayers on mica surfaces [38,42–44]. However we have unequivocally demonstrated here that bilayers on mica can be formed by vesicle fusion in pure water. Such systems may be useful for example when trying to exclude ion mediated interactions.

In Figure 2 we illustrate the delicate, fluidic nature of DOPC. We first prepared a sparsely covered sample of DOPC on mica, and then scanned successively over the same area. As can be seen, the tip drags the lipid together, forming successively larger patches, with the space between them growing. The effect is most obvious when the boundary between a previously scanned area and an untouched area is observed, such as at the corner. By carefully flattening an image of such patches, we are able to observe two distinct populations, representing the mica surface and the top surface of the bilayer. By fitting two Gaussian distributions and then subtracting the difference between the peak heights, we find that the bilayer depth is 5 nm. The precision of this measurement is very high (standard error of the mean < 0.02%), however the largest uncertainty in this measurement is due to the force setpoint. By scanning with a higher force, it is easy to compress the bilayer, measuring slightly less than the "true" depth, and by scanning with a low force, it is possible that longer range repulsive forces play a role, resulting in an overestimation. We believe a quantitative approach to depth measurement of soft bilayer systems can only be achieved by analysing force distance curves.

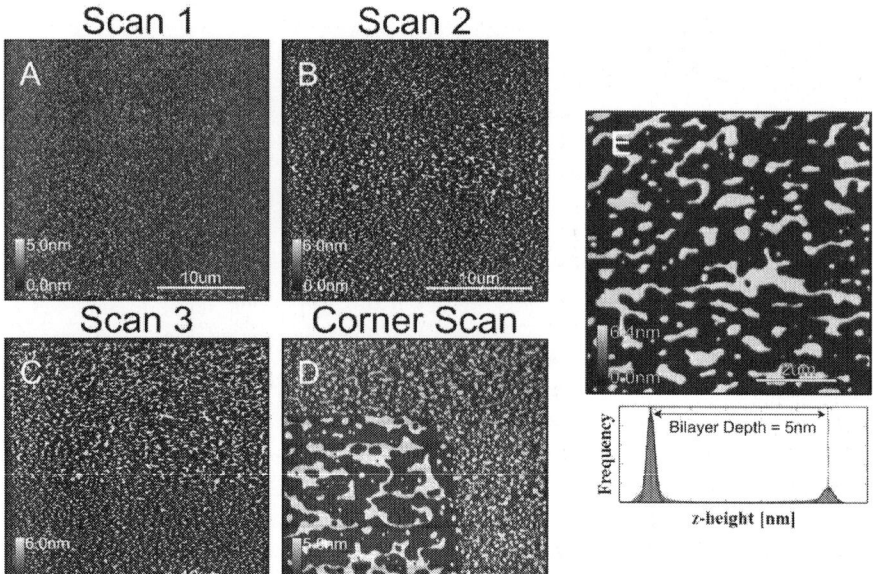

Figure 2. Tip induced DOPC gathering effects on sparsely covered samples indicating both the fluidity of DOPC and the weak attraction of DOPC with mica. (**A**) DOPC sample prepared with a very sparse coverage of DOPC; (**B**) and (**C**) successive scans of sample in A cause the lipid to be drawn together, making larger patches; (**D**) The effect is most clearly seen when a corner at the boundary between a previously scanned region, and an untouched region is scanned; (**E**) The distribution of pixel heights shows two distinct populations, representing the mica surface and the surface of the bilayer. The bilayer depth is found by subtracting the Gaussian fitted peaks of the two populations. The value of 5 nm obtained here agrees well with expected values.

Another useful technique for testing a bilayer surface is the de-wetting approach. In Figure 3A we observe a predominantly flat sample, except for some protrusions. It is difficult from this single image to be certain as to whether we are observing a sample predominantly consisting of uncoated mica but with a few vesicles adsorbed to the surface, or whether there is a complete bilayer covering the mica but with a few vesicles either adsorbed or existing in a trapped state [6,43,45]. By simply removing some of the water from the fluid cell so that the surface de-wets for a few seconds and then replacing the water quickly, we observe many holes as shown in Figure 3B. This clearly demonstrates that the original surface was in fact a complete bilayer. Again the fluidity of this DOPC sample is demonstrated when we scan some time later, over exactly the same area, as in Figure 3C, where we observe the holes to have decreased in size. In contrast to the de-wetting technique, we observe that increasing the strength of the washing stream or the volume used to wash the sample has no perceivable effect.

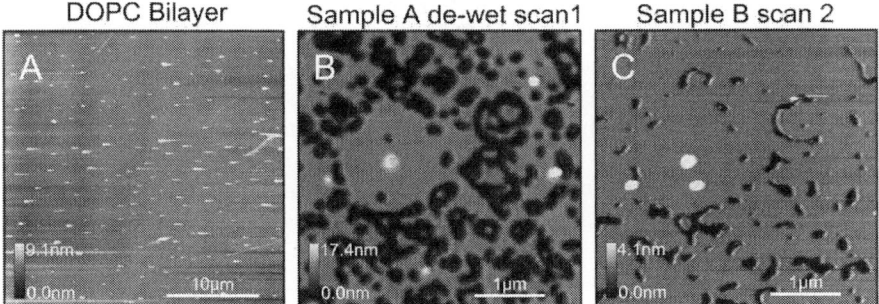

Figure 3. Illustration of the dewetting test used to confirm bilayer coverage. (**A**) A sample is prepared using DOPC liposomes. Apart from some protrusions the sample is defect-free, making a confirmation of the state of the sample (mica, single bilayer, multilayer) difficult; (**B**) Sample instantaneously dewetted and then rehydrate. The holes confirm the presence of a single bilayer; (**C**) After a short period the holes begin to close, demonstrating again the fluidic nature of DOPC. All images were taken in pure water at room temperature.

2.2. DPPC Bilayers from Vesicles in Water

Whereas liposomes of DOPC in solution have a transition temperature of -16.5 °C [39], liposomes of DPPC have been determined to have a transition temperature of 41.3 °C [40]. This means that DOPC is in the fluid phase at room temperature, whereas DPPC is in the gel phase. When bilayers are attached to a mica surface, the proximal and distal leaflets are decoupled due to the strong interaction of the proximal leaflet with the mica surface [46] leading to a broadening of the transition temperature to between 41 °C and 46 °C. AFM based studies demonstrated that there is a transition temperature width of 10 °C [47] for single bilayers on mica. Even with substrate induced broadening of the transition temperature, it is clear that DPPC is in the gel phase at room temperature. The dynamics of vesicle fusion are completely different for lipids in the gel phase compared with the fluid phase. Typically it is reported that DPPC vesicles should be heated above the transition temperature (50–60 °C) in order for them to fuse with the mica surface and form planar bilayers [23,38,48]. We typically observe that DPPC vesicles in water deposited at room temperature either partially fuse, or only form a vesicle layer. However, we demonstrate later that for high concentrations of DPPC solution deposited for short time periods before washing, the vesicles will fuse to mica at room temperature.

We prepared DPPC liposomes using the general protocol described in the experimental section, with the lipid at 0.3mgml^{-1}. The solution was then added to a fluid cell containing mica maintained at 60 °C using the heating stage. After 5min the sample was washed with water and allowed to cool at room temperature (~ 5 °Cmin^{-1}). Previous experiments (data not shown) with low lipid concentrations ($< 0.5 \text{mgml}^{-1}$) demonstrated that the vesicles would not fuse to mica at room temperature. However, as shown in Figure 4A, when heating in-situ, the vesicles are able to fuse. We observe complete bilayer coverage across the mica surface, however we very distinctly observe ~ 2 nm high domains. We then reheated the sample to 60 °C, which was maintained for 5min, and then cooled the sample (5 °Cmin⁻¹) to room temperature. We now observe holes in the bilayer surface exposing mica, due to lipid loss to the water [29], allowing us to clearly see the three levels correlating to bare mica, a low DPPC domain, and a high DPPC domain. The highest domain appears to be about 6 nm in height, consistent with DPPC in its gel phase [25,47,49]. Furthermore, we see that the higher domains have become larger in size and more uniform at their edges, suggesting a re-organization has occurred due to the extra heating step. Similar domains have been reported before [23,29,47,50] at room temperature although the domains were usually much smaller and experiments were carried out in buffer with salt. It is not clear exactly what they are due to, however they are usually attributed to either interdigitation, or tilting of the lipids. These experiments have been repeated using a liposome solution that was prepared straight from powdered DPPC, so as to be completely certain that there are no trace solvents in the solution. The results were qualitatively identical to those presented here.

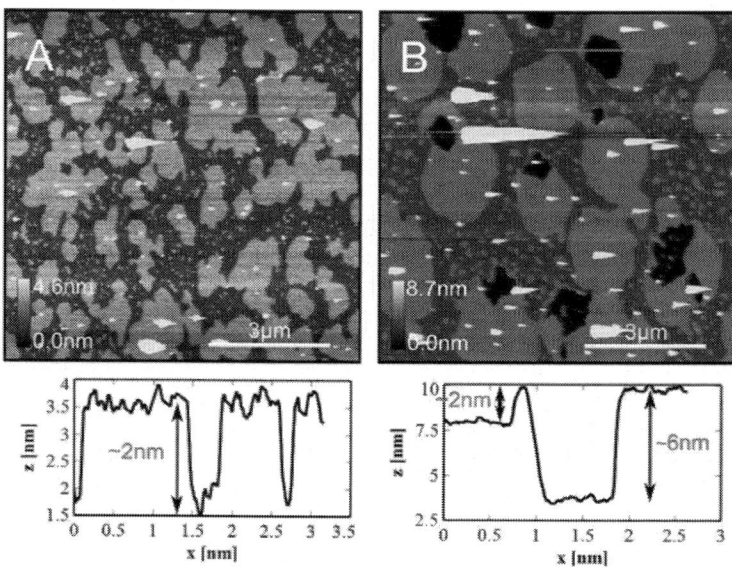

Figure 4. DPPC domains observed at room temperature. (**A**) DPPC liposome solution in pure water was added to a fluid cell at 60 °C. After 5min incubation the sample was washed and allowed to cool with the heater off ~ 5 °Cmin⁻¹. Lipid completely covers the surface with two different domains with a height difference of ~ 2 nm; (**B**) The same sample was reheated to 60 °C, held at that temperature for 5min and cooled at 5 °Cmin⁻¹. Holes in the lipid exposing mica appear after re-heating, which are thought to be due to lipid loss into the liquid.

As shown in the large area (~ 30 μm^2) scans of Figure 5, we investigated the effect of cooling rate on sample topography. Three separate samples were prepared at cooling rates of 5 °Cmin^{-1}, 1 °Cmin^{-1} and 0.5 °Cmin^{-1}. There is a clear progression observed in which higher domains cover an increasingly larger proportion of the mica surface relative to the lower domains with decreasing cooling rate. We also observe regions of bare mica in the samples cooled at 1 °Cmin^{-1} and 0.5 °Cmin^{-1}, which is a reflection of the increased time spent at higher temperatures as compared with the sample cooled at 5 °Cmin^{-1}. As shown in Figure 5C we sometimes observe three different types of domains, with the lowest two being 1.1 nm and 0.6 nm below the upper domain in this case. It is not clear what the lower domains are due to, however there is likely a decoupling effect due to the proximal leaflet interacting with the mica surface. For all samples we tend also to see some vesicles that are not removed by washing and are likely to be trapped [6,43,45]. However, with the sample cooled at 0.5 °Cmin^{-1} we do see extended regions several micrometers square of defect-free bilayer that could be used to test interactions with other molecules either by surface imaging or force spectroscopy.

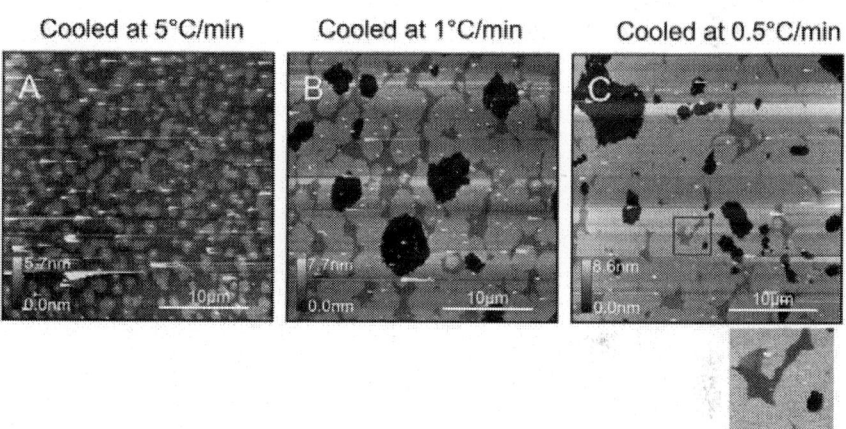

Figure 5. Effect of cooling rate on DPPC bilayers in water. All samples were prepared by incubation with DPPC liposome solution in water at 60 °C and incubation for 5min followed by cooling to room temperature at (**A**) 5 °Cmin^{-1}; (**B**) 1 °Cmin^{-1}; (**C**) 0.5 °Cmin^{-1}. The region enclosed by the blue box has been enlarged and is shown below. For this image of DPPC at room temperature, we see three different types of domains with the lowest two being 1.1 nm and 0.6 nm below the upper domain.

We also tested the effect of changing the temperature of the mica in the fluid cell. When only 39 °C is maintained, we see a dramatic reduction in the number of lower domains compared with a sample prepared at 60 °C as shown in Figure 6. The trend of decreasing lower domains with decreasing temperature continues down to 33 °C where the domains are almost completely eliminated. However, at 30 °C we start to observe unfused vesicles. This suggests that although the transition temperature for a DPPC liposome solution is 41 °C, they will fuse with mica in pure water at and above 33 °C. Thus it seems the minimum temperature for complete vesicle fusion is well below the lipid transition temperature. However, for these samples prepared between 33 °C and 39 °C, we observe many more protrusions across the surface, suggesting that trapped vesicles are more likely to form at these lower temperatures. Reimhult et al. [45] made the same observation with eggPC (transition temperature ~ -15 °C) using the Quart Crystal Microbalance technique. We also tested a sample prepared at 33 °C and cooled at 0.5 °Cmin^{-1}, for which we observe the lower domains completely eliminated (data not shown as very similar to Figure 6E).

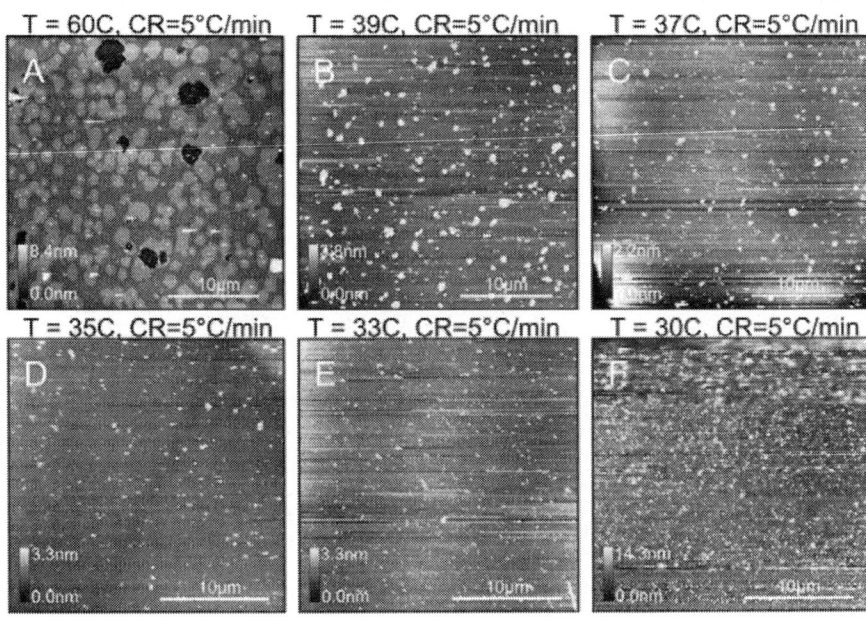

Figure 6. Effect of mica temperature during deposition. All samples were prepared with a cooling rate of 5 °Cmin^{-1}. Mica temperature during deposition was: (**A**) 60 °C; (**B**) 39 °C; (**C**) 37 °C; (**D**) 35 °C; (**E**) 33 °C; (**F**) 30 °C. Lower domains decrease dramatically as a proportion of total lipid coverage between 60 °C and 39 °C, after which a slow decrease is observed to 33 °C. At 30 °C we mostly observe unfused vesicles. Protrusions thought to be trapped vesicles are seen for all samples but dominate for temperatures below 60 °C.

Although we mentioned earlier that the protrusions are likely to be trapped vesicles, it is actually very difficult to be certain about the exact form that they take. As shown in Figure 7 we suggest that they could be either trapped vesicles, adsorbed vesicles or partially fused vesicles. In order to gain insights into this process, we prepared a new liposome solution at 1mgml^{-1} and diluted it to create samples at 0.5mgml^{-1}, 0.33mgml^{-1} and 0.17mgml^{-1} and deposited at 33 °C. Higher resolution images of these samples are shown in Figure 8. At 0.5mgml^{-1}, we observe a complete bilayer, but with partially fused bilayer patches adsorbed on top. When the solution is diluted to 0.33mgml⁻¹, we see a decrease in the quantity of the adsorbed bilayer patches, and some holes in the first bilayer that expose the mica surface. When the solution is diluted to 0.17mgml⁻¹, we see the mica is only partially covered, and that although there are some flat bilayer patches, a good proportion of them appear to exist in the partially fused state. It seems that preparing bilayers at these lower temperatures in water results in partially fused vesicles that lead to the protrusions observed. Although lower domains and holes can be eliminated at lower temperatures, the drawback is that partially fused vesicles are inevitable. Such a surface is not ideally suited as a model test surface. Samples created at higher temperatures, although lacking vesicle type protrusions, are also not ideal as the tip can more easily get contaminated at the edge of holes in the bilayer.

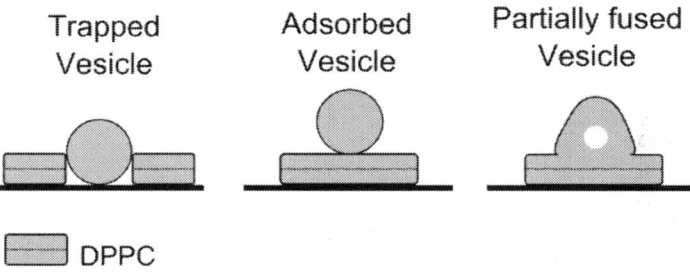

Figure 7. Assigning identity to protrusions observed in bilayer samples. The schematics illustrate three possible variants that may lead to the protrusions seen in DPPC bilayer samples: trapped vesicle, adsorbed vesicle and partially fused vesicle.

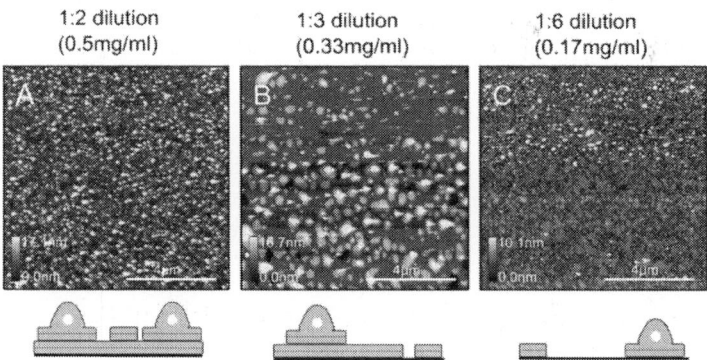

Figure 8. Dilution test for samples prepared at 33 °C. DPPC samples in water were prepared at (**A**) 0.5mgml⁻¹ ; (**B**) 0.33mgml⁻¹ and (**C**) 0.17mgml⁻¹ . By diluting the liposome solution to the point where bare mica is seen in the samples, we are able to see the very initial stages of vesicle fusion, which indicate protrusions are partially fused vesicles. Below: Schematics illustrating proposed generalised models for respective samples.

Interestingly, we observe a very similar trend when varying the incubation time of a 1mgml^{-1} DPPC solution. As shown in Figure 9, we vary the incubation time from 0min (sample immediately washed after deposition, actual incubation time < 2 sec) to 3min. At the shortest incubation time, we observe patchy coverage across the mica surface, including partially fused vesicles. The coverage increases in a time-dependent manner, and at the longest incubation time we again observe a complete bilayer but with lipid patches and partially fused vesicles across the surface. Based on this, it seems that an increase in incubation time is equivalent to an increase in concentration. Another important point to note is that in this case these samples were prepared at room temperature. It seems therefore that although vesicles are not able to completely fuse when at lower concentrations (~ 0.3mgml^{-1}) and below 33 °C, vesicles will completely fuse even at room temperature when the concentration is high enough. Therefore, this suggests that there is interdependence between the concentration and the temperature at which vesicles will fuse.

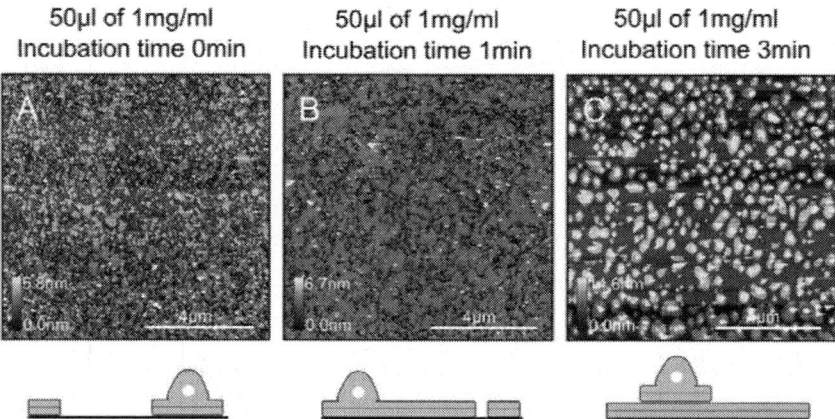

| 50µl of 1mg/ml | 50µl of 1mg/ml | 50µl of 1mg/ml |
| Incubation time 0min | Incubation time 1min | Incubation time 3min |

Figure 9. Variation of short incubation times when using a DPPC solution of high concentration (1mgml^{-1}). Incubation times of (**A**) 0min; (**B**) 1min and (**C**) 3min. The increase in surface coverage with increasing incubation time is qualitatively the same as seen for experiments increasing the concentration.

In Figure 10 we demonstrate the dewetting test for DPPC. In this case we removed the liquid from the fluid cell carefully until the surface was instantaneously dewetted, after which water was immediately replenished. The surface was re-imaged, and then the dewetting step repeated again followed by a final image being taken. As can be seen, holes are formed after the initial dewetting, which become larger after the second dewetting step. As with DOPC, this is a good test used to verify that the surface was completely covered with a first layer of lipid. In addition we can see that dewetting seems to preferentially remove the bilayer patches from the top as opposed to the vesicle structures. As with DOPC, we also tested the effect of various washes with increased force but again observed no change in the bilayer topography. We also tested the effect of washing with a buffer at a different ionic strength (PBS) followed by again washing with water, for which we also observe no change in the surface topography.

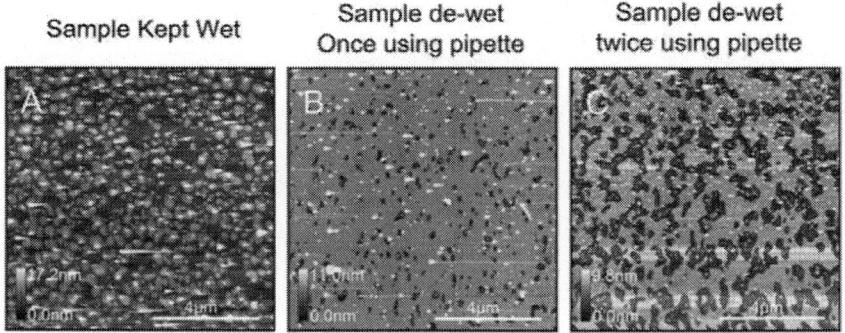

Figure 10. Illustration of the dewetting test for DPPC. The sample bilayer in (**A**), which is very similar to the bilayer in Figure 9C, was dewetted instantaneously and immediately rehydrated (**B**); The dewetting step was repeated a second time (**C**). The holes in the bilayer become progressively larger after each successive wash. The test proves that the sample in (**A**) was a complete bilayer with partially fused lipid patches on top.

2.3. DPPC Bilayers from Vesicles in Buffered Salt Solutions

We see in the literature that bilayers can be prepared both in pure water and in solutions containing buffers and salts of various types and concentrations. It is also reported in the literature that divalent cations such as Ca^{2+} or Mg^{2+} are very important in aiding fusion of vesicles to mica surfaces [42,43,51]. We therefore tested two buffers, one containing only NaCl and HEPES buffer (HEPES-NaCl: 10mM HEPES, 150mM NaCl, pH 7.4), and another in addition containing 20mM Mg^{2+} (HEPES-NaCl-Mg:10mM HEPES, 150mM NaCl, 20mM $MgCl_2$, pH 7.4).

We see in Figure 11 that vesicles do not fuse to mica unless the deposition temperature is raised to 70 °C. Below this temperature, we see a combination of small fused bilayer patches and unfused vesicles.

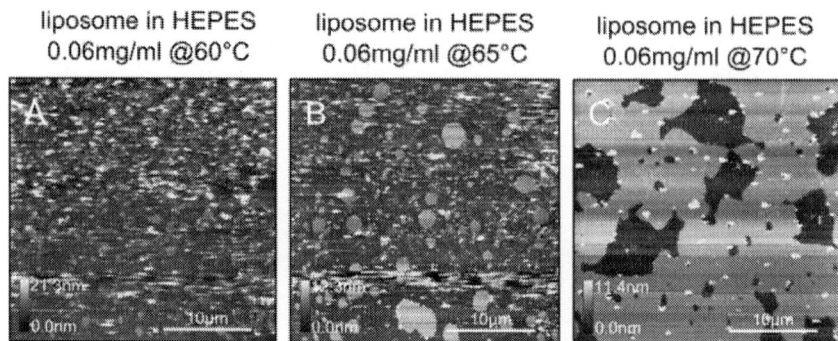

liposome in HEPES
0.06mg/ml @60°C

liposome in HEPES
0.06mg/ml @65°C

liposome in HEPES
0.06mg/ml @70°C

Figure 11. DPPC bilayers prepared from a HEPES-NaCl liposome solution. Bilayers prepared at (**A**) 60 °C; (**B**) 65 °C and (**C**) 70 °C. We observe continuous fusion only when the sample plate is maintained at 70 °C; below this temperature mostly observed are vesicles with occasional fused patches.

Samples were prepared using liposomes in HEPES-Mg at exactly the same concentration (0.06mgml^{-1}) and deposition time (2min) as for the HEPES-NaCl samples shown in Figure 11. We see from Figure 12 that at 50 °C a vesicle layer is formed. By increasing the force whilst scanning, we are able to fuse the vesicles to some extent as shown by the contrasting horizontal regions of fused and unfused regions. From 55 °C to 70 °C we clearly see that the bilayer has fused, however quite a lot of holes still appear. The change in minimum temperature for complete vesicle fusion of DPPC liposomes at 0.06mgml^{-1} between HEPES-NaCl buffer and HEPES-NaCl-Mg buffer from ~ 70 °C to ~ 55 °C is thought to be due to the increased attraction between positively charged liposomes and negatively charged mica. It has been shown that for DMPC liposomes, which have the same head group as DPPC, the charge on the liposomes can vary depending on the ionic strength of the medium [48]. In the case of pure water, they observe a strong repulsion due to negative charges, whereas in 100–150mM NaCl the liposomes are close to neutral. With 150mM NaCl plus 20mM MgCl$_2$, they observe a repulsion due to positive charges.

Figure 13A shows another DPPC sample prepared in HEPES-Mg, similar to the samples shown in Figures 12B–D except that it was incubated at 60 °C and for 60min as opposed to just 2min. The sample is very similar to those prepared for shorter incubation times, indicating that increasing the time further has little effect on the quality of the bilayer. We note that by increasing the force that the tip images with, we are able to superficially make the sample appear smoother, with less trapped vesicles. Although some of the vesicles were dislodged and moved away whilst scanning, many of them still remain and the surface mainly looks much cleaner because the tip is tracking across the surface of the vesicles better with the higher force. At these higher forces however, damage can occur to the sample. We see a higher resolution scan in Figure 13C. A few lines of this area were then scanned with much increased force (Figure 13D), which then causes holes to appear with the second pass at a much lighter force. In addition to the formation of holes, we see that some holes fill in even when scanning relatively lightly. Although the extent of tip induced changes is far less than with fluidic DOPC, we still see that DPPC is quite easily deformed by the tip. We believe that the quality of the bilayer shown in Figure 13D is comparable to DPPC bilayers presented by several other groups, which used their DPPC samples as test systems to assess changes due to other proteins/peptides or drug interactions [16,23,25,30,47,50,52].

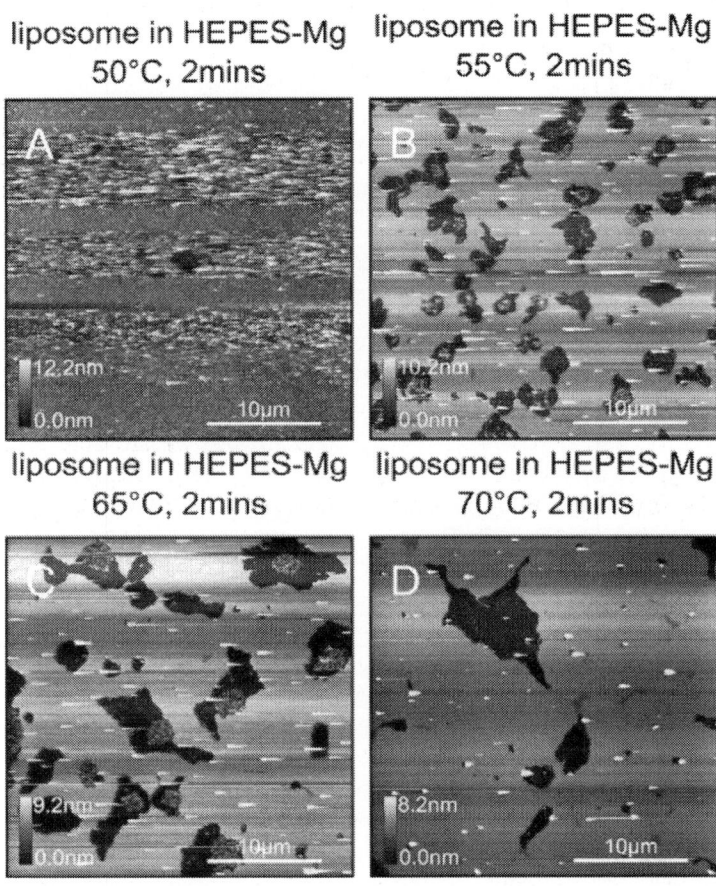

Figure 12. DPPC bilayers prepared from a HEPES-NaCl-Mg liposome solution. Bilayers prepared at (**A**) 50 °C; (**B**) 55 °C; (**C**) 65 °C; (**D**) 70 °C. When the sample plate is maintained at 50 °C we only observe a vesicle layer, which can be partially fused by the tip as indicated by the horizontal streaks in A. For 55 °C, 65 °C and 70 °C deposition temperatures, we observe extended regions of fused bilayer. More domains are observed for the highest temperatures. Samples were prepared using liposomes in HEPES-Mg at exactly the same concentration (0.06mgml^{-1}) and deposition time (2min) as for the HEPES-NaCl samples shown in Figure 11.

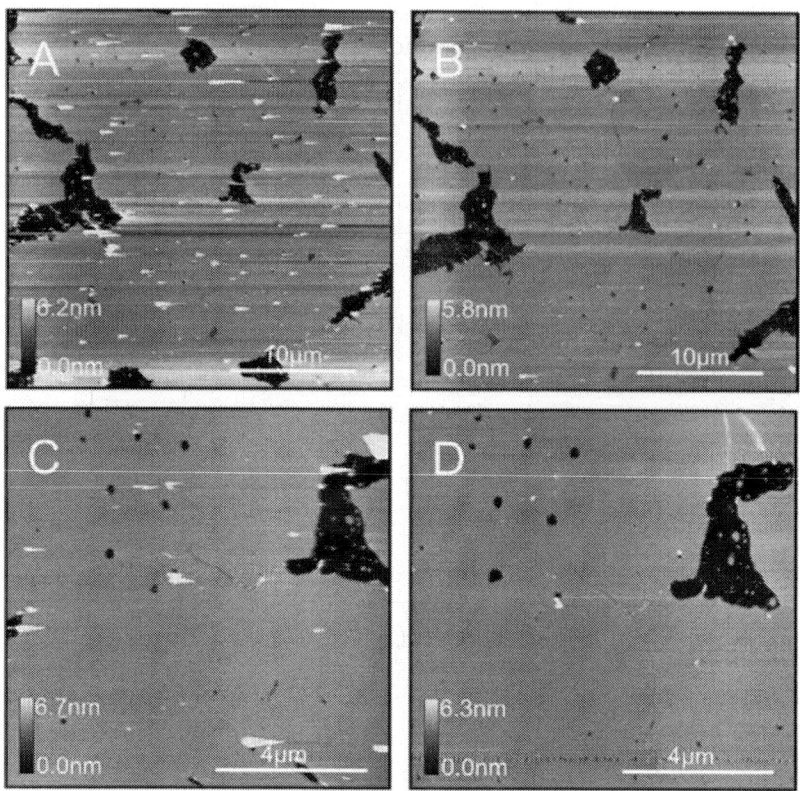

Figure 13. DPPC sample prepared from HEPES-NaCl-Mg liposome solution and incubated at 60 °C for 60min. (**A**) Light force scanning highlights vesicles; (**B**) Higher force scanning makes the sample appear more continuous than it really is. Although some vesicles were dislodged and swept away, the smoother appearance of the surface is mostly due to the tip tracking across the vesicles better; (**C**) Extended regions of defect-free bilayer are observed; (**D**) A few lines of (**C**) were scanned with a high force and then scanned lightly again (area enclosed by red box). We see holes due to the hard scanning. We also see holes disappearing even with relatively light scanning (area enclosed by red circle), indicating that the DPPC bilayers are delicate and dynamic, although much less so than DOPC.

Overall we do not observe any significant improvement in the quality of the DPPC bilayer surface (extent of flat regions, absence of holes, domains or trapped vesicles) when comparing bilayers prepared in water (Figure 5C), HEPES-NaCl (Figure 11C) or HEPES-NaCl-Mg (Figure 12D). However, these samples were prepared with slightly different experimental conditions; pure water: $C_l = 0.3\text{mgml}^{-1}$, $t_d = 5\text{min}$, $T_d = 60$ °C, $R_c = 0.5$ °Cmin^{-1} ; HEPES-NaCl: $C_l = 0.06\text{mgml}^{-1}$, $t_d = 2\text{min}$, $T_d = 70$ °C, $R_c = 5$ °Cmin^{-1} ; HEPES-NaCl-Mg: C_l

$= 0.06 \text{mgml}^{-1}$, $t_d = 2\text{min}$, $T_d = 55\ °\text{C}$, $R_c = 5\ °\text{Cmin}^{-1}$ (C_l, t_d, T_d, R_c are lipid concentration, deposition time, deposition temperature and cooling rate respectively).

Generally we observe that DPPC bilayers prepared in either HEPES-NaCl or HEPES-NaCl-Mg form fused bilayers with less concentrated solutions and with less time than compared with vesicles prepared in pure water alone. In order to prepare fluid phase DOPC bilayers in pure water, we need higher concentration (> 0.5mgml^{-1}) and longer incubation time ($\sim$ 15min). Generally we are able to use much faster cooling rates for DPPC samples prepared in either buffer and see fewer domains compared with samples prepared in pure water, for which a high proportion of the sample surface contains lower domains unless very slow cooling rates are used. Since DOPC is in the fluid phase at room temperature, we never observe domain formation during sample preparation. We also observe interdependence between lipid concentration and the minimum temperature at which vesicles fuse for DPPC. We observe that although vesicles are not able to fuse when at lower concentrations ($\sim$ 0.3mgml^{-1}) and below 33 °C, vesicles will fuse even at room temperature when the concentration is high enough (1mgml^{-1}).

2.4. Force Spectroscopy of DOPC Bilayers

During a typical force spectroscopy experiment, the tip successively approaches and leaves the surface in a cyclic manner. The force experienced by the cantilever is detected and then plotted against the z-piezo displacement or tip-sample separation. During each cycle the x- and y-coordinates are typically fixed, however in order to get a more reliable measure of the surface properties, the tip is typically moved laterally between cycles, in an approach often referred to as "force volume mapping" [53]. This is illustrated schematically in Figure 14C. Collecting data at several different points across a surface is preferable over measurements at a single point, since slight deviations in bilayer properties can be averaged, giving a more representative measurement. However, as seen in the image, if the bilayer is full of holes, the tip can easily become contaminated at the mica–lipid edges, resulting in misleading results. It is imperative therefore that the bilayer be continuous for such measurements.

Figure 14 illustrates the most important features of a typical force-distance curve exhibiting a rupture event associated with a fluid-like lipid bilayer. Typically, the rupture force (F_B), bilayer depth (z_{A-B}) and Young's modulus (E) may be determined from the approach curve. In addition, the force of adhesion (F_{adh}) and work of adhesion (W_{adh}) may be determined from the retract curve. For the current work however, we focus just on the rupture force and depthmeasurements. All of the significant discontinuities in the approach curve have been labeled (A–D). The physical significance of these points and their transitions is interpreted as follows; (A) First contact of the tip with the top surface of the bilayer; (A–B) elastic compression of the bilayer; (B) Rupture of the upper surface of the bilayer; (B–C) Rapid tip transition through the central portion of the bilayer; (C) On-set of increased repulsion associated with compression of proximal head groups, water layer [54] and other trapped material; (C–D) Compression of trapped material; (D) Tip in direct contact with mica surface.

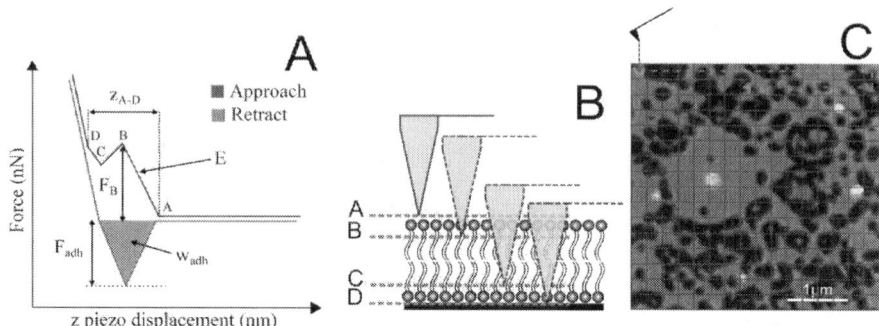

Figure 14. (**A**) Schematic force-distance plot highlighting the most significant features of a typical rupture event associated with a fluid-like lipid bilayer. From the approach curve (blue), the rupture force F_B, the bilayer depth z_{A-D} and the Young's modulus E may be determined. From the retract curve (red), the maximum force of adhesion F_{adh} and the work of adhesion W_{adh} may be determined; (**B**) Schematic illustrating tip penetration through the bilayer (not to scale); (**C**) A 16 × 16 grid illustrating "force volume mapping", where force-distance curves are conducted at each point on a grid across a surface (in this case 4 μm × 4 μm).

We are able to get several different measures of the bilayer depth. The full depth of the bilayer may be considered to be the distance between points A and D (z_{B-D}). It should be noted however that several factors can lead to over- or under-estimation of the bilayer depth using this approach. Point A is not always well-defined, for example, due to long range electrostatic repulsion leading to a curved region around A, and a similar problem may be observed around D. The nature of the curve about these discontinuities is dependent on the tip chemistry, the ionic strength of the medium and lipid composition and phase. Typically ill-defined contact regions have been observed when a bilayer has formed on the tip, creating strong repulsion for example due to hydration effects [37,55]. For these reasons, the depth z_{B-D} is typically quoted in the literature and often referred to as the "jump depth".

In Figure 15 we show a single representative force distance curve for the rupture of DOPC in pure water. We have converted the z-piezo displacement to tip-sample separation so as to more accurately reflect the tip dynamics. We observe that the tip first senses the upper surface of the bilayer at a distance z_{A-D} = 6.8 nm from the mica surface and the rupture depth occurs at z_{B-D} = 4.9 nm. The rupture occurs at F_B = 3.0 nN.

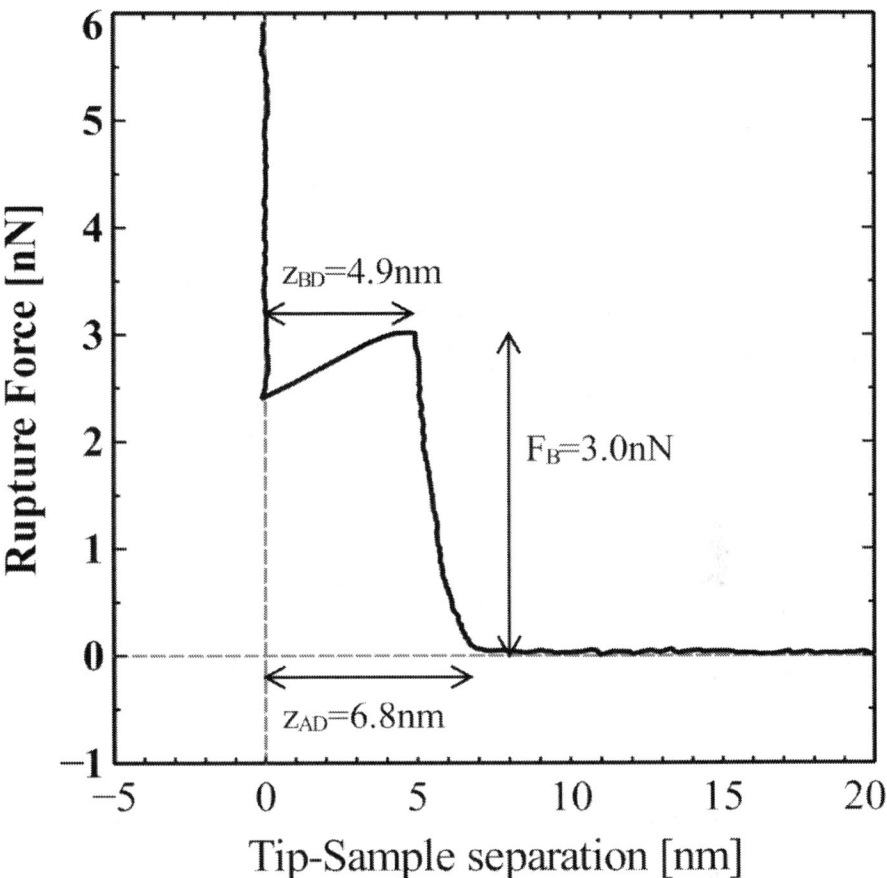

Figure 15. A representative force–distance curve for the rupture of DOPC in water. The rupture occurs at F_B = 3.0 nN. The tip first senses the bilayer at a distance z_{A-D} = 6.8 nm from the mica surface. The rupture depth occurs at z_{B-D} = 4.9 nm. The bilayer depth for DOPC is in good agreement with values reported by X-ray diffraction [54].

In order to have a more accurate reflection of the bilayer properties, it is necessary to repeat the force-distance measurements several times. As shown in Figure 16 we have collected data for 136 approach–retract cycles for a single tip. The mean values are F_B = 3.1 ± 0.3 nN, z_{A-B} = 2.4 ± 0.3 nm, z_{B-D} = 4.6 ± 0.2 nm, z_{A-D} = 7.0 ± 0.3 nm.

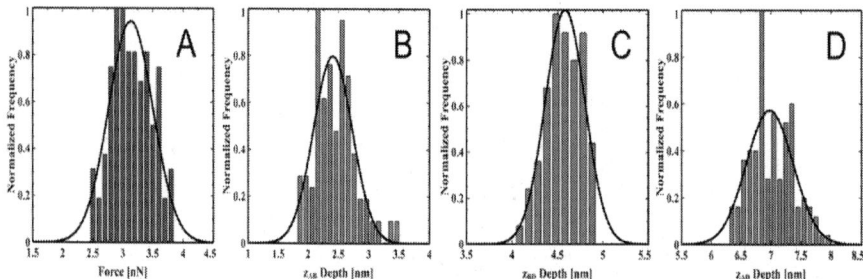

Figure 16. DOPC rupture force and depth distributions from a representative data set. (**A**) Rupture force, F_B; (**B**) z_{A-B} depth; (**C**) z_{B-D} depth; (**D**) z_{A-D} depth. Mean values are $F_B = 3.1 \pm 0.3$ nN, $z_{A-B} = 2.4 \pm 0.3$ nm, $z_{B-D} = 4.6 \pm 0.2$ nm, $z_{A-D} = 7.0 \pm 0.3$ nm (errors are standard deviations, number of data points per distribution is 136). Black lines are Gaussian fits to the distributions.

The rupture force for DOPC is comparable to other reported values [27,34]. The z_{B-D} depth is in good agreement with the value of 5.0 nm reported here from the earlier image analysis of bilayer patches and agrees with other reports in the literature using the same approach where values of 5.5 nm [47] and 4.1 ± 0.2 nm were observed [56]. As mentioned before however, determining bilayer depths for soft DOPC samples using the imaging approach is typically unreliable due to the deformation that the tip causes and the strong susceptibility to small changes in the setpoint force, which is confounded by tip contamination by the soft lipid molecules. We have seen depth values ranging from ~ 3 nm to ~ 7 nm for DOPC using standard image analysis of DOPC patches on mica.

It is difficult to assign a "true depth" to the bilayer because that depends on from where depth is measured for these measurements. The value $z_{A-D} = 7.0 \pm 0.3$ nm from the force spectroscopy analysis is undoubtedly an overestimate, since the tip may sense electrostatic interactions before it actually makes contact, and the term "contact" is also ill-defined because the head groups are typically hydrated, so the tip may be first sensing those water molecules. The depth $z_{B-D} = 4.6 \pm 0.2$ nm is taken at a point when the bilayer is in compression, and so is likely an underestimation. Interestingly, an X-ray diffraction study [54] reports a fully hydrated DOPC bilayer thickness of $D = 6.31$ nm, which is most similar to the depth $z_{A-D} = 7.0 \pm 0.3$ nm. They observe the thickness only due to the lipids as being $D'_B = 4.53$ nm, which is comparable to the depth $z_{B-D} = 4.6 \pm 0.2$ nm. They also report a water layer of $D'_W = 1.79$ nm and comparable to the depth $z_{A-B} = 2.4 \pm 0.3$ nm. We believe that presenting the constituent depth contributions, especially z_{A-D}, z_{B-D} and z_{A-D}, is a much more robust way of characterising a bilayer as compared with the standard image analysis approach.

3. EXPERIMENTAL SECTION

3.1. Materials and Instrumentation

Stock ampules (25mg) of 1,2-dioleoyl-sn-glycero-3-phosphocholine (DOPC, purity > 99%) in chloroform were purchased from Avanti Lipids and stored at −20 °C immediately after receipt. Powdered 1,2-dipalmitoyl-sn-glycero-3-phosphocholine (DPPC, purity ≥ 99%) was purchased from Sigma-Aldrich and stored at −20 °C. Chloroform (purity > 99.8%) and sodium chloride (purity > 99%) were purchased from EMD chemicals (USA). Methanol (purity > 99.8%), sodium hydroxide (purity > 97%) and magnesium chloride (purity > 99%) were purchased from Caledon Laboratories (Georgetown, Ontario, Canada). HEPES (purity > 99.5%) was purchased from Sigma-Aldrich. Millipore water (resistivity > 18Mcm) from a Synergy UV-system was used throughout. A Branson 1510 sonicator bath was also used. Muscovite mica (grade V-4, 22mm diameter, 0.15mm thick circular discs) was purchased from SPI Supplies (West Chester, PA, USA). Silicon MAC-2 cantilevers (nominal spring constant 2.8Nm^{-1}) were purchased from Agilent Technologies. DNPS silicon nitride cantilevers (4 levers, nominal spring constants range in 0.06Nm^{-1} − 0.350Nm^{-1}) and gold coated NPG levers (4 levers, nominal spring constants range in 0.06Nm^{-1} − 0.350Nm^{-1}) were purchased from Bruker. 1-Undecanethiol (purity > 98%) was purchased from sigma.

3.2. General Handling of Lipids

An important point to note when handling DOPC is that due to the double bond in its tail, it is particularly susceptible to hydrolysis or oxidation [57]. The lipid powders are extremely hygroscopic and so we prefer to purchase DOPC dissolved in chloroform and layered with argon. Upon receipt, we store at −20 °C and use one ampoule at a time, which can be divided into aliquots, layered with nitrogen/argon and again stored at −20 °C until required. Degassing water with nitrogen/argon to displace oxygen before liposome preparation and always keeping solutions of DOPC layered with nitrogen/argon when not in use are also recommended.

3.3. DOPC Bilayer Preparation

Stock Solution Preparation

- Open the ampule containing 25mg DOPC in chloroform.

- Transfer the entire ampule directly to a 20ml glass vial. Add an extra 1000 μl of chloroform to the ampule and wash out to the glass vial so as to remove all lipid.

- Evaporate the solvents under a stream of nitrogen until visibly dry (approximately 10–15 min).

- Add 6ml chloroform/methanol (2:1) and split into even aliquots to 6 small glass vials (capacity
- mL each). Each stock vial therefore contains 4.17mg DOPC.
- Top with nitrogen and store at −20 °C.

Liposome Preparation

- Transfer one of the 1000 µl stock aliquots to a clean 20ml glass vial. Wash out the small vial with an extra 1600 µl chloroform/methanol (2:1) and add this to the 20ml vial as well to make 2mM and ensure no lipid is lost.
- Evaporate solvents with a continuous stream of nitrogen. After ~ 15min all should be visibly dry. Continue for an extra 15min to ensure all solvents are evaporated (30min total).
- Bubble nitrogen gas through ~ 20ml of millipore water for ~ 15min to remove the oxygen.
- Add de-oxygenated millipore water to make 0.5mgml^{-1} . The lipid should quickly start to swell, separate from the glass vial and form an inhomogeneous suspension of cloudy material. Also add a stir bar and top the vial with nitrogen.
- Stir using magnetic stirrer for 30min at 1100 rpm (the solution should appear homogeneous and milky).
- Place the solution at 4 °C and allow to swell for 1 hour.
- Stir using magnetic stirrer at 1100 rpm for 30min at room temperature.
- Place the vial in the middle of the sonicator where cavitation is greatest. The most powerful region of any bath sonicator can be found easily by placing a sheet of aluminium foil on the surface of the water and sonicating for ~ 10min—a large hole should appear at the region of most intense power. Sonicate for 30min, during which we observe slight heating of the water (start temperature is ~ 23 °C, finish temperature is ~ 31 °C. The solution should appear completely clear after sonication. If using higher concentrations of lipid, the solution may take longer to go to clarity.
- Store the liposome solution at 4 °C until needed.

Vesicle Fusion

- Remove the liposome solution from the fridge and stir at 1100 rpm for ~ 45 s.
- Place 300 µl of cold solution directly into the AFM fluid cell containing freshly cleaved mica.
- Wait 15min.

- Wash with 10ml water through a syringe by slowly allowing the fluid cell to overflow so as to prevent sample de-wetting.

Optimization

Typically the above protocol will give a continuous bilayer of DOPC. However, slight decreases in the concentration (e.g., a little extra lipid lost during preparation of the liposome solution) may mean that a complete bilayer is not formed. In this case we have found that it is best to prepare several samples that have been incubated for slightly longer and shorter periods of time and image these using the AFM. By capturing the state of bilayer coverage at various points in time, it is then possible to be certain that for a given incubation time a complete bilayer was formed. The time series test is described in more detail in the main text. In some cases, a complete bilayer may not form even for extended incubation times. In this case, another approach can be taken whereby a higher concentration of lipid (1mgml^{-1}) liposome solution is prepared. This solution can then be sequentially diluted to find the optimum concentration to form a complete bilayer.

3.4. DPPC Bilayer Preparation

Liposome Preparation

- Weigh out powdered DPPC and add chloroform/methanol (2:1) to make 2mM.

- Evaporate solvents with a continuous stream of nitrogen for ~ 30min or until visibly dry.

- For liposomes in water: Add water to make 0.3mgml^{-1} (this concentration was used for the sample shown in Figure 5). For liposome in buffer: Add HEPES buffer (10mM HEPES, 150mM NaCl, 20mM MgCl$_2$, pH 7.4) to make 1mgml^{-1} (this liposome concentration was used for samples shown in Figure 13 but diluted later). Also add a stir bar and top the vial with nitrogen.

- Stir using magnetic stirrer for 30min at 1100 rpm and at room temperature.

- Place the solution at 60 °C and allow to swell for 1 hour.

- Stir using magnetic stirrer for 30min at 1100 rpm.

- Sonicate for 45min at 60 °C. The solution should appear completely clear after sonication. However, if using higher concentrations of lipid, the solution may take longer to go to clarity.

Vesicle Fusion

- Stir the liposome solution for ~ 45 s.

- Prepare the fluid cell with freshly cleaved mica at 60 °C.

- For liposomes in water: Add 300 µl and wait 5min. For liposomes in buffer: dilute to ~ 0.06mgml^{-1} in a small centrifuge tube, add 300 µl to the fluid cell and wait 2min.

- For liposomes in water: Cool at a rate less than 0.5 °Cmin^{-1} . For liposomes in buffer: Cooling at 5 °Cmin^{-1} will produce a few domains, and a lower cooling rate will produce even fewer.

- Wash with 10ml water through a syringe by slowly allowing the fluid cell to overflow so as to prevent sample de-wetting.

3.5. Atomic Force Microscopy

All AFM imaging was conducted in pure water or buffer either using an Agilent 5500 AFM (Agilent technologies Inc, Santa Clara CA, USA) equipped with a MAC mode 3 control box, or a JPK nanowizard-2 (JPK Instruments AG, Berlin, Germany). Temperature controlled experiments were performed using the Agilent system equipped with a LakeShore 325 temperature controller and fluid cell. Imaging was performed in intermittent contact mode or MAC mode (Agilent AFM only) using MAC-II cantilevers (nominal k = 2.8Nm^{-1}). MAC mode allows for excellent control of the cantilever in liquid environments and is ideally suited for imaging soft supported bilayer surfaces. Since the mode is an intermittent contact (or AC) technique, cantilevers of slightly higher spring constant than those used for contact mode can be used with great effect. We have also imaged in contact mode using DNPS (Bruker) levers with a nominal spring constant of 0.06–0.350Nm^{-1} and find the image quality comparable. However, MAC mode is less sensitive to cantilever drift (as the cantilever deflection drifts the force set-point must be continually adjusted to maintain a constant contact force) and so is often more convenient.

3.6. Atomic Force Spectroscopy

For force spectroscopy experiments presented here, we used gold coated probes from Bruker (NPG, nominal k = 0.06Nm^{-1}). Cantilevers were first functionalised with > 2mM 1-undecanethiol for > 12 h to provide a more homogeneous surface. As with Loi et al. [58] we observe improved consistency with this approach rather than using uncoated levers. Several different samples and cantilevers have been tested, all showing qualitatively the same results. We have also replicated experiments using uncoated MAC-2 silicon cantilevers from Agilent.

Cantilevers were first calibrated in air using the thermal tune method [59], which is implemented in both the Agilent PicoView software and the JPK software. Briefly, a thermal noise spectrum is first recorded in air. The cantilever is then pressed against a hard material such as mica to obtain the cantilever sensitivity in mV^{-1} . The cantilever spring constant is then calculated from these two measurements as detailed in the reference.

Force curves were obtained by repeatedly approaching and retracting the cantilever to the surface (tip velocity was 715 nms^{-1}) whilst simultaneously recording the tip deflection. The tip deflection data V_d (in Volts) was converted to force units (Newtons) using $F = kV_dS$, where S is the normal cantilever sensitivity in mV^{-1}. The normal sensitivity is the inverse of the gradient of the linear region of the force-distance curve when the cantilever is in hard contact. The force calibration was carried out for each force curve separately to ensure the most accurate value for the sensitivity in case of drift. Care was taken to ensure that high enough loads were reached such that the sensitivity could be found from a linear region. The z-piezo displacement values, z_d, were converted to tip-sample separations, z_{ts}, using $z_{ts} = |z_p| - |z_d|$, where z_d is the tip deflection in nanometers. Typically force curves are taken at points on a grid of 16 by 16 points across a 2 μm square area. Several different areas have also been tested and give qualitatively the same results. Results were analysed using a script developed in our laboratory with MATLAB 7.4.0.

4. CONCLUSIONS

We have developed robust protocols to reproducibly generate extended regions of DOPC bilayer surfaces on mica in water using the vesicle fusion approach. Such test surfaces are typically defect-free and thus ideally suited for studying the influence of effector molecules such as proteins/peptides or drugs either through AFM imaging or by force spectroscopy. Such surfaces are especially suitable for force spectroscopy studies, in which the tip can easily get contaminated at edges of defects in the bilayer. We also demonstrate that DPPC bilayers can be prepared either in pure water or in buffer solutions with additional ions. These surfaces typically have defects either in the form of lower domains or holes exposing bare mica. However, such defects are reported by several other groups and there are still regions ~ 5 μm² that are suitable for further testing.

The large number of different experimental parameters that can influence both the preparation of the liposome solution and their fusion with mica, together with the interdependence of some of those parameters, make it very difficult to understand and control the whole process. There are a staggering number of laboratories that have their own unique "recipe" for preparing bilayers on surfaces, and the complexity of bilayer preparation demonstrated here undoubtedly contributes to such diversity. We have focused here on just two examples of lipids containing phosphocholine (PC) head groups in their fluid and gel phases. It is expected that different lipids will require different conditions in order to consistently prepare bilayers suitable as model systems, however the general themes should apply to all. We hope that by highlighting the most critical experimental parameters, and by providing their general protocols, we provide crucial information to aid scientists, especially those new to the field.

We have also demonstrated force spectroscopy studies of DOPC bilayers prepared using our vesicle fusion protocol. Both the depth measurements and rupture force data are comparable to literature values. We also demonstrate an approach for the assessment of the bilayer thickness characteristics. We subdivide the force-distance plots into depths about the discontinuities and find that these compare well to hydrated and unhydrated lipid values obtained by X-ray diffraction. Without a robust and reliable protocol for bilayer preparation, much time can be wasted with inconsistent and misleading results. With the protocol that we present here however, force spectroscopy experiments can be quickly used to assess lipid surfaces in a reliable manner.

ACKNOWLEDGEMENTS

The authors acknowledge funding from Canadian Foundation for Innovation (CFI), Ontario Research Fund (ORF) and Natural Science and Engineering Council of Canada (NSERC) to Z.L. as well as Ontario Graduate Scholarship and Waterloo Institute for Nanotechnology (WIN) Fellowship to Y.V.C.

REFERENCES

1. Voet, D.; Voet, J.G. Biochemistry; ACS: Washington, DC, USA, 1990.
2. Mouritsen, O.; Jorgensen, K. A new look at lipid-membrane structure in relation to drug research. Pharm. Res **1998**, 15, 1507–1519.
3. Phillips, R.; Ursell, T.; Wiggins, P.; Sens, P. Emerging roles for lipids in shaping membrane-protein function. Nature **2009**, 459, 379–385.
4. Dufrene, Y.; Barger, W.; Green, J.; Lee, G. Nanometer-scale surface properties of mixed phospholipid monolayers and bilayers. Langmuir **1997**, 13, 4779–4784.
5. Dufrene, Y.; Boland, T.; Schneider, J.; Barger, W.; Lee, G. Characterization of the physical properties of model biomembranes at the nanometer scale with the atomic force microscope. Faraday Discuss **1998**, 111, 79–94.
6. Richter, R.; Berat, R.; Brisson, A. Formation of solid-supported lipid bilayers: An integrated view. Langmuir **2006**, 22, 3497–3505.
7. El Kirat, K.; Dufrene, Y.F.; Lins, L.; Brasseur, R. The SIV tilted peptide induces cylindrical reverse micelles in supported lipid bilayers. Biochemistry **2006**, 45, 9336–9341.
8. Goksu, E.I.; Vanegas, J.M.; Blanchette, C.D.; Lin, W.C.; Longo, M.L. AFM for structure and dynamics of biomembranes. Biochim. Biophys. Acta-Biomembr **2009**, 1788, 254–266.
9. Alessandrini, A.; Seeger, H.M.; Di Cerbo, A.; Caramaschi, T.; Facci, P. What do we really measure in AFM punch-through experiments on supported lipid bilayers? Soft Matter **2011**, 7, 7054–7064.

10. Picas, L.; Milhiet, P.; Hernandez-Borrell, J. Atomic force microscopy: A versatile tool to probe the physical and chemical properties of supported membranes at the nanoscale. Chem. Phys. Lipids **2012**, 165, 845–860.

11. Lesniewska, E.; Milhiet, P.; Giocondi, M.; Le Grimellec, C. Atomic Force Microscope Imaging of Cells and Membranes. In Atomic Force Microscopy in Cell Biology; Academic Press Inc: San Diego, CA, USA, 2002; Volume 68, pp. 51–65.

12. Seantier, B.; Giocondi, M.C.; Le Grimellec, C.; Milhiet, P.E. Probing supported model and native membranes using AFM. Curr. Opin. Colloid & Interface Sci **2008**, 13, 326–337.

13. Petty, M. Langmuir-Blodgett Films: An Introduction; Cambridge University Press: Cambridge, UK, 1996.

14. Rinia, H.; Demel, R.; van der Eerden, J.; de Kruijff, B. Blistering of Langmuir-Blodgett bilayers containing anionic phospholipids as observed by atomic force microscopy. Biophys. J **1999**, 77, 1683–1693.

15. Picas, L.; Suarez-Germa, C.; Teresa Montero, M.; Hernandez-Borrell, J. Force spectroscopy study of langmuir-blodgett asymmetric bilayers of phosphatidylethanolamine and phosphatidylglycerol. J. Phys. Chem. B **2010**, 114, 3543–3549.

16. Leonenko, Z.; Cramb, D. Revisiting lipid-general anesthetic interactions (I): Thinned domain formation in supported planar bilayers induced by halothane and ethanol. Can. J. Chem.-Revue Can. Chimie **2004**, 82, 1128–1138.

17. Leonenko, Z.; Finot, E.; Cramb, D. AFM study of interaction forces in supported planar DPPC bilayers in the presence of general anesthetic halothane. Biochim. Biophys. Acta-Biomembr **2006**, 1758, 487–492.

18. Merino, S.; Domenech, O.; Diez, I.; Sanz, F.; Vinas, M.; Montero, M.; Hernandez-Borrell, J. Effects of ciprofloxacin on Escherichia coli lipid bilayers: An atomic force microscopy study. Langmuir **2003**, 19, 6922–6927.

19. Berquand, A.; Mingeot-Leclercq, M.; Dufrene, Y. Real-time imaging of drug-membrane interactions by atomic force microscopy. Biochim. Biophys. Acta-Biomembr **2004**, 1664, 198–205.

20. Leonenko, Z.; Carnini, A.; Cramb, D. Supported planar bilayer formation by vesicle fusion: The interaction of phospholipid vesicles with surfaces and the effect of gramicidin on bilayer properties using atomic force microscopy. Biochim. Biophys. Acta-Biomembr **2000**, 1509, 131–147.

21. Choucair, A.; Chakrapani, M.; Chakravarthy, B.; Katsaras, J.; Johnston, L.J. Preferential accumulation of A beta(1-42) on gel phase domains of lipid bilayers: An AFM and fluorescence study. Biochim. Biophys. Acta-Biomembr **2007**, 1768, 146–154.

22. Legleiter, J.; Fryer, J.D.; Holtzman, D.M.; Kowalewski, T. The modulating effect of mechanical changes in lipid bilayers caused by apoe-containing lipoproteins on a beta induced membrane disruption. Acs Chem. Neurosci **2011**, 2, 588–599.

23. Sheikh, K.; Giordani, C.; McManus, J.J.; Hovgaard, M.B.; Jarvis, S.P. Differing modes of interaction between monomeric A beta(1-40) peptides and model lipid membranes: An AFMstudy. Chem. Phys. Lipids **2012**, 165, 142–150.

24. Quist, A.; Doudevski, I.; Lin, H.; Azimova, R.; Ng, D.; Frangione, B.; Kagan, B.; Ghiso, J.; Lal, R. Amyloid ion channels: A common structural link for protein-misfolding disease. Proc. Natl. Acad. Sci. USA **2005**, 102, 10427–10432.

25. Yarrow, F.; Kuipers, B.W.M. AFMstudy of the thermotropic behaviour of supported DPPC bilayers with and without the model peptide WALP23. Chem. Phys. Lipids **2011**, 164, 9–15.

26. Hui, S.; Viswanathan, R.; Zasadzinski, J.; Israelachvili, J. The structure and stability of phospholipid-bilayers by atomic-force microscopy. Biophys. J **1995**, 68, 171–178.

27. Dekkiche, F.; Corneci, M.C.; Trunfio-Sfarghiu, A.M.; Munteanu, B.; Berthier, Y.; Kaabar, W.; Rieu, J.P. Stability and tribological performances of fluid phospholipid bilayers: Effect of buffer and ions. Colloids Surf. B-Biointerfaces **2010**, 80, 232–239.

28. Benz, M.; Gutsmann, T.; Chen, N.; Tadmor, R.; Israelachvili, J. Correlation of AFM and SFA measurements concerning the stability of supported lipid bilayers. Biophys. J **2004**, 86, 870–879.

29. Fang, Y.; Yang, J. The growth of bilayer defects and the induction of interdigitated domains in the lipid-loss process of supported phospholipid bilayers. Biochim. Biophys. Acta-Biomembr **1997**, 1324, 309–319.

30. Garcia-Manyes, S.; Redondo-Morata, L.; Oncins, G.; Sanz, F. Nanomechanics of lipid bilayers: Heads or tails? J. Am. Chem. Soc **2010**, 132, 12874–12886.

31. Mueller, H.; Butt, H.; Bamberg, E. Adsorption of membrane-associated proteins to lipid bilayers studied with an atomic force microscope: Myelin basic protein and cytochrome c. J. Phys. Chem. B **2000**, 104, 4552–4559.

32. Mueller, H.; Butt, H.; Bamberg, E. Force measurements on myelin basic protein adsorbed to mica and lipid bilayer surfaces done with the atomic force microscope. Biophys. J **1999**, 76, 1072–1079.

33. Garcia-Saez, A.J.; Chiantia, S.; Salgado, J.; Schwille, P. Pore formation by a bax-derived peptide: Effect on the line tension of the membrane probed by AFM. Biophys. J **2007**, 93, 103–112.

34. Chiantia, S.; Ries, J.; Kahya, N.; Schwille, P. Combined AFM and two-focus SFCS study of raft-exhibiting model membranes. ChemPhysChem **2006**, 7, 2409–2418.

35. An, H.; Nussio, M.R.; Huson, M.G.; Voelcker, N.H.; Shapter, J.G. Material properties of lipid microdomains: Force-volume imaging study of the effect of cholesterol on lipid microdomain rigidity. Biophys. J **2010**, 99, 834–844.

36. Sullan, R.M.A.; Li, J.K.; Hao, C.; Walker, G.C.; Zou, S. Cholesterol-dependent nanomechanical stability of phase-segregated multicomponent lipid bilayers. Biophys. J **2010**, 99, 507–516.

37. Pera, I.; Stark, R.; Kappl, M.; Butt, H.; Benfenati, F. Using the atomic force microscope to study the interaction between two solid supported lipid bilayers and the influence of synapsin I. Biophys. J **2004**, 87, 2446–2455.

38. Mingeot-Leclercq, M.P.; Deleu, M.; Brasseur, R.; Dufrene, Y.F. Atomic force microscopy of supported lipid bilayers. Nat. Protoc **2008**, 3, 1654–1659.

39. Ulrich, A.; Sami, M.; Watts, A. Hydration of dopc bilayers by differential scanning calorimetry. Biochim. Biophys. Acta-Biomembr **1994**, 1191, 225–230.

40. Biltonen, R.; Lichtenberg, D. The use of differential scanning calorimetry as a tool to characterize liposome preparations. Chem. Phys. Lipids **1993**, 64, 129–142.

41. Thompson, G.A.J. In Form and Function of Phospholipids; Elsevier: Amsterdam, The Netherlands, 1973.

42. Leckband, D.; Helm, C.; Israelachvili, J. Role of calcium in the adhesion and fusion of bilayers. Biochemistry **1993**, 32, 1127–1140.

43. Reviakine, I.; Brisson, A. Formation of supported phospholipid bilayers from unilamellar vesicles investigated by atomic force microscopy. Langmuir **2000**, 16, 1806–1815.

44. Jass, J.; Tjarnhage, T.; Puu, G. From liposomes to supported, planar bilayer structures on hydrophilic and hydrophobic surfaces: An atomic force microscopy study. Biophys. J **2000**, 79, 3153–3163.

45. Reimhult, E.; Hook, F.; Kasemo, B. Temperature dependence of formation of a supported phospholipid bilayer from vesicles on SiO2. Phys. Rev. **2002**, 66, 051905:1–051905:4.

46. Yang, J.; Appleyard, J. The main phase transition of mica-supported phosphatidylcholine membranes. J. Phys. Chem. B **2000**, 104, 8097–8100.

47. Leonenko, Z.; Finot, E.; Ma, H.; Dahms, T.; Cramb, D. Investigation of temperature-induced phase transitions in DOPC and DPPC phospholipid bilayers using temperature-controlled scanning force microscopy. Biophys. J **2004**, 86, 3783–3793.

48. Garcia-Manyes, S.; Oncins, G.; Sanz, F. Effect of ion-binding and chemical phospholipid structure on the nanomechanics of lipid bilayers studied by force spectroscopy. Biophys. J **2005**, 89, 1812–1826.

49. Leidy, C.; Kaasgaard, T.; Crowe, J.; Mouritsen, O.; Jorgensen, K. Ripples and the formation of anisotropic lipid domains: Imaging two-component double bilayers by atomic force microscopy. Biophys. J **2002**, 83, 2625–2633.

50. Rinia, H.; Kik, R.; Demel, R.; Snel, M.; Killian, J.; van der Eerden, J.; de Kruijff, B. Visualization of highly ordered striated domains induced by transmembrane peptides in supported phosphatidylcholine bilayers. Biochemistry **2000**, 39, 5852–5858.

51. Shinozaki, Y.; Siitonen, A.M.; Sumitomo, K.; Furukawa, K.; Torimitsu, K. Effect of Ca2+ on vesicle fusion on solid surface: An in vitro model of protein-accelerated vesicle fusion. Jpn. J. Appl. Phys **2008**, 47, 6164–6167.

52. Marques, J.T.; Viana, A.S.; de Almeida, R.F.M. Ethanol effects on binary and ternary supported lipid bilayers with gel/fluid domains and lipid rafts. Biochim. Biophys. Acta-Biomembr **2011**, 1808, 405–414.

53. Soussen, C.; Brie, D.; Gaboriaud, F.; Kessler, C. Modeling of Force-Volume Images in Atomic Force Microscopy. Proceedings of the 2008 IEEE International Symposium on Biomedical Imaging: From Nano to Macro, Paris, France, 14–17 May 2008; 1–4.

54. Tristram-Nagle, S.; Petrache, H.; Nagle, J. Structure and interactions of fully hydrated dioleoylphosphatidylcholine bilayers. Biophys. J **1998**, 75, 917–925.

55. Abdulreda, M.H.; Moy, V.T. Atomic force microscope studies of the fusion of floating lipid bilayers. Biophys. J **2007**, 92, 4369–4378.

56. Picas, L.; Rico, F.; Scheuring, S. Direct measurement of the mechanical properties of lipid phases in supported bilayers. Biophys. J **2012**, 102, L1–L3.

57. Grit, M.; Desmidt, J.; Struijke, A.; Crommelin, D. hydrolysis of phosphatidylcholine in aqueous liposome dispersions. Int. J. Pharm **1989**, 50, 1–6.

58. Loi, S.; Sun, G.; Franz, V.; Butt, H. Rupture of molecular thin films observed in atomic force microscopy. II. Experiment. Phys. Rev. E **2002**, 66, 031602:1–031602:7.

59. Hutter, J.; Bechhoefer, J. Calibration of atomic-force microscope tips. Rev. Sci. Instrum **1993**, 64, 1868–1873.

Index